Francis Wolle

Diatomaceæ of North America

Illustrated with Twenty-Three Hundred Figures from the Author's Drawings...

Francis Wolle

Diatomaceæ of North America
Illustrated with Twenty-Three Hundred Figures from the Author's Drawings...

ISBN/EAN: 9783337280895

Printed in Europe, USA, Canada, Australia, Japan

Cover: Foto ©berggeist007 / pixelio.de

More available books at **www.hansebooks.com**

OF

NORTH AMERICA,

ILLUSTRATED WITH

TWENTY-THREE HUNDRED FIGURES

FROM THE

AUTHOR'S DRAWINGS

ON

ONE HUNDRED AND TWELVE PLATES.

BY THE REV. FRANCIS WOLLE,

Author "Desmids of the United States," "Fresh Water Algæ of the United States"

To
Professor H. L. Smith,
OF HOBART COLLEGE, GENEVA, N. Y.
THIS WORK IS
Dedicated
WITH GREAT RESPECT AND ESTEEM
BY HIS GRATEFUL FRIEND,
THE AUTHOR.

PREFACE.

THE Diatoms have perhaps been more thoroughly studied than the other Algæ, especially within the past fifty years. Among the noted investigators of American Diatoms was Dr. G. C. Ehrenberg, of Berlin, Germany, who in his several visits to this country, made careful explorations all the way from the Atlantic to the Pacific coast. His labors resulted in his celebrated work on *Microgeologie* or Microscopic Organisms Diatomaceae) published 1854-55, besides numerous short papers in the Journals of the Academy of the Natural Science of Berlin. Previous to Ehrenberg's publications, Prof. J. N. Bailey, of West Point, N. Y., turned his attention to the Diatoms of the United States, and frequently contributed papers on this subject to the American Journal of Sciences and Arts (Siliman's Journal), New Haven. In 1850 the Smithsonian Institute published his Microscopical observations in South Carolina, Georgia and Florida. Somewhat later, A. M. Edwards published several papers on the United States Diatomaceae; and F. W. Lewis, M.D., contributed between 1861-65, notes on "New and Rarer Species of Diatoms of the United States Sea Board;" "On New and Singular Forms," etc. From this time contributions to the literature of Diatoms became more numerous. One of the most valuable of them, appeared in *The Lens*, Chicago, 1872, by Prof. H. L. Smith, entitled "A Conspectus of the Families and Genera of the Diatomaceae." Among other prominent investigators are Prof. J. D. Cox, of the Law School of Cincinnati; Dr. H. H. Chase, of Geneva, N. Y., and later of Michigan; C. S. Peticolas, of Richmond, Va., an industrious collector of the fossil Diatoms in his vicinity; Messrs. Briggs and Thomas, of Chicago, whose names are associated with the local Diatomaceae; Kain, of Philadelphia, Pa., familiar with the forms about Atlantic City, N. J.; Shultze, whose field is Staten Island and the neighborhood of New York City; Vorce, of Cleveland, Ohio; Newcomber, of Indianapolis, Ind.; M. A. Booth, Longmeadow, Mass., and many others, all of whom have rendered valuable service as collectors, as preparers of specimens or as students of the structure, habits, growth and development of Diatoms.

Among foreign writers who have described many North American forms are Ch. Stodder, who described the diatomaceous earth found at Randolph, Mass., Richmond, Va., and at Santa Monica, Cal.; B. K. Greville, F. Kitton, C. Janisch, L. Rabenhorst, A. Grunow and others are associated with species found in western territories and waters from Vancouvers Sound and southward across the Continent to Campeachy Bay and the Gulf of Mexico.

Upwards of eight thousand species of Diatoms are recorded in Chase's edition of Habirshaw's Catalogue of the Diatomaceæ of the world, 1888. Of these, less than two thousand are North American species, but the figures and descriptions of these are scattered through many and rare volumes of Journals of Scientific Societies, magazines, reports of expeditions, beside special works; for which reason students find the literature of the Diatoms very difficult and costly to obtain. In the present volume is collected the cream of what has been already written on the subject, as well as the figures of all known North American species together with the most approved system of classification.

I follow the example of Adolph Schmidt in his *Atlas der Diatomacean Kunde* in letting the figures suffice for a description of the species. The Conspectus of the Families and Genera of the Diatomaceæ in this work, is that of Prof. H. L. Smith, of Hobart College, Geneva, N. Y., originally published in *The Lens*, January, 1872, and which Dr. H. Van Heurck has also adopted in his *Synopsis der Diatomees de Belgique*, a few subdivisions of genera being added.

I gratefully acknowledge the very valuable assistance freely extended to me by that veteran authority, Prof. H. L. Smith, to whom I submitted the plates of this work prior to their being photolithographed. I am also greatly indebted to Dr. H. H. Chase, of Michigan, through personal correspondence and his invaluable catalogue of the Diatomaceæ of the world. Nor can I omit to mention Prof. J. D. Cox, of Cincinnati, Ohio, B. N. Thomas, of Chicago, or Kain, of Philadelphia, Pa., all willing friends in time of need.

BIBLIOGRAPHY,

OR

ABBREVIATIONS OF NAMES OF AUTHORS AND THEIR WORKS CONSULTED IN PREPARING THIS MONOGRAPH.

A. J. M.—American Journal of Microscopy, N. Y., 1876–1880.
A. Q. J. M.—American Quarterly Journal of Microscopy, N. Y., 1878–1879.
BAIL. SOUND.—Microscopic Examination of Soundings made by U. S. Atlantic Coast Survey, J. W. Bailey, Washington, 1851.
BAIL. N. SP.—Notes on the New Species of Microscopic Organisms, J. W. Bailey, M.D., Washington, 1854.
BAIL. M. O.—Micro-observations in South Carolina, Georgia and Florida, J. W. Bailey, Washington, 1850.
BAIL. AND HARV.—U. S. Exploring Expedition, 1838, 1842, under command of Ch. Wilkes, Philadelphia, 1862 (1874).
B. J. N. H.—Boston Journal of Natural History.
BREB. ALG. FAL.—Algues des environs de Falaise, par M. M. de Breb. et Godey, 1855.
CASTRACANE, 1873–1876.—Exploring voyage of Challenger.
CLEVE. N. L. K. D.—On some New and Little Known, D. R. S. A., 1881.
DONKIN. B. D.—Natural History of British Diatoms, London, 1870.
EDWARDS.—Sketch of Natural History of Diatomaceae, Concord, N. H., 1874.
E. ABH.—Ehrenberg in Abhandlungen der König. Academie der Wissenschaften zu Berlin, 1830, etc.
E. BER.—Ehrenberg in Monats Berichten der K. A. der Wis. zu Berlin.
E. INF.—Die Infusions-Thierchen, von G. C. Ehrenberg, Leipsic, 1833.
E. MIK.—Mikrogeologie von Ehrenberg, 1856.
GIRARD.—Des Diats. fossils par J. Girard, Paris, 1867.
Etudes Sur Guano Comptes rends, 1868.
GREGORY, W.—New forms of Diatomaceae, British Association Reports, 1856.
GREV. CAL.—Description of New Species and Varieties of Naviculeae, etc., observed in California, Guano; R. K. Greville, Edinburgh. New Phil. Journal, 1859.
GRUN. 1860.—Ueber neue oder ungenügend gekannte Algen, von M. Grunow, Verh. K. K. Zoöl. Bot. Gesell. in Wien, 1860.

HEIB.—Conspectus Criticus Diatomacearum, per P. A. Heiberg, Copenhagen, 1862.
JAN. GUAN.—Zur Characteristic des Guanos Abh'd. der Schleisch Gesel. von C. Janisch, Breslau, 1861, 1862.
JAN. AND RAB.—Ueber Meeres Diatomeen von Honduras, von C. Janisch and Dr. L. Rabenhorst, 1862.
J. Q. C.—Jour. of the Quekett Mic. Club, London, 1868.
KITTON.—(F.) Diate of Norfolk Trans. N. H. S., of Norwich, 1875.
K. B.—Die Kieselschaliger Bacillarien. F. T. Kützing Nordhausen, 1844.
LENS.—The Lens, a Quarterly Journal of Mic., Chicago, 1872, 1873.
LEWIS, N. AND I. FORMS.—On some new and Singular Intermediate Forms of Diatæ by F. W. Lewis, Proc. Phila. Acad. Nat. Sciences, 1863.
LEWIS, N. AND R. SP.—Notes on new and rare Sp. of Diatæ from the U. S. Sea Board, F. W. Lewis, Phila., 1861.
LEWIS, W., M. D.—On extreme and exceptional variations of Diatæ in White Mountains, F. W. Lewis, Philadelphia Academy of Science, 1855.
M. D.—Micrographic Dictionary, Griffith F. Henfrey, London, 1875.
M. J.—Quarterly Journal of Microscopic Science, London.
M. M. J.—Monthly Microscopic Journal, London, 1869.
NITZ.—Beitrag zur Infusiorienkunde, von Dr. C. L. Nitzsch, Halle, 1819.
O. M. I. D.—Report on the Irish Diatæ by Rev. Eug. O'Meary. Proc. Royal Irish Academy. Vol. 2, 1876.
PETIT, LISTE.—Liste des diatomees, etc., environs de Paris, Bot. de France, Paris, 1877.
RAB. S. D.—Süsswasser Diatm, Dr. L. Rabenhorst, Leipsic, 1853.
R. M. S—Journal Royal Microscopic Society, London, 1841–1880.
SCHM. AT.—Atlas der Diatomaceen-Kunde, von Adam Schmidt. Plates 1-140. Ascherleben, 1874–1890.
SCHUMAN, T. Beitrage zur Nat. Diatomeen, Wien, 1869.
SILL. JOUR.—American Journal of Science and Art, New Haven.
S. B. D. Synopsis of the British Diatæ by the Rev. Wm. Smith, London, 1853.
SM. SP. T. Species Diatomacearum Typical Studies, H. L. Smith, Geneva, N. Y.
TOR. B. CLUB.—Torrey Botanical Club, Bulletin, New York.
T. M. S. Transactions of the Royal Microscopic Society, London.
V. H. Synopsis des Diatomes de Belgique, Dr. Henri von Heurck, 1884.
W. AND C. Notes on new and rare species, Walker and Chase, Series I.

INTRODUCTION.

The Diatomaceae comprise an order of microscopic Algae remarkable for their silicious epiderm and their singular beauty. The silex of their epidermal covering renders all their forms, excepting a few which affect brackish waters, indestructible by the ordinary agency of decomposition. They belong to both salt and fresh waters, though the denizens of the one are not to be found in the other. The marine forms abound in the sea depths, in marshes which are flooded at high tide, in shallow inlets and the muddy bottoms underlying the sandy surface of the seashore. The fresh water forms are plentiful on the mossy stones of mountain streamlets, pools bordering rivulets, dripping rocks, in ponds, creeks and rivers; every sluggish stream has a bed of them of more or less thickness.

Owing to their indestructible covering, vast quantities of Diatoms have formed fossiliferous deposits in many States of the Union, notably Virginia. The city of Richmond is built upon such a deposit from twenty to eighty feet thick and several miles in extent; the Church Hill tunnel was cut through it three-quarters of a mile.

The silicious epiderm although characteristic of the Diatomaceae, is not without exceptions; in a few it is wholly wanting; others develop it irregularly from almost nothing to nearly perfect silicification.

All living Diatoms possess a gelatinous envelope, which, owing to its transparency, can be readily detected only by means of coloring matter added to the surrounding fluid; it is secreted by the Diatom and is necessary to its existence; sometimes it forms tubes or stalks or stipes which have misled observers into basing genera upon them; nevertheless these are features which should not be ignored.

The end of these tubes, stalks or stipes attach themselves to stones, wood, and other adjacent objects to prevent the Diatom from being swept away by currents and waves. When the Diatom is about to propagate, the frustule is often immersed in large masses of this transparent gelatinous envelope.

STRUCTURE OF DIATOMS.

In the pill-box-like silicious structure of Diatoms, the top and bottom constitute the *valves;* the sides are known as the *connecting membranes* or *sutural zones* and when detached are termed *hoops;* in some, the valves do not fold over but merely rest against each other edge on edge. The line of junction forms a suture, along which the valves readily separate. H. L. Smith divides Diatoms into three groups, according to the presence or absence of a raphé or median line, viz., Raphidieæ, Pseudo-Raphidieæ and Crypto-Raphidieæ. The first has a distinctly true raphé or median line which usually has median and terminal enlargements or nodules; the second has no raphé but a pseudo-median line without nodules or enlargements; in both of these groups, the valves are much longer in the direction of the raphé or pseudo-raphé and therefore are more or less bacillar. The third group has the raphé concealed and the pseudo-raphé absent, the valves circular, often angular and sometimes broadly elliptical.

The raphé is a true cleft which in the first group divides the valves equally and is supposed to be the means by which the contents of the frustule communicate with the outer world. In one or the other of these three groups, all known forms of the Diatom may be included.

Besides the simple frustule and envelopes, diatoms often present hairs, horns or bristles, which are usually silicious. They are most abundant on group three, less so on group two and are rarely found in group one. They are mostly marginal or sub-marginal, and are sufficiently persistent to have led to the naming of genera and species in accordance with their presence or absence.

The mere adherence of the frustules gives rise to what are termed *filamentous* forms; thus, in group I, we have *Diadesmis*, a genus founded alone on this characteristic; *Fragilaria*, in group II, is but an association of *Synedra* in the form of a straight filament, and in group III *Melosira* is a straight filament of frustules no way different from *Cyclotella* or even *Coscinodiscus* except in cohering more or less firmly after the self-division of the frustules.

The typical forms of the three groups are regular in outline; thus, a *Navicula* from group I is in side view (the valve) more or less inflated and rectangular in front; the side view in *Synedra* of group II presents straight, parallel margins, the ends drawn more or less together; in the front view, however, the margins are straight to the ends; the side view of the *Coscinodiscus* in group III is circular and the front view rectangular. Though these serve as types, yet each group has many exceptional forms; thus the frustules of *Cymbella* appear more or less swollen in a front view, while the side view shows one valve more swollen than the other. The frustules of *Gomphonema* are cuneate on side view and more or less irregularly swollen in front view. The frustules of *Meridion* and *Rhipidophera* are also cuneate in a greater or less degree.

The absence of the raphé or median line from one of the valves, is another departure from the normal form in group I. The frustule is then curved with the raphé and central nodule, on the concave side (valve) as in *Achnanthes*, *Achnanthidium* and *Cocconeis*. The obliteration, partial or entire, of the central nodule as in *Stauroneis*, is another variation in group I. In some of the Naviculeae the median line is more or less sigmoid, not straight; of these, Grunow made the genus *Scoliopleura*; also, the Pleurosigma differ from the Naviculeae in the sigmoid median line as well as in the sigmoid form of the frustules.

As a general rule, free forms belong to group I, the flat filamentous to group II, the cylindrical and most of the irregular forms to group III.

CONTENTS OF DIATOMS.

The colored internal portions of a Diatom are for the same genus, usually arranged in the same manner. Any departure therefrom, is accompanied by differences in the build of the frustule; for which reason, microscopists have been disposed to classify Diatoms agreeably to the arrangement of the endochrome; which, however, has nothing to commend it especially as regards the immense masses of fossil Diatoms which can be described only by their frustules.

The contents of the Diatom are enclosed in a membrane that is firmly attached to the frustule, and possesses a higher vitality than the external membrane; for when the contents thrust the halves of the frustule apart, the internal membrane elaborates the silex for the inside edge of the sutural zone and the zones for the new valves.

Within the membranous sac is a dimly colored band more or less granular, attached to the inner surface of the internal membrane directly beneath the valves and along the zones; it is contractile and apparently serves to hold the valves together besides equally dividing the contents of the Diatom, by means of extremely thin membranous walls which issue from it.

Possibly this structure pertains to all diatoms, though it can be detected only in the larger specimens.

MOVEMENTS OF DIATOMS.

These have excited the curiosity of all observers and given rise to a belief with many, that the Diatom belongs to the animal rather than to the vegetable world; but though the latter opinion now prevails, yet the cause of the movement of Diatoms has not thus far been satisfactorily accounted for even by those who have, like H. L. Smith, devoted much attention to the subject. This gentleman, at the tenth annual meeting of the American Society of Microscopists, presented "*A Contribution to the Life History of the Diatomacea*," in which he says: "I am disposed to consider" (for reasons he had previously stated at length) "that the motion of the Naviculeae is due to injection and expulsion of water, and that these currents are caused by different tension of the" (internal) "membranous sac in the two halves of the frustule. In those Diatoms which do not have the central band thus binding the frustules, as for example, the *Synedra*, the *Fragilaria* and the circular and angular forms, no motion is to be observed, or at best but a slight trembling, as in this case, the tension is more nearly uniform over the whole surface of the internal sac." More recently, Mr. Cornelius Onderdonk, in *The Microscope* for August, 1890, ascribes the movements of Diatoms to "a thin fluid mass in rhythmical motion," which covers the surface of the Diatom. "We cannot conceive that the fluid has power to drive itself, for it is unformed material. What then remains except that the motive force is on the walls of the cell . . . . ."

REPRODUCTION OF DIATOMS.

Diatoms reproduce themselves by three methods of conjugation.

1st. By two frustules uniting their undifferentiated endochrome, thereby producing a single sporangium about double the size of the parent frustules, which finally develops into a frustule with ends more rounded than the parent frustules and the valves not so wedge-shaped. *Surisella splendida* is a good example of this method, see Plate I.

2d. By two frustules uniting their differentiated endochrome and producing *two* sporangial cells; sometimes one of these proves abortive, but that has no influence on the growth and development of the other. *Navicula amphirhynctus* is propagated in this manner.

3d. By the differentiated contents of a *single* frustule producing one sporangium, the most common mode of propagation as well as the lowest in the scale, as in the free form, *Cymbella cuspidata*.

Besides reproduction by conjugation there is multiplication by division, and as diatoms do not grow these divisions tend to make the resultant diatoms smaller and smaller until some become so minute as in the *Nitzschia*, *Navicula* and *Amphora*, that while living they can go through the pores of filtering paper; a single frustule by constant division and subdivision of its progeny, would give rise in twenty-five repetitions of this process, to more than thirty million individuals, did they all survive.

As regards the longevity of diatoms, it may be said that dried specimens can not be revived, but they have been known to survive nearly a quarter of a century in their natural element even though kept for long periods in the dark, and at times frozen in solid ice. Their silicious covering is almost indestructible, resisting the strongest acids and passing unscathed through very high degrees of heat.

There are very few microscopic objects of more interest than the Diatom or which return more ample rewards to the patient, enthusiastic student.

CONSPECTUS OF THE FAMILIES AND GENERA

OF THE

DIATOMACEÆ

SYNOPTICALLY ARRANGED BY PROF. H. L. SMITH
IN "THE LENS," FOR JANUARY, 1872.

CLASS CRYPTOGAMIA.

SUB-CLASS ALGÆ.   NATURAL ORDER DIATOMACEÆ.

[s. v. *Side View;* f. v. *Front View.*]

Plant a FRUSTULES (*silicious box*), bivalve, pseudo-unicellular, and with an external, more or less apparent, gelatinous envelope, and an internal investing membrane, with endochrome. GEMMIPAROUS INCREASE by SELF-DIVISION. REPRODUCTION by CONJUGATION.

TRIBE I. RAPHIDIEÆ.

FRUSTULES mostly bacillar in s. v.; sometimes broadly oval;

always with — a distinct raphé and nodules on one or both valves; central nodule rarely wanting or obscure; valves simple or complex; raphé generally prominent in s. v., occasionally in f. v., especially when constricted, with nodules at the constrictions;

without — teeth, spines, awns, or processes.

TRIBE II. PSEUDO-RAPHIDIEÆ.

FRUSTULES generally bacillar in s. v., sometimes broadly oval; or suborbicular —very rarely orbicular. Frustules with or without nodules;

always with either — a pseudo-raphé (simple line or blank space) on one or both valves; or longitudinal septa or vittæ in f. v., or valves fusiform, sigmoid, beaked, or alate, or with numerous transverse ribs, plicae, costae, striae, or rows of granules on one or both valves, rarely regularly radial; costae sometimes showing in f. v.

without — processes, teeth, spines, awns, or true raphé on the valves; except spines very rarely among the Surirelleæ, or Tabellarieæ, when the character is already sufficiently indicated by the above.

never — angular in s. v.

rarely — hyaline, unstriate, or much developed in f. v. unless longitudinally septate.

Tribe III. CRYPTORAPHIDIEÆ.

FRUSTULES generally circular, sub-circular or angular in s. v., more rarely elliptical oval, or bacillar.

- either ... much developed in f. v., and filamentous; *or* with processes, teeth, spines, or awns; *or* more or less hyaline or irregular; *or* transversely septate (or costate) in f. v. with a central *linear* blank space or true raphé, *on* the valves, except a raphé or pseudo-raphé in Raphidodiscus. v. microscope, May, 1889, p. 135.
- never ...

ANALYSIS OF THE FAMILIES.

Tribe I. RAPHIDIEÆ.

Lettered figures refer to Families named in last paragraph of each section. Plain figures refer to the corresponding numbers on the left of each section.

	Frustules with valves alike......................	2
	Frustules with valves unlike......................	1
1.	Valves cuneate......................	8C
	Valves not cuneate......................	7
2.	Valves symmetrically divided by the raphé......................	1
	Valves not so divided......................	3
3.	Valves alate, or obliquely striate......................	4
	Valves not as above, more or less arcuate or cymbiform......................	8A
4.	Valves with central nodule equally distant from ends.	8B
	Valves not as above......................	5
5.	Valves with central nodule obscure or wanting......................	8B
	Valves not as above......................	6
6.	Valves with central nodule unequally distant from ends......................	8C
	Valves not as above......................	7
7.	Frustules genuflexed, nodule or stauros on one valve, usually on the concave margin (at the constriction), valves rarely broadly oval......................	8D
	All others, valves generally broadly oval, rarely bent	8E

	A .... CYMBELLEÆ......	(I.)
	B .... NAVICULEÆ......	(II.)
8.	C .... GOMPHONEMEÆ.	(III.)
	D .... ACHNANTHEÆ...	(IV.)
	E .... COCCONIDEÆ.....	(V.)

TRIBE II. PSEUDO-RAPHIDIEÆ.

	Frustules compound; really or apparently longitudinally septate or vittate; septa or vittae showing distinctly in f. v..............	1
	Frustules not as above, or only seen in s. v.............	2
1.	Arcuate in f. v. (apparently septate?), valves alike, or differing only by a pseudo-nodule at ends of concave valve..............	9A
	All others.............	9B
2.	Valves circular, sub-circular, very broadly oval, or differently costate	3
	Valves not as above.............	4
3.	Valves mostly hyaline, with a few pervious costae (scalariform, or arcuate f. v., with valves differently costate, really septate in f. v.............	9B
	All others.............	9C
4.	Valves fusiform, sigmoid, or beaked, or one margin more strongly marked than the other..............	9C
	Valves not as above.............	5
5.	Valves transversely undulate (undulations conspicuous in f. v., with transverse shaded bands........	9C
	Valves not as above.............	6
6.	Frustules in f. v. with beaded margins (especially on one side), not ends of costae; or carinate, or alate....	9C
	Frustules not as above.............	7
7.	Valves pervious or dimidiate, costate or striate, or transversely or irregularly dotted (not carinate or alate).............	9A
	Valves not as above.............	8
8.	Frustules with a row of marginal sub-capitate processes, or alate, or carinate.............	9C
	Frustules not as above.............	9A
9.	A .... FRAGILARIEÆ...	(VI.)
	B .... TABELLARIEÆ...	(VII.)
	C .... SURIRELLEÆ.....	(VIII.)

Tribe III. CRYPTO-RAPHIDIEÆ.

1. { Frustules cylindrical, or flattened, valves alike, terminated by a calyptra; pointed with a bristle...... 18A
{ Frustules not as above 1

1. { Frustules with valves unlike, or mostly smooth; and furnished with awns, horns (elongated processes, spines, or setae, which are sometimes absent or imperfect in fossil forms; frequently imperfectly silicious, valve not ribbed radically, or cellulose... 18A
{ Frustules not as above... 2

2. { Frustules imperfectly silicious, connecting zone more or less turgid; connected in distant series; valve angular, with long central spine 18A
{ Frustules not as above... 3

3. { Valves with a single pseudo-nodule............. 18G
{ Valves not as above.... 4

4. { Valves lunate, not transversely costate, or septate... 18G
{ Valves not as above...... 5

5. { Valves somewhat hispid; with sinuato-reticulate lines not rayed 18G
{ Valves not as above...... 6

6. { Valves circular or angular, not much developed in f. v., with (obscurely) reticulate center, and conspicuous pore-like puncta............. .... 18G
{ Valves not as above............. 7

7. { Valves alike, smooth (hyaline), with radiating lines (linear rays not terminating in a spine); rays definite (few)........... .... 18F
{ Valves not as above 8

8. { Frustules cuneate in f. v., or with decided ocelli, processes, or tubercles, generally few and prominent in f. v. (not spines alone) 11
{ Frustules not as above... 9

10. { Frustules cohering; generally much developed in f. v., and cylindrical; firmly silicious; valves rarely hyaline, unlike, or elliptical; without median line; sometimes apiculate, or conical, or with a peculiar central nodule (spine; radiate punctate, or cellulate; and frequently with marginal or sub-marginal spines. Frustules cohering by smooth sutural lines, or by marginal spines or teeth, or by a central spine 18B
{ Frustules and valves not as above.. 12

9.	{ Frustules traversely septate, or costate; cuneate, angular, or sub-angular...............	11
	(Frustules not as above...........	10
11.	{ Frustules not much developed in f. v. (free), rarely angular, neither lunate nor cuneate.........	18D
	{ All others, filamentous, generally much developed in f. v......	18C
12.	{ Valve disc more or less undulate, divided into regular compartments, usually alternate light and dark; mostly with marginal or sub-marginal spines or teeth......	18E
	{ Valves not as above..........	13
13.	{ Valves hyaline, with umbilical lines............	18F
	(Valves not as above............	14
14.	{ Valves with definite, irregular, flexuose, or bifurcate rays; not hispid, nor with marginal spines........	18F
	(Valves not as above.......	15
15.	{ Valves hyaline, rays definite, not reaching margin...	18F
	(Valves not as above.....	16
16.	{ Valves with spathulate, cordate, or deltoid rays, their bases frequently forming a hyaline central area.....	18F
	(Valves not as above	17
17.	{ Valves with large marginal hyaline spaces, which are neither circular nor hexagonal.........	18F
	(All others.....	18G
18.	{ A.............CHÆTOCEREÆ...... (IX.)
	{ B............MELOSIREÆ....... (X.)
	{ C...........BIDDULPHIEÆ...... (XI.)
	{ D...........EUPODISCEÆ....... (XII.)
	{ E...........HELIOPELTEÆ.....(XIII.)
	{ F..........ASTEROLAMPREÆ.(XVI.)
	{ G...........COSCINODISCEÆ... (XV.)

ANALYSIS OF THE GENERA.

Names of the Genera printed in Italics are not North American.

TRIBE I. RAPHIDIEÆ.

FAMILY I. CYMBELLEÆ.

{ Frustules sometimes complex; frequently hyaline, or inflated or constricted in f. v.; with central nodules approximate, and touching the connecting zone; median line often inflexed. Valves frequently with a transverse line (stauros); "connecting membrane" often longitudinally striate or punctate......................AMPHORA 1.

{ Frustules not as above....................1

1. {
   Frustules simple, arcuate in s. v., linear in f. v., raphé and central nodule marginal in s. v., valve sometimes inflated in the center of ventral margin .... CERATONEIS II.
   All others, raphé frequently curved......... CYMBELLA III.

FAMILY II. NAVICULEÆ.

1. {
   Frustules compound,—valves with loculi (marginal cells)................................MASTOGLOIA IV.
   Frustules not as above...1
   Frustules compound (each valve with two plates): one (superior) with transverse ribs, the other (inferior) striate, with raphé and nodules........ .... STICTODESMIS V.
   Frustules not as above..2

2. {
   Valves with a conspicuous, transverse, smooth silicious band (stauros) not alate ....STAURONEIS VI.
   Valves not as above 3

3. {
   Valves sigmoid, or arcuate; or frustules alate, or raphé inflexed or reflexed............................5
   Valves and frustules not as above........................4

4. {
   Valves with a central nodule................NAVICULA VII.
   Valves without central nodule (or obscure)............. .... AMPHIPLEURA VIII.

5. {
   Valves symmetrically divided by the raphé; frustules not alate, rarely constricted in f. v............. .... PLEUROSIGMA IX.
   Valves or frustules not as above...........................6

6. {
   Valves not symmetrically divided by the raphé, which is arcuate, ends reflexed, and marginal..... .... TOXONIDIA X.
   All others, alate, usually constricted in f. v .... AMPHIPRORA XI.

FAMILY III. GOMPHONEMEÆ.

{ Frustules curved in f. v., nodule on concave valve... .... RHOIKOSPHENIA XII.
  All others............................... GOMPHONEMA XIII.

FAMILY IV. ACHNANTHEÆ.

Free or stipitate ....ACHNANTHES XIV.

FAMILY V. COCCONIDEÆ.

{ Valves with marginal cells....................Orthoneis XV.
  Valves not as above 1

	Valves not symmetrically divided by the raphé........	
1.		*Anortheis* XVI.
	All others............... ....COCCONEIS XVII.	

TRIBE II. PSEUDO-RAPHIDIEÆ.
FAMILY VI. FRAGILARIEÆ.

	Frustules arcuate in f. v., valves with interrupted transverse striae or costae; and one or both valves with pseudo-nodules (blank spaces) at the ends. ... GEPHYRIA XVIII.
	Frustules and valves not as above............ 1
1.	Frustules more or less arcuate in s. v., with transverse (often granulate) ribs (canaliculi, *W. S.*); which in f. v. frequently cause the margin or submargin to appear beaded, or dentate.....EPITHEMIA XIX.
	Frustules not as above 2
2.	Frustules arcuate in s. v.; not ribbed; valves transversely striate, without median line or nodule, and with pseudo-nodules at the ends, on concave margin ....EUNOTIA XX.
	Frustules not as above... 3
3.	Valves with a pseudo-raphé, and transverse rows of granules within square cells (clathrate); central and terminal nodules distinct ........GLYPHODESMIS XXI.
	Valves not as above. 4
4.	Valves cruciform; with interrupted transverse striae (not clathrate), central nodule very distinct. Frustules in f. v., with terminal vittae?............... *Omphalopsis* XXII.
	Valves not as above 5
5.	Valves with central and terminal nodules (or blank spaces), latter prominent in f. v.; and pseudo-median line (not always distinct); valves frequently constricted or inflated at the middle; frustules cohering.. ....*Diadesmis* XXIII.
	Valves not as above 6
6.	Valves with a central (generally transverse) blank space, and a central pseudo-ocellus; or with two or more (few) strong pervious costae in the middle, which are prominent in f. v.; and transverse (generally moniliform interrupted) striae or costae, or square cellules; and terminal nodules........ .... PLAGIOGRAMMA XXIV.
	Valves not as above 7

CONSPECTUS OF THE FAMILIES AND GENERA. 23

7. { Frustules cohering; quadrangular in f. v., valves without central nodule; striae interrupted by a smooth median line or a blank space; valves inflated or constricted; terminal nodules present, and generally prominent in f. v......DIMEREGRAMMA XXV.
   Frustules not as above...8

8. { Frustules with a serrated suture; valves without a median line, and with transverse conspicuous rows of pores or dots..................................Terebraria XXVI.
   Frustules not as above...9

9. { Frustules in f. v. narrow, linear; valves lanceolate or inflated; with conspicuous moniliform (generally somewhat radiate), transverse striae; and a median line or black space (frequently obscure or wanting); nodules absent..................RAPHONEIS XXVII.
   Frustules not as above...10

10. { Frustules sessile, solitary, or in twos, elongated, linear; slightly cuneate; valve finely striate, constricted at one end, and without median line........
     ....Peronia XXVIII.
     Frustules not as above...11

11. { Frustules linear in f. v.; somewhat hyaline, and enlarged at one end; valves striate, without median line or nodule; cuneate, and constricted at one end; united in a stellate or zigzag manner...........
     ....ASTERIONELLA XXIX.
     Frustules not as above...12

12. { Frustules much elongated; valves with a smooth median line or blank space, sometimes obscure; frequently a central pseudo-nodule; transversely striate, never costate; sometimes slightly cuneate, or bent, sessile, filamentous, or attached end to end...SYNEDRA XXX.
     Frustules not as above...13

13. { Frustules very much elongated, straight or undulate; valves inflated at the middle, slender (awn-like), somewhat irregularly punctate in s. v., without median line, or nodules...................Toxarium XXXI.
     Frustules not as above...14

15. { Frustules cuneate; margin smooth; valves hyaline, or finely striate, with a median line....................
     ....LICMOPHORA XXXIII.
     Frustules not as above...16

14. { Valves finely striate, punctate, or more or less hyaline; never costate; median line wanting or obscure; frustules narrow in f. v., sometimes inflated (undulato-arcuate) or constricted; margins smooth, cohering, forming a straight, sometimes zigzag, filament..FRAGILARIA XXXII.
Valves and frustules not as above..................................15

16. { Frustules cuneate; valves transversely costate or distinctly granular striate, with a median line ....
PODOCYSTIS XXXIV.
Frustules not as above...17

17. { Frustules compound; valves costate or ribbed; ends of costae prominent, sub-marginal and capitate in f. v.; frustules not cuneate ....DENTICULA XXXV.
All others; valves costate; linear, or cuneate in f. v.; cohering; filament zigzag, curved, or straight
DIATOMA XXXVI.

FAMILY VII. TABELLARIEÆ.

Frustules linear, or cuneate, and in f. v., with moniliform vittae (ends of costae); valves divided into chambers by transverse ribs (scalariform); outer valve finely pervious striate, without median line.
CLIMACOSPHENIA XXXVII.
Frustules not as above...1

1. { Frustules curved in f. v.; valves costate; dissimilar; septae rudimentary............................*Entopyla* XXXVIII.
Frustules not as above 2

2. { Frustules in f. v. with straight vittae, which are usually alternate; cohering, forming a zigzag filament; valves transversely striate, and inflated at center and ends.................................TABELLARIA XXXIX.
Frustules not as above...3

3. { Frustules with straight vittae, in pairs, and interrupted at the ends and center in f. v..........*Diatomella* XL.
Frustules not as above..4

4. { Frustules with vittae in pairs, which are straight or undulate in f. v.; not interrupted or enlarged at the ends; cohering, forming a zigzag filament......
GRAMMATOPHORA XLI.
Frustules not as above..5

5. { Frustules filamentous; vittae in f. v. somewhat enlarged at the ends (clavate); valves pervious costate; costae few, showing in f. v......*Gomphogramma* XLII.
Frustules not as above..6

6. { Valves without pervious costae, quite smooth, or *very* finely striate; often with fine median line; frustules hyaline in f. v.................7
Valves and frustules not as above.................9
7. { Frustules with spines (bristles) at the angles in f. v.. *Attheya* XLIII.
Frustules without spines.................8
8. { Septae (or vittae) in f. v., interrupted, alternate...... TESSELLA XLIV.
Septae not interrupted...................STRIATELLA XLV.
9. { Frustules compound; valves with a median line, and generally blank ends; costate or striate. Septae in f. v. connected by transverse striae (latticed); frustules cohering, forming a flat filament............ RHABDONEMA XLVI.
Valves mostly hyaline, with a few pervious costae (scalariform); linear, orbicular, or inflated......... BIBLARIUM XLVII.

FAMILY VIII. SURIRELLEÆ.

{ Frustules compound; valves broadly oval, unlike; one costate, the other sieve-like (cribrose), punctate; without nodules.................CAMPYLONEIS XLVIII.
Frustules not as above.................1
1. { Valves transversely undulate, undulations conspicuous in f. v.; striate, and with a few transverse shaded bands..................CYMATOPLEURA XLIX.
Valves not as above.................2
2. { Frustules in f. v. showing a marginal row of short, sub-capitate processes; valves transversely costate (scalariform).................. *Clavularia* L..
Frustules not as above.................3
3. { Frustules bent in s. v., margins finely dotted, valves striate, without median line, and with an unequally notched inflation at one end; cohering in a stellate manner.................ACTINELLA LI.
Frustules not as above.................4
4. { Frustules linear in f. v.; arcuate, with rounded ends in s. v.; margins punctate; valves finely striate, without median line, terminal nodules on the concave margin. Frustules cohering in a zigzag manner, or in tables. ....*Desmogonium* LII.
Frustules not as above.................5

5. { Frustules (inconspicuously) alate; valves elliptical or linear; not cuneate; parallel (dimidiate, marginal, or wavy) striate; rarely costate; median line, if present, simple, and not beaded or conspicuously marked; valves sometimes bent along the longer axis (apparently).. ....*Tryblionella* LIII.
Frustules not as above 6

6. { Frustules alate (sometimes inconspicuously); valves cuneate, reniform, oval, or subcircular; rarely linear, frequently twisted; with simple median line, or more or less linear blank space; center sometimes blank, or finely dotted; margins or submargins strongly marked (somewhat radiate); costate or plicate (canaliculate, *W. S.*). .....*Surirella* LIV.
All others usually, finely striate, linear, constricted, inflated, sigmoid, beaked, fusiform or carinate; free, adherent, or in tubes.....................*Nitzschia* LV.

Tribe III. CRYPTO-RAPHIDIEÆ.

Family IX. CHÆTOCEREÆ.

Frustules annulate; cohering; elongate; ends alike, calyptriform; tipped with a spine or mucro; often imperfectly silicious......... ....*Rhizosolenia* LVI.
Frustules not as above.................. 1

1. { Frustules with horns (elongated processes; not spines simply); valves frequently unlike............2
Frustules with spines or bristles; or awns; or smooth.3

2. { Frustules compressed, the sutural portion narrow; horns frequently branching or bifurcate; sometimes mucronate; valves and horns sometimes with short scattered spines; horns sometimes obtuse, short (mammae)........ ....*Dicladia* LVII.
Frustules elongated; horns mucronate; valves dissimilar; generally one horn (or process) or one valve, two on the other ....*Syringydium* LVIII.

3. { Spines on one valve only; frequently long, and sometimes branching........ ......*Syndendrium* LIX.
Frustules not as above 4

4. { Valves angular; spine central; sutural portion more or less turgid (imperfectly silicious); valves with a radiate series of dots; frustules connected in a distant series ....*Ditylum* LX.
Valves not as above. 5

5. {
- Spines (setae?) marginal; on both valves; valves dissimilar, mostly hyaline............... HERCOTHECA LXI.
- All others, valves frequently awned ; or with minute scattered spines ; or dissimilar; hyaline, or imperfectly silicious; or frustules compound; awns often absent in fossil forms; leaving valves entirely smooth ....CHÆTOCEROS LXII.

FAMILY X. MELOSIREÆ.

{
- Frustules apiculate (drawn out at the extremities or margins to a point)..1
- Frustules not as above..2

1. {
- Frustules cylindrical ; apiculate in f. v. ; valves unlike...PYXILLA LXIII.
- Frustules not cylindrical ; apiculate in s. v.; valves alike..Peponia LXIV.

2. {
- Valves with a central spine, or coronal or scattered spines ; not ribbed ; frustules cohering by the spines..STEPHANOPYXIS LXV.
- Valves and frustules not as above............................3

3. {
- Frustules cylindrical ; with somewhat large regular marginal teeth, and peculiar central clasping spine. Syndetocystis LXVI.
- Frustules not as above..4

4. {
- Valves elliptical or constricted; with marginal spines or teeth ; and central, peculiar nodule................. RUTILARIA LXVII.
- Valves not as above..5

5. {
- Frustules cylindrical ; ends constricted and finally expanded into a connecting nodule................... Strangulonema LXVIII.
- Frustules not as above..6

6. {
- Frustules cylindrical ; with a border of much elongated cells at junction of margins in f. v............. Skeletonema LXIX.
- Frustules not as above..7

7. {
- Valves circular ; with curved marginal rays, and minute marginal teeth......DISCOSIRA=MELOSSIRA LXXII
- Valves not as above..8

8. {
- Valves dissimilar ; somewhat conical or inflated in f. v. ; with radiating lines or ribs ; not branching, nor bifurcate at extremities ; apex truncated ; usually spinous ; interspaces punctate.................. STEPHANOGONIA LXX.
- Frustules not as above..9

9. {
- Frustules marked with spiral or crossed bands in f. v. .... *Liparogyra* LXXI.
- All others...................................MELOSIRA LXXII.

FAMILY XI. BIDDULPHIEÆ.

1. {
- Frustules with one neck-like process; generally oblique; cohering irregularly, valves unlike............ ISTHMIA LXXIII.
- Frustules not as above 1
- Frustules transversely costate in f. v.; costae more or less capitate, resembling music notes; valves transversely costate; without spines or median line.......................................TERPSINOE LXXIV.
- Frustules not as above.................................2

2. {
- Frustules transversely costate, or scalariform; costae (septae) showing in f. v., not capitate, valves often lunate, connecting zone hyaline or striate............ ANAULUS LXXV.
- Frustules not as above 3

3. {
- Frustules mostly hyaline, or imperfectly silicious, forming a straight or curved filament; processes obscure, or absent.........................EUCAMPIA LXXVI.
- Frustules not as above 4

4. {
- Processes generally straight on the outer margin in f. v.; and tipped with a spine or mucro, which is sometimes obscure.......................HEMIAULUS LXXVII.
- All others.........................BIDDULPHIA LXXVIII.
- Angular forms..............................TRICERATIUM.

FAMILY XII. EUPODISCEÆ.

2. {
- Valves with plumose rays or dots about the flat mastoid processes (or ocelli); rarely obscure; sometimes with a sub-quadrate central portion; or a radiant cellulation interrupted by a linear series terminating in the ocelli...............AULISCUS LXXIX.
- Valves not as above...................................1
- Valves circular or oval; with the ocelli, or pseudo-openings, in compartments............*Craspedoporus* LXXXI.
- Valves discoid with a central thickening or obscure nodule, and an interrupted raphé terminated by minute spines or spiniform nodules somewhat within the margin of the disc; central portion of the disc naviculoid, depressed, its ends terminating at the spines. Striae radiate moniliform, extending from raphé to the margin of the valve...RAPHIDODISCUS.
- Valves not as above......................................3

CONSPECTUS OF THE FAMILIES AND GENERA.   29

1. { Valves with decided ribs, rays, or furrows, connecting the (usually large) processes or tubercles ..... AULACODISCUS LXXX.
    Valves not as above..2

3. { Valves circular; with a radiating series of minute punctae, and marginal tubercles............ *Perithyra* LXXXII.
    Marginal tubercles smaller, valves radiate-granulate, circular or oval............................. CESPODISCUS LXXXII².
    All others, ocelli (tubercles, or processes) generally quite large, and few, usually sub-marginal; cellules or granules, rarely radial, or minute............
                                                EUPODISCUS LXXXIII.

FAMILY XIII. HELIOPELTEÆ.

    Valves with marginal spines obsolete; or if present, few, and in alternate compartments........... ACTINOPTYCHUS LXXXIV.
    Valves not as above...1
    Valves with a hyaline stellate umbilicus, with marginal spines or teeth, connected by a radial rib ....
                                                Halionyx LXXXV.
1. { Valves with numerous marginal spines or teeth; and a hyaline umbilicus; often hyaline spaces at the base (angles) of each compartment........................
                                                HELIOPELTA LXXXVI.

FAMILY XIV. ASTEROLAMPREÆ.

    Valves hyaline, angular or circular; with straight rays or ribs not expanded at margin or center, and not reaching the margin....... LISOTEPHANIA LXXXVII.
    Valves not as above 1

1. { Disc radiate-punctate, cellulate or granulate; with several well-defined linear blank spaces (ribs), from the margin inwards; center granulate, not stellate.
                                                ACTINODISCUS LXXXVIII.
    Valves not as above 2

2. { Valves inflated, hyaline, or punctate, center sometimes stellate; rays linear; more or less bifurcating, and somewhat irregular; interspaces blank, or with curved or sinuose lines CLADOGRAMMA LXXXIX.
    Valves not as above 3

3. { Valves hyaline; with a broad margin divided by simple rays; center hyaline, or granulate, reticulate, or minutely punctate...................MASTOGONIA XC.
All others..................................ASTEROLAMPRA XCI.

FAMILY XV. COSCINODISCEÆ.

1. { Disc with a circle of large marginal or intra-marginal cellules; and radiate, or scattered cellules or punctae ....Heterodictyon XCII.
Valves not as above 1
{ Disc with an interior ring of cellules separating the center from the broad marginal rim; cellulation of center, curved or spiral..................Brightwellia XCIII.
Valves not as above 2

2. { Disc very convex, or conical in f. v.; with a conspicuous, central-pseudo-opening.. ....Porodiscus XCIV.
Valves not as above 3

3. { Disc cellulose; large, with a broad border of a different structure, separated by a well-defined margin. CRASPEDODISCUS XCV.
Valves not as above................... 4

4. { Disc hyaline; with distinct umbilicus, and (very) finely marked: with rayed or decussating lines... HYALODISCUS XCVI.
Probably often valves of PODOSIRA=MELOSIRA LXXII.
Valves not as above 5

5. { Disc cellulose; with a narrow (somewhat dentate) rim; connecting zone cellulose .... ENDICTYA=MELOSIRA LXXII.
Valves not as above................. 6

6. { Disc without marginal spines, teeth, or pseudo-nodule; usually of small or medium size; and generally with an outer ring-like portion either smooth or striate; center often bullate, smooth, or granulate; granules equal, scattered, or rayed, or disc hyaline (or finely punctate), with strong linear straight rays................................CYCLOTELLA XCVII.
Valves not as above...... 7

7. { Frustules complex? disc circular, generally with a marginal or submarginal pseudo-nodule (sometimes absent); frequently with minute marginal spines or teeth; and with a single or double series of radiating dots or punctate often subulate (!) blank spaces.. ....................ACTINOCYCLUS XCVIII.
Frustules not as above... 8

CONSPECTUS OF THE FAMILIES AND GENERA. 31

8. { Frustules cuneate in f. v., lunate in s. v.................9
   { Frustules not as above.................................11

9. { Valves cellulose, center blank, margin veined........
   {   *Hemidiscus* XCIX.
   { Valves not as above 10

10. { Valves with indistinct umbilicus, finely punctate
    {   with radiating lines, dorsal and ventral margins
    {   with minute teeth or spines...............*Palmeria* C.
    { All others, dorsal margin without spines, ventral
    {   frequently with pseudo-nodule............*Euodia* CI.

11. { Disc with a radiating series of small, equal or sub-
    {   equal granules; and generally with a granular
    {   umbilicus or center; marginal teeth or spines
    {   rarely absent............*Stephanodiscus* CII.
    { Valves not as above.....................................12

12. { Valves circular; much inflated. Frustules in f. v.
    {   with the longitudinal axis much longer than the
    {   transverse; not ribbed, nor cellulose; sutural por-
    {   tion ("connecting zone") narrow; sometimes
    {   minute marginal teeth................*Pyxidicula* CIII.
    { Valves not as above....................................13

13. { Valves elliptic, circular, or sub-angular; with a
    {   prickly aspect hispid; often with minute spines,
    {   and with sinuato-reticulate rays or lines...*Liradiscus* CIV.
    { Valves not as above....................................14

14. { Disc circular or angular; with conspicuous punctae;
    {   and divided into more or less plicate compart-
    {   ments, often obscurely, by radiating, often dicho-
    {   tomizing, lines or blank spaces; center sometimes
    {   bullate, or more or less distinctly reticulate......
    {   *Stictodiscus* CV.
    { Valves not as above 15

15. { Frustules compound. Disc circular, with numerous,
    {   strong, straight, radial ribs; and a hyaline center;
    {   ribs connected by concentric lines, or rows of gem-
    {   maceous granules without spine or teeth............
    {   *Arachnoidiscus* CVI.
    { Frustules not as above.................................16

16. { Disc with a circle of well-defined marginal or intra-
    {   marginal subulate spines; cellules in parallel
    {   rows............................*Systephania* CVII.
    {   (Cellules not in parallel rows, valves of
    {   *Creswellia* — *Stephanopyxis*? LVIII.
    { Valves not as above....................................17

17. { Disc very convex, and strongly cellulose, without marginal teeth or spines.................... *Dictyopyxis* CVIII.
    { Valves not as above...18

18. { ⌠ Disc without rays; frequently hyaline; and with
      scattered spines............................XANTHIOPYXIS CIX.
    { All others, without strong linear rays, or large spines
      or teeth.................................... ...........COSCINODISCUS CX.

INDEX

TO NAMES OF FOURTEEN HUNDRED NORTH AMERICAN DIATOMS ILLUSTRATED IN THIS WORK WITH TWENTY-THREE HUNDRED FIGURES.

[*Roman numerals indicate the Plates—common figures, the Illustrations on each Plate.*]

ACHNANTHES.
    arenicola, xxix, 9.
    brevipes, xxvii, 29, 30.
    exilis,   "   25.
    Hudsonis, "   28.
    longipes,   "   22-24.
    microcephalum, xxvii, 27.
    minutissima, xxix, 10, 11.
    subsessilis, xxvii, 31-34.
ACTINELLA.
    diodesmis? xxxvii, 9.
    punctata,   "   6-10.
ACTINISCUS.
    Sirius, lxix, 9.
ACTINOCYCLUS.
    alienus, lxxxv, 14.
    crassus, ciii, 1.
    curvatulus, xciv, 14.
    Ehrenbergii, lxxxv, 9, 12.
    ellipticus,   "   1.
    moniliformis, ciii, 5.
    Niagarae,   "   4.
    Ralfsii, lxxxv, 13.
    tenuissimus, ciii, 6.
    triradiatus, lxix, 8.
ACTINODISCUS.
    Atlanticus, lxxviii, 11.
ACTINOPTYCHUS.
    amblyoceros, lxxxv, 5.
    areolatus,   "   4.
    biseptinarium, cviii, 9.
    Bismarkii, ciii, 3.
    elegans, lxxiv, 14.
    excellens, xcii, 8.
    glabratus, lxxxv, 16, 17, 19.

ACTINOPTYCHUS.
    Grundleri, lxxiv, 11; xcii, 3, 7.
    heliopelta, xcii, 1; ciii, 2.
    incertus, lxxxv, 8.
    interpunctatus, lxxvii, 5, 6.
    irregulare, cxii, 3.
    laevigatus, lxxxv, 10.
    nitidus,   "   6.
    Pfitzeri, xcii, 2.
    praeter, lxxxv, 3.
    pulchellus, ciii, 7.
    quinarius, cviii, 10.
    Racanus, lxxvii, 7.
    socius, lxxxv, 2.
    spiniferus, lxxxv, 18.
    splendens, xcii, 9-12.
    subtilis, lxxxv, 7.
    triangulus, xcii, 13.
    trigonus, lxxxv, 11.
    undulatus, lxxiv, 12; lxxviii,
      13; xcii, 4, 5, 6.
    vulgaris, lxxxv, 15, 20.
AMPHICAMPA.
    ? ——, xlviii, 29-31.
AMPHIPLEURA.
    Lindheimerii, xxxi, 2.
    maxima, cxii, 1.
    Oregonica, xxxi, 1.
    pellucida,   "   3-5.
    Weissflogii, "   6.
AMPHIPRORA.
    alata, ii, 20, 21.
    calumetica, xlviii, 22-24.
    conserta, cvii, 3, 4.
    conspicua, ii, 16.

AMPHIPRORA.
constricta, ii, 14, 15.
costata, cvii, 5.
discussata, ii, 19.
elegans, v, 3.
hyalina, ii, 7 8.
lepidoptera, ii, 6.
Lindheimerii, ii, 1.
maxima, ii, 17, 18.
nervis, cvii, 1, 2.
ornata, ii, 12, 13; xlviii, 21.
paludosa, ii, 22, 23.
pulchra, ii, 4, 5.
quadrifasciata, ii, 9-11.
vitrea, v, 13-15.

AMPHITETRAS.
antediluviana, lxiv, 39.
elegans,   "   35.
minuta,   "   40.
ornata, cviii, 4.

AMPHORA.
affinis, ix, 13.
arcuata, ix, 11.
areolata, iv, 15.
bigibba, iv, 7, 8.
binodes, ix, 9.
cingulata, iv, 27.
Clevia, iv, 21.
coffeaeformis, iv, 11, 19, 20; ix, 1.
contracta, ix, 21.
costata,   "   3.
crassa, iv, 36, 37.
cymbifera, iv, 10, 17, 18.
Delphinea, iv, 23, 24.
dubia, iv, 25, 26.
egregia, iii, 20, 21; iv, 1.
Eulensteinii, iv, 22.
excisa, iii, 4.
exsecta, iii, 15.
farcimen, iv, 2.
flexuosa, ix, 10; cxii, 12.
formosa, iv, 33, 34.
fusca, ix, 7.
gibba, ix, 18.
gigantea, ix, 20.
gigas, v, 4.
granulata, iii, 6; iv, 35.
Grevilliana, iv, 28.
Gründlerii, iv, 38, 39.
hyalina, ix, 2.
inflata, iv, 12, 13.

AMPHORA.
laevissima, ix, 4; iv, 3-6.
lanceolata, iv, 16; xxxi, 11.
larvis, iii, 22.
libyca, iii, 14; iv, 14.
lineata, ix, 5, 15-17.
Mexicana, iii, 24, 25.
micans, iii, 26.
micromata, xxxi, 14.
Nova-Caledonica, ix, 12.
obtecta, iii, 16, 28; ix, 6.
obtusa, iii, 29, 30.
ovalis, iv, 29-32.
pellucida, iii, 9, 17.
plicata, xxii, 25.
proteus, iii, 10, 11; ix, 19.
rectangularis, ix, 14.
Richardtiana, iv, 9.
rimosa, iii, 23.
robusta, iii, 1-3.
salina, ix, 8.
Schmidtii, iii, 5.
spectabilis, iii, 7, 8.
sulcata, iii, 28.
truncata, iii, 27.
turgida, iii, 18, 19.
veneta, iii, 12, 13.

ANAULUS.
birostratus, lxiv, 12-14.

ARACHNOIDISCUS.
Ehrenbergii, xci, 1-4.
Grevilleanus, " 5.
Indicus,   " 6, 7, 10, 11.
ornatus,   " 8, 9.

ASTERIONELLA.
Bleakeleyi, xlvi, 7.
formosa, xxxi, 16; xlvi, 1, 2, 4-6, 8.
notata, xlvi, 3.

ASTEROLAMPRA.
Brookii, xciii, 3.
Brebissoniana, xciii, 7.
Darwinii,   " 8, 9.
elegans,   " 11.
Grevillei, lxxxi, 5.
Hiltoniana, lxxxi, 1.
Marylandica, xciii, 1, 4-6.
Moroniensis, lxxxi, 2.
Ralfsiana,   " 3.
rotula, xciii, 10.
variabilis, xciii, 2.

ASTEROMPHALUS.
   arachne, lxxxi, 4.
   flabellatus, lxxxi, 6.

AULACODISCUS.
   argus, lxxx, 11.
   Brownei, lxxxviii, 10.
   circumdatus, " 7.
   crux, " 1.
   decorus, lxxv, 2.
   Kinkeri, lxxxviii, 11.
   Kittonii, lxxxii, 5; lxxxviii, 2.
   margaritaceus, lxxxii, 1, 2.
   Mölleri, lxxv, 1.
   Oregonus, lxxxviii, 4-6.
   Petersii, lxxxii, 3, 6, 7.
   probabilis,lxxx,12; lxxxviii,8.
   pulchir, lxxi, 1.
   Rogersii, lxxx, 10.
   sollitianus, lxxxviii, 9.
   sparsus, lxxxii, 4.
   Thumii, lxxxviii, 3.

AULISCUS.
   Americanus, lxxx, 9.
   Biddulphia, " 3, 4.
   caelatus, lxxxix, 6, 9, 10.
   Caribaeus, lxxx, 7.
   Clevei, lxxix, 8, 9, 10.
   confluens, lxxxix, 8, 11.
   Grunowii, lxxix, 11, 12.
   Hardmanianus, lxxix, 1.
   incertus, lxxx, 8.
   intestinalis, " 5.
   Johnsonii, " 1.
   Macraeanus, lxxxix, 2.
   mirabilis, " 12, 13.
   mutabilis, cviii, 8.
   Peruvianus, lxxxix, 4.
   pruinosus, lxxix, 13, 14; cxi, 5.
   punctatus, " 2, 3.
   racemosus, lxxxix, 7.
   radiatus, " 3.
   rectienlatus, lxxix, 5, 6, 7.
   Schmidtii, lxxxix, 5.
   sculptus, " 1.
   speciosus, lxxx, 2.
   spinosus, lxxix, 4; cxi, 6.
   Stockhardtii, lxxxix, 14.
   textilis, lxxx, 6.

BACILLARIA.
   paradoxa, lxviii, 32, 34.

BACTERIASTRUM.
   curvatum, lxviii, 20.
   furcatum, lxvii, 1.
   varians, " 2, 3, 4.

BERKELEYA.
   micans, xxxi, 15.

BIBLARIUM.
   clypeus, lxii, 25, 26.
   ellipticum, cxi, 13, 14.
   stylobiblium, xxii, 5, 6.

BIDDULPHIA.
   angulata, cx, 6, 7.
   aurita, xcvi, 9, 10, 11.
   Baileyi, xcviii, 1, 2, 3.
   Brittoniana, xcviii, 18, 19.
   Californicus, xcvii, 9.
   circinus, lxiv, 1.
   Cookiana, xcviii, 10.
   decipirus, " 5, 6.
   Edwardsii, xcvi, 4, 5.
   elegantula, xcviii, 7, 8, 9.
   gigas, ciii, 11, 12.
   Johnsonianus, xcvii, 6.
   laevis, xcvii, 1, 2; ciii, 8-10.
   longierucis, xcvi, 6.
   longispina, xcvi, 12.
   lunata, xxii, 29.
   membranacea, cx, 9, 10.
   Mobilensis, xcv, 9.
   occidentalis, lxiv, 11.
   ovalis, xcvii, 8, 10.
   polymorphus, xcvii, 5, 11.
   Porpeia, xcviii, 15, 16, 17.
   pulchella, xcvi, 1, 2, 3.
   quadricornis, lxiv, 8, 9.
   reticulata, xcvi, 13, 14.
   rhombus, xcv, 7, 10.
   Roperiana, xcv, 8, 11-13.
   seticulosa, xcvi, 7, 8.
   suborbicularis, xcv, 5, 6.
   tenuis, xcviii, 11, 12.
   trinacria, xcviii, 13, 14.
   Tuomeyii, xxii, 1; xcv, 1-4.
   turgidus, xcvii, 3, 4, 7.
   Weissflogii, lxxxviii, 3-5, 6.
   Woolmanii, xcviii, 4.

CAMPYLODISCUS.
   adornatus, lxxii, 3.
   ambiguus, lxxiii, 2.
   Americanus, lvi, 7.
   argus, lxx, 5, 6, 7.

CAMPYLODISCUS.
　bicostata, ex, 3, 4.
　bifurcatus, lxxii, 2.
　bimarginatus, lxxxiii, 8.
　Castilii, lvi, 9.
　clypeus, lxxiii, 3.
　concinnus, lxxii, 5, 6.
　costatus, lxix, 13, 14.
　crebrostriatus, lxxxiii, 9.
　cribrosus, lxx, 1, 2.
　ecclesianus, lxxxiii, 7.
　Echeneis, lxxiii, 5, 7.
　Ehrenbergii, lvi, 6.
　Greenleafianus, lxxxiii, 1.
　Gründerii,　　　"　　4.
　hibernicus, lxxi, 2, 3.
　Hodgsonii, lxx, 3.
　Humboldtii, lvi, 12; lxx, 12.
　imperialis, lxxii, 1.
　intermedius, lxxxiii, 6.
　latus, lvi, 10.
　limbatus, lxx, 10, 14.
　marginatus, lxx, 15.
　Mexicanus, " 13.
　Mulleri, lxxiii, 1.
　noricus, lxxi, 5, 7.
　parvulus, lxx, 8, 9.
　Phalangium, lxxxiii, 14, 15.
　punctulatus,　　"　　5.
　radiosus, lxx, 11.
　Ralfsii, lxxii, 4; lxxxiii, 10.
　rotula, lxxxiii, 11.
　Samoensis, lxx, 4, 16.
　Sauerbekii, lxxxiii, 12, 13.
　Schmidtii, lxxii, 6.
　simulans, lxxxiii, 16.
　stellatus, " 2.
　tabulatus, lxxiii, 4.
　triumphans, lxxxiii, 3.

CAMPYLONEIS.
　argus, xxxiii, 24, 25.
　Grevillei, " 28, 29, 30.
　regalis, " 39, 40.

CERATONEIS.
　arcus, xxx, 24, 25.

CHAETOCEROS.
　bacillaria, lxiii, 9, 10.
　boreale, lxv, 11.
　Californicum, lxv, 13.
　distans, lxii, 29.

CHAETOCEROS.
　didymus, lxv, 12, 14 16;
　　lxxvii, 16.
　incurvata, lxv, 9, 10.
　monicae, " 22.

CLIMACOSPHENIA.
　elongata, xlviii, 32, 33.
　moniligera, xxix, 7, 8.

COCCONEIS.
　ambigua, lxiii, 35, 36.
　Americana, xxxiii, 32.
　borealis, lxiii, 18.
　Californica, xxxiii, 19, 20.
　costata,　　　"　9, 10.
　decussata,　　"　31.
　dirupta,　　　"　15, 16.
　elongata, lxiii, 37.
　finnica, xxxiii, 36.
　gemmata, v, 7.
　Grevillei, xxxiii, 8; xxxiv, 8,
　　9, 10.
　interrupta, xxxiii, 5, 11.
　lineata, xxxiii, 6, 7; xxxiv, 4.
　marginata, xxxiii, 38.
　Mormonorum, xxxvii, 23.
　nitida, xxxiv, 3, 5.
　nitidulans, xxxiii, 37.
　oblonga, " 34.
　Oceanica, " 35.
　pediculus, " 21, 27, 33.
　pinnularia, " 41.
　placentula, " 17, 18.
　pseudo-marginata, xxxlii, 1.
　regalis, v, 11; xxxiv, 6, 7.
　rhombia, lxiii, 19.
　rhombifera, v, 8, 9.
　scutellum, xxxiii, 12, 13, 14,
　　22, 23.
　striata, xxxiii, 26.
　sulcata, v, 12.
　Thwaitesii, xxxiii, 3, 4.

COCCONEMA.
　asperum, xxxiv, 14.
　Janischii, vi, 5.
　lanceolata, vi, 14-16.
　lanceolatum, vi, 1-3, 11.
　Mexicanum, " 4.
　parvum,　　" 8, 9.

COLECTONEMA.
　vulgare, xxii, 28.

COSCINODISCUS.
  apiculatus, lxxxvi, 9; xciv, 13.
  argus, lxxxi, 8.
  armatus, lxxxvi, 2.
  asteromphalus, lxxxvii, 4.
  biangulatus,   " 1.
  borealis,   " 1.
  Californica, cviii, 7.
  contralis, lxxxvii, 6; xciv, 12.
  cocconeiformis, xciv, 4, 5.
  concavus, xciv, 3.
  Crassus, lxxiv, 1; lxxxvii, 2.
  denarius, cxii, 6.
  diorama, lxxxvii, 3.
  elegans, xciv, 1.
  excavatus, xc, 1, 2.
  excentricus, xc, 10.
  Floridulus,   " 5.
  gigas, lxxxi, 9.
  heteroporus, lxxxv, 6; xc, 12.
  impressus, xc, 9.
  incretus, cx, 1.
  intermedius, lxxiv, 13.
  leptopus, xc, 6.
  Lewisianus, xciv, 18.
  lineatus, lxxxvii, 10.
  liocentrum, lxxiii, 30.
  marginato-lineatus, xciv, 16.
  marginatus, cxii, 8; xciv, 21.
  marginulatus, xciv, 25, 26, 27.
  minor, xciv, 20.
  nitidulus, xciv, 10, 15.
  nitidus, xciv, 22, 23.
  noduliter, xciv, 7.
  Nottinghamensis, xciv, 24.
  Omphalanthus, lxxxvii, 8.
  patina, lxxiv, 6, 7.
  perforatus, lxxxvii, 9.
  punctatus, cviii, 11, 12.
  radiatus, lxxxi, 7; xc, 3.
  radiolatus, xciv, 6.
  radiosus,   " 11.
  robustus, lxxxvii, 5, 11.
  rotula, xciv, 17.
  sol, lxxiv, 2, 4.
  stelliger, xciv, 19.
  subconcavus, lxxxi, 10.
  subcelatus, xc, 11.
  subtilis, xciv, 8, 9.
  suspectus, lxxxvii, 7.
  symmetricus, xciv, 2.

COSCINODISCUS.
  velatus, xc, 7, 8; cxi, 12.
  Woodwardia, xc, 4.

COSMIODISCUS.
  tenuis, lxxv, 12.

CRASPEDODISCUS.
  Coscinodiscus, lxxxvi, 3, 8.
  elegans, lxxvii, 1, 2; lxxxvi, 7.
  Isoporus, lxxiv, 5.
  microdiscus, lxxiv, 3.
  Oculis-Iridis, lxxxvi, 4.
  rhombicus, lxxxvi, 5.

CYCLOTELLA.
  antiqua, lxvi, 18, 19.
  compta, lxiii, 30, 31; lxvi, 20, 21; cx, 5.
  Kützingiana, lxvi, 8, 9.
  Meneghiniana, lxvi, 13-15.
  operculata,   " 22-24.
  physoplea, lxii, 20, 22.
  rotula, lxvi, 10-12.
  striata,   " 16, 17.

CYMATOPLEURA.
  angulata, lx, 16.
  apiculata,   " 9, 12.
  Campylodiscus, lxxi, 9.
  elliptica, lx, 5-7.
  ellipticum, cviii, 5, 6.
  Hibernica, lx, 8, 10, 11.
  marina,   " 14, 15.
  solea,   " 1-4, 13.

CYMBELLA.
  affinis, vii, 31, 32.
  Americana, vii, 24.
  amphicephala, vii, 15.
  anglica,   " 33, 34.
  cistula, vi, 6, 7, 12; vii, 29, 35.
  curta, vii, 30.
  cuspidata, vii, 16, 20.
  Cymbiformis, v, 10; vii, 28.
  Ehrenbergii, vii, 3, 21, 25.
  excisa,   " 9, 10.
  Gastroides,   " 4, 5, 12, 13, 15.
  gracelis,   " 26.
  helvetica,   " 6.
  Heteropleura,"   1, 2.
  Kamtschatica,"   11.
  leptoceros,   " 8.
  maculata,   " 22.
  naviculiformis, vii, 7, 14.
  parva,   " 19.

CYMBELLA.
    rotundata, cxi, 19.
    Sporangial growth, i, 10-12.
    Stodderi, lxiii, 34.
    stomataphora, vii, 17, 18.
    turgidula,   " 27.
    ventricosa, vi, 13.
CYMBOSIRA.
    Agardhii, xxix, 5, 6.
DIMEREGRAMMA.
    fossile, lxiii, 27
    fulvum, v, 23.
    marinum, xlv, 4, 5.
    Novæ Cæsareæ, v, 16, 22.
DENTICULA.
    antillaria, lxiii, 32, 33.
    lauta, xlvi, 10; lxviii, 16, 17.
    thermalis, xlvi, 9; lxviii, 14, 15.
    valida, lxviii, 9-11.
DIATOMA.
    anceps, xlvi, 11, 17, 18.
    Ehrenbergii, xlvi, 16, 21, 22.
    elongatum,   " 15, 26, 27.
    pectinale,   " 19, 20.
    tenue,   " 23-35.
    vulgare,   " 12, 14.
DICLADIA.
    capriolus, lxiv, 5, 7.
    clathrata,   " 2.
    mitra,   " 3, 7.
DICTYOCHYA.
    ? xlviii, 25-28.
DISCLOPEA.
    Oregonica, cxii, 14.
DISCUS.
    porcelaineus, lxxiv, 17.
    unbenant, lxxviii, 15.
ENCYONEMA.
    Auerwaldii, viii, 21.
    caespitosum,   " 20.
    gracile,   " 16, 17.
    paradoxum,   " 23.
    prostratum,   " 24.
    triangulatum,"   22.
    ventrico,   " 18, 19.
ENTOMOGASTER.
    Woodwardii, v, 25.
EPITHEMIA.
    argus, xxxv, 14-17; xxxvi, 1-3.
    constricta, xxxv, 20, 21.
    gibba,   " 1-3, 8, 9.

EPITHEMIA.
    gibberula, xxxv, 26, 28.
    granulata,   " 4, 5.
    Hyndmanii, "   6, 7.
    marina, xxii, 23, 24.
    musculus, xxxv, 18, 19; lxviii, 27.
    ocellata, xxxv, 22, 23.
    sorex, xxxiv, 13; xxxv, 24, 25.
    succincta, xxxiv, 12.
    tentricula, xxxv, 34
    turgida,   " 10-13.
    ventricosa, i, 7, 8.
    zebra, xxiii, 31, 32; xxxv, 29-33.
EUCAMPIA.
    Virginica, lxvi, 32.
    Zodiacus,   " 34, 35.
EUNOTIA.
    Americana, lxxviii, 12.
    Bactriana, xxxviii, 10.
    biceps, lxviii, 23.
    bidentula, xxxviii, 9.
    cygnus, xxxvii, 5.
    decaodon, xxxvi, 14.
    declives,   " 23.
    depressa, xxxvii, 3.
    diodon, xxxvi, 5, 6, 21, 22.
    dizyga, xxxviii, 7.
    formica,   " 20, 21.
    gibbosa,   " 8.
    heptodon, xxiii, 23.
    impressa, xxxviii, 22.
    incisa,   " 24.
    lunaris,   " 16-19.
    major,   " 1-4, 14.
    monodon, xxxvi, 4, 12.
    mosis, xxxviii, 30-32.
    parallela,   " 15.
    pectinalis, xxxvi, 15-20; xxxviii, 12, 13.
    pentaglyphis, xxxviii, 29.
    polyglyphis,   " 11.
    polydon, xxxvii, 1.
    praerupta, xxxviii, 5, 6.
    robusta, xxxvi, 9-13.
    sella, xxiii, 29.
    serra, xxxvii, 2.
    serrulata, xxxviii, 33, 34.
    tridentula,   " 25, 26.
    triodon, xxxvi, 7, 8.
    undinaria, xxxviii, 35.
    zygodon,   " 27, 28.

EUODIA.
  gibba, lxviii, 26.
  Janischii, cv, 19, 20, 21.
EUNOTOGRAMMA.
  debilis, lxxxv, 16.
  laevis,   "   17-19.
EUPODISCUS.
  argus, lxxvi, 1, 2, 4.
  Californicus, lxxvi, 7.
  oculatus,   "   5.
  radiatus, lxxvi, 6, 11; lxxvii, 8.
  Rogersii, " 3; cxi, 17.
FRAGILARIA.
  amphicephala, xlvii, 21.
  biceps,   "   9.
  binodes,   "   8.
  brevistriata, "   15.
  Californica, "   22.
  capucina,   "   1-3.
  construens,   "   13, 14, 20.
  Dimeregramma, xxxvii, 4.
  entomon, xlvii, 26.
  Harrisonii, "   12.
  inflexa, v, 21.
  mutabilis, xlvii, 18, 19.
  Newberyii, v, 20.
  Pacifica, xlvii, 23-25.
  paradoxa,   " 16, 17.
  parasitica, " 10.
  Tremontii, v, 17.
  turgens, xlvii, 11.
  venter, v, 19.
  verescens, xlvii, 4-7.
GEPHYRIA.
  constricta, lxi, 3.
  gigantea, " 1, 2.
  Japonica, xxxiv, 15, 16.
  media, lxi, 4, 5.
GLYPHODESMIS.
  distans, xlv, 21, 22.
  Williamsonii, xlv, 23, 24.
GLYPHODISCUS.
  Grünowii, lxxvi, 9.
  stellatus, " 8, 10.
GOMPHONEMA.
  acuminatum, xxviii, 22-27.
  Americanum, xxiii, 22, 28.
  augur, xxviii, 17, 18.
  capitatum, xxii, 19; xxvi, 6, 7;
    xxviii, 16, 20, 21.

GOMPHONEMA.
  constrictum, xxvi, 12-15;
    xxviii, 15, 19.
  cristatum, xxvii, 3, 11.
  dichotomum, xxvi, 16, 17.
  elegans, xxvii, 8.
  geminatum, xxvi, 1-4, 10, 11.
  gracile, xxvii, 6, 7.
  intricatum, xxvii, 4.
  mammilla, xxvi, 5.
  maximum, " 8, 9.
  Mexicanum, xxvii, 12.
  olevaceum, " 14, 15.
  olor, xxix, 19.
  Oregonicum, xxii, 8, 9.
  sarcophagus, xxvii, 10.
  semapertum, " 16.
  sphaerophorum, xxvii, 1, 2.
  tenellum,   " 13.
  turgidum,   " 9.
  vibrio,   " 5.
GONIOTHECIUM.
  didymum, lxiv, 31.
  monodon, " 30.
  obtusum, " 33.
  odontella, " 17, 18.
  Rogersii, xxxiv, 17, 18, 19; xxv,
    20; lxiv, 32.
GRAMMATIPHORA.
  angulosa, xlix, 11-13.
  Caribaea, " 19.
  gibba,   " 6, 7.
  marina,   " 8-10, 14-16.
  maxima,   " 26.
  Mexicana, " 24, 25.
  serpentina, xlix, 1, 20-23.
  stricta,   " 17, 18.
  undulata, xlix, 3-5.
HANTZSCHIA.
  Amphioxys, li, 15-17.
HELIOPELTA.
  nitida, lxxi, 4.
HEMIAULUS.
  affinis, lxiv, 15, 16.
  bifrons, lxiv, 26, 28, 29.
  Californicus, lxiv, 34.
  polycestenorum, lxiv, 19, 20,
    24, 25, 27.
  polymorphum, xxv, 23, 24, 25.
HERCOTHECA.
  mammilaris, lxiv, 22, 23.

HOMOCLADIA.
    capitata, xxix, 13, 15.
    filiformis, xxix, 16-18.
HYALODICTYA.
    Dianæ, lxviii, 31.
HYALODISCUS.
    laevis, lix, 2.
    maximus, lix, 7, 8.
    retienlatus, cx, 8.
    Stelliger, lix, 6.
    subtiles,   " 3.
    Whitneyi, " 5.
HYDROSERA.
    triquetra, cv, 9-12.
ISTHMIA.
    enervis, cix, 6-9.
    nervosa, " 1-5.
LICMOPHORA.
    Californica, xlvii, 22, 32a.
    flavellata, xxix, 1, 2; xlvii, 27-29.
    gracilis, xlvii, 30, 31.
    Jurgensii, xxix, 12; xlvii, 33, 34.
    papeana, xxii, 26, 27.
    tincta, xlvii, 35, 37.
LIOSTEPHANIA.
    compta, lxii, 21.
LIRADISCUS.
    minitus, lxiii, 14-16.
LITHODESMIUM.
    contractum, lxii, 28.
MASTOGLOIA.
    angulata, cvii, 10.
    apiculata, v, 24.
    elegans, cvii, 8.
    exigua, xxx, 15, 16.
    Grevillei, " 17, 18.
    Kinsmanii, xxv, 26, 27.
    lanceolata, xxx, 19.
    Smithii,   " 21, 23.
    submarginata, cvii, 11.
MASTOGONIA.
    crux, lix, 1.
    heptagonia, lix, 9.
    sexangulata, lxii, 18.
MELOSIRA.
    Americana, lvii, 10.
    arenaria, lviii, 5-7; cxii, 2.
    Baileyi, lxxiv, 9, 10.
    Borreri, lviii, 8-11.

MELOSIRA.
    clavigera, lviii, 16.
    crenulata, lvii, 16-20.
    crotonensis, lvii, 6.
    distans, lvii, 20-32, 36.
    granulata, lvii, 7-9, 21, 22, 33.
    horologium, lxiii, 20, 21.
    Huttonia, ciii, 13.
    nummuloides, lvii, 1-5.
    orichalcea, lviii, 17, 18.
    scalaris, lvii, 23-25, 28.
    sculpta,   " 34, 35; lviii, 3, 4.
    sol, lviii, 1, 2.
    spiralis, lvii, 26, 27, 29.
    sulcata, lviii, 12-15.
    undulata, cxii, 4, 5.
    varians, lvii, 11-15.
MERIDION.
    circulare, xxxvii, 24, 25, 27, 28; cxi, 15, 16.
    intermedium, xxxvii, 26, 29, 30.
NAVICULA.
    acrosphaeria, xxi, 8, 9, 11.
    acuminata, xvi, 15; xx, 5.
    affinis, xiv, 1, 2.
    amphigomphus, xviii, 10; xix, 5.
    amphionys, xxiii, 24, 25.
    Americana, x, 51.
    amphirhyncus, xiv,3,4; xix,13.
    amphisbaena, xix, 23.
    angelorum, xi, 3.
    Anglica, x, 1.
    angusta, x, 34.
    apiculata, xiv, 13, 14.
    apis, xvi, 19.
    approximata, xvi, 18.
    aspera, xx, 4.
    Bacillum, x, 29, 30.
    Baileyana, xxiv, 3, 4.
    binoides, xvii, 23.
    bioculata, xii, 13.
    Bleischii, xviii, 9.
    Boeckii, xxv, 8, 9.
    Bohemica, ix, 46.
    Bomboides, xii, 17.
    Bombus, x, 47, 50; xxiii, 3.
    Borealis, ix, 23, 24.
    brasiliensis, xiv, 21.
    Braunii, lxviii, 13.
    Brebissonii, x, 26; xix, 16-19.
    brevis, x, 7, 17.

NAVICULA.
Californica, xiv, 17, 24; xvi, 1.
campylodiscus, xii, 15.
carassius, xiv, 20.
cardinalis, x, 45; xii, 18.
Caribaea, xvi, 54.
carinifera, " 15.
Chersonensis, xv, 19.
circumsecta, ix, 27.
cluthensis, xi, 10; xvi, 27.
coarctata, xv, 3, 11.
cocconeiformis, x, 14.
columnaris, xviii, 1, 6.
commutata, xvii, 18.
confecta, xv, 12.
Comperii, xvi, 7; xvii, 2.
costata, xxiv, 6, 7.
crabro, x, 33; xv, 7.
crabromiformis, xv, 9.
cruciata, xi, 6.
cryptocephala, x, 13.
cruciformis, xxiv, 19.
cuspidata, xii, 16.
Cynthia, ix, 25, 26.
dactylus, xxi, 1, 2, 3, 4.
Dariana, xvii, 5.
decurrens, ix, 38, 39; xxv, 3.
denta, xv, 15.
De Wittians, xxii, 18.
dicephala, xvii, 21; xx, 22.
didyma, xii, 12; xix, 24, 25.
diffluens, xvi, 21.
diffusa, " 11.
digitus, xxiv, 1.
dilatata, xviii, 5, 7.
diomphala, xxiv, 8.
diplosticha, xii, 3.
dirhynchus, xi, 18, 19;
  xxv, 6.
dirrhombus, xii, 14.
discrepans, x, 4.
disphenia, xxiii, 9.
distenta, xvi, 10.
divergens, xix, 20, 21.
Division of Diatoms, i, 9–13,
  14, 15.
Donkinii, xv, 17.
dubia, x, 27; xvii, 22.
duplicata, xxiii, 8.
elegans, ix, 22; xxiv, 9.
Elginensis, xx, 22.
elliptica, xiv, 18, 19.
elongata, ix, 37.
4

NAVICULA.
entomon, xii, 6; lxiii, 1.
esox, xxiii, 14.
excavata, xiv, 15, 27.
excentrica, xviii, 2.
excenta, xv, 4.
exilis, x, 40.
Eugenia, x, 10.
eudonia, xiv, 6.
Febegerii, xi, 8.
firma, xix, 9, 10.
Fischeri, xiv, 11.
Floridana, xi, 7.
forcipata, xvi, 13, 20.
formica, xxiv, 12.
formosa, xix, 2, 11.
fusca, xii, 25.
fusidum, xxiii, 10.
futilis, xvii, 8.
gastrum, x, 2.
gemina, xi, 12, 13; xii, 8.
gemmata, x, 46.
gibba, xx, 9, 10, 11; lxviii, 8.
gibberula, x, 35, 38; xxii, 13.
Giebilli, xv, 8.
gigas, xiii, 5.
globiceps, ix, 33.
gracilis, x, 22, 23.
Greenlandica, xi, 5.
granulata, xvii, 14.
Gründleri, cxi, 8, 9.
hemiptera, xvii, 19, 20;
  xx, 6.
Hennedyi, xiv, 28, 29, 30.
hexapla, xix, 4.
Hitchcockii, xix, 30, 31.
humerosa, xvii, 12, 15.
impressa, xiv, 25, 26.
incomperta, xxv, 2.
Indica, xvii, 11.
inflata, xvi, 23; xix, 29.
inflexa, x, 25.
interposita, xi, 14.
interrupta, x, 31, 32;
  xii, 4.
iridis, xviii, 4.
irrorata, xvi, 2, 17.
Isocephela, xxiii, 6.
Johnsonii, " 13.
Kutzingii, x, 48.
lacrimans, xii, 28; xv, 1.
lacunarum, x, 39.
lanceolata, ix, 32, 42.

NAVICULA.
  lata, xxiii, 26, 27.
  latissima, xvii, 24, 25.
  leptostigma, xxiv, 14.
  Lewisiana, " 2.
  liber, xxiii, 5, 11.
  linearis, xix, 6.
  littoralis, x, 11, 12.
  longa, xii, 23, 24.
  lyra, xvi, 6, 9, 14, 26.
  maculata, xvii, 1; exi, 1, 2.
  marginata, xi, 16; xxiii, 2; exi, 7.
  marginulata, xi, 10.
  marina, xiv, 7, 23.
  mesogongyla, xxv, 1.
  mesolepta, xx, 3, 13, 14, 16.
  Moesta, xix, 26.
  Mormonorum, xix, 7.
  multicostata, xv, 2.
  musceaeformis, xii, 7.
  nitescens, xiv, 8, 22.
  nobilis, xiii, 2, 4, 6.
  nodosa, xix, 14.
  nodulosa, xxiv, 16.
  notabilis, x, 49.
  nummularia, ix, 40, 43; x, 41, 42, 44.
  oblonga, xxi, 10.
  obtusa, xx, 19, 20.
  oculata, x, 36.
  Oregonoea, xi, 15.
  ornata, xv, 20.
  oscitans, xvii, 7.
  palpebralis, x, 6, 16.
  pandura, xv, 5.
  papula, ix, 11.
  parca, x, 5.
  parva, x, 3.
  Pelagi, xiv, 16.
  pelliculosa, ix, 31.
  pennata, xiv, 9, 10.
  Pensacola, xi, 9.
  peregrina, xii, 20, 22.
  permagna, xi, 2; xvii, 3, 4.
  placentula, ix, 35.
  polysticta, ix, 28.
  polyonca, xi, 17; xxiv, 13; lxiii, 2.
  Powelli, xi, 4.
  praestes, xix, 28.
  prestiophora, xxiii, 1.
  pretexta, xx, 1, 7; exi, 3.

NAVICULA.
  prisca, xv, 6, 13.
  probabilis, xix, 1.
  producta, " 3, 12, 15.
  puella, xv, 10, 16, 18.
  pumila, xxiv, 15.
  pupula, xxiii, 19, 20.
  pusilla, x, 18, 19.
  pygmaea, xvi, 12.
  quinquenodis, x, 8, 9.
  radiosa, xxi, 6, 7.
  rectangulata, ix, 36.
  Reinhardtii, xxiii, 17.
  retusa, xvii, 16, 17.
  rhombica, xxiv, 17, 18.
  rhomboides, xvii, 9, 10; xxii, 21.
  rhyncocephela, ix, 34; x, 21.
  Robertsiana, xvi, 8.
  salva, xii, 9.
  saugeri, x, 43.
  Schmidtiana, xii, 19, 21.
  Schultzei, xxiv, 5.
  Schumanniana, xxii, 14.
  scintillans, xvi, 22.
  scoliopleura, xii, 27.
  sculpta, ix, 44, 45.
  sejuncta, x, 24.
  semen, xx, 21.
  separabilis, xix, 27.
  serians, x.
  serratula, xiv, 5.
  Smithii, xii, 11; xiv, 12; xx, 8.
  sigma, xxiv, 10, 11.
  simulans, xxiii, 4.
  Sillimanorum, xi, 1; xxiii, 12.
  singularis, xx, 2, 18.
  solaris, xvii, 6.
  spectabilis, xvi, 3.
  sphaesophora, xvii, 13.
  splendida, xii, 5; xv, 14.
  Stauntonii, xxv, 13, 14, 15.
  stauroneiformis, xxv, 10.
  stauroneis, xx, 4.
  subinflata, xxii, 11.
  suborbicularis, x, 37.
  subnuda, xii, 2.
  suspecta, " 26.
  tabellaria, x, 28; xii, 1; xx, 12, 17; xxi, 5.
  tenella, xxii, 15, 16.

NAVICULA.
   termis, ix, 29, 30; xxiv, 21, 22.
   trinodis, lxviii, 12.
   truncata, xxiii, 15, 16.
   tumescens, xviii, 3.
   tumidula, xxiv, 23.
   tuscula, xxii, 10.
   undetermined, xiv, 20.
   undosa, xxiii, 18.
   undulata, xxii, 12.
   vacillans, xii, 10; xix, 22.
   velox, xxiii, 21.
   veneta, " 7.
   viridis, xiii, 1, 7-10.
   viridula, x, 20.
   vulpina, cxi, 10.
   Weissflogii, xvi, 25.

NITZCHIA.
   amphionys, li, 20.
   angularis, xliii, 24-26.
   angustata, xliv, 18, 19.
   biblobata, xliii, 5, 6, 15; cxii, 15.
   Brebissonii, xlii, 6, 8.
   Campeachiana, xliii, 2, 3.
   circumsuta, xliv, 7.
   closterium, xl, 10-13, 18.
   coarctata, " 10.
   communis, " 24, 25.
   cursoria, xxii, 17; xliii, 18.
   denticula, xliii, 31-33.
   dissipata, xlii, 19, 20.
   dubia, xliv, 5, 6, 11.
   epithemioides, xliii, 12-14.
   fasciculata, xl, 14-17.
   Febigerii, xliii, 1.
   fluminensis, xliii, 20.
   frustulum, xl, 22, 23.
   granulata, xliii, 11.
   Hungarica, xliv, 12, 13.
   Kittonii, xliii, 30.
   lanceolata, xl, 8, 9.
   levidensis, xliv, 8.
   limnicola, xliii, 10.
   linearis, xlii, 17, 22.
   littorea, xliii, 16, 17.
   longissima, xl, 2, 3.
   major, xlii, 4, 5.
   majuscula, xliii, 19.
   marina, xliv, 16, 17.
   obtusa, xlii, 7, 18.
   palia, xl, 26, 27.

NITZCHIA.
   panduriformis, xliv, 3, 4, 9.
   paradoxa, xliii, 27-29.
   plana, xliv, 1, 2.
   punctata, xliv, 20-22.
   scalaris, xxxiv, 1; xliii, 9.
   scaligera, xliii, 4.
   scutellum, xliv, 24.
   sigma, xl, 4-7; xlii, 10, 11.
   sigmoidea, xliii, 9.
   sinuata, xliii, 21.
   spathalata, xliii, 7, 8.
   spectabilis, xl, 1.
   tabellaria, xliii, 22, 23.
   tennis, xl, 19, 20.
   tryblionella, xliv, 23.
   valida, xlii, 1-3.
   vermicularis, xlii, 12-15.
   virgata, xxv, 16, 17.
   vitrea, xlii, 16, 21, 23.
   vivax, xliv, 14, 15.

ODONTELLA.
   obtusa, lxii, 23, 24.

ODONTIDIUM.
   hyemale, xlviii, 5, 6.
   mesodon, " 1-4.
   mutabile, " 7-12.
   tabellaria, " 15-20.

ODONTODISCUS.
   subtilis, lxxxiv, 6.

PERIPTERA.
   capra, lxiv, 36.
   chlamydophora, lxvii, 5.
   tetracladia, " 17-19.

PERISTEPHANIA.
   Baileyi, lxxiv, 15.
   eutycha, lxiv, 18.

PLAGIOGRAMMA.
   Californicum, xlv, 16, 17.
   Gregorianum, " 10, 11.
   obesum, " 12-14.
   ornatum, " 20.
   pulchellum, " 1, 2, 3.
   pygmaeum, " 6, 7.
   tessellatum, " 18, 19.
   validum, " 8, 9.
   ? lxvi, 31.

PLECROSIGMA.
   aestuarii, xxxii, 10.
   angulatum, xxx, 5, 9.

PLEUROSIGMA.
    attenuatum, xxx, 2.
    Balticum, xxxii, 5, 6.
    decorum, " 4.
    delicatulum, xxii, 22.
    elongatum, xxxii, 2.
    eximium, " 14.
    Fasciola, " 11.
    formosum, " 8, 9.
    Hippocampus, xxxii, 7.
    Kützingii, " 13.
    Macrum, " 1.
    nubecula, xxx, 4.
    obscurum, " 1; xxxii, 12.
    obtusatum, " 13, 14.
    pulchrum. xxxii, 3.
    quadratum, xxx, 10, 11.
    rigidum, " 3.
    sciotense, " 12.
    Spenceri, " 6, 8.
    virginiacum, xxxi, 7.
PODOCYSTIS.
    Americana, li, 14, 18.
    Adriatica, li, 19.
PODOSIRA.
    argus, lix, 4.
    Febigerii, lix, 10, 12, 13.
    hormoides, lix, 14-16.
    maculata, lxix, 4, 5.
    Montagnei, lix, 17, 18.
    stellulifera, " 11.
    variegata, lxix, 3.
PODOSPHENIA.
    Baileyi, lxix, 1, 2.
    papeana, xxii, 26, 27.
PORPEIA.
    quadrata, lxvi, 33.
    quadriceps, lxvi, 25, 30.
PSEUDO-AULISCUS.
    radiatus, lxxxiv, 6.
PYXIDICULA.
    compressa, lxii, 17.
    cristata, " 9.
    cruciata, " 7, 11.
    gigas, " 30.
    globata, xxii, 2, 3, 4.
    lens, lxii, 10.
    limbata, lxii, 8.
    urceolaris, lxii, 19.
PYXILLA.
    Americana, lxv, 8, 17, 18.

PYXILLA.
    boreale, lxv, 20, 23.
    dubia, " 24.
    Kittoniana, lxv, 21.
    subulata, " 19.
RAPHIDODISCUS.
    Christiana, lxxxiv, 2.
    Febigerii, " 3, 4.
    Marylandica, " 1.
RAPHONEIS.
    affinis, lxiii, 25.
    amphiceros, xxxvii, 18 22.
    Archeri, lxiii, 29.
    Belgica, lxviii, 3.
    cocconeis, xxxvii, 11.
    fluminensis, lxviii, 5.
    fuscus, lxiii, 39.
    gemmifera, xxxvii, 17.
    linearis, lxiii, 28.
    Oregonica, lxviii, 4.
    Petropolitana, lxiii, 26.
    pretiosa, xxxvii, 15.
    scalaris, " 16.
RHABDONEMA.
    adriaticum, li, 2, 3, 4.
    arcuatum, li, 5, 6, 7.
    Atlanticum, lxiii, 17.
    frustule, xxxiv, 20.
    minutum, li, 8-11.
RHIPIDOPHORA.
    paradoxa, xxix, 3, 4.
RHIZOSOLENIA.
    Americana, lxv, 3-5.
    calypta, " 6.
    Eriensis, " 2.
    gracilis, " 1.
    ornithoglossa," 7.
    pileolus, lxiv, 21.
    styliformis, lxviii, 28, 29.
RHOICOSPHENIA.
    curvata, xxvii, 17, 18, 19, 20, 21.
RUTILARIA.
    epsilon, lxviii, 1.
    hexagona, " 2.
SCEPTRONEIS.
    caduceus, xxxvii, 13.
    gemmata, " 14.
    nitzschioides," 12.
SCHIZONEMA.
    Americanum, xxviii, 13.
    comoides, " 11, 12.

SCHIZONEMA.
　cruciger, xxviii, 9, 10.
　divergens, " 14.
　Grevillei, " 5, 6.
　implicatum, " 7, 8.
　ramosissima," 1, 2.
　Smithii, " 3, 4.
　vulgare, xxii, 20.

SCOLIOPLEURA.
　antillarum, xxxi, 12.
　Jennerii, " 13, 17.
　latestriata, " 8.
　tumida, " 9, 10.

STAURONEIS.
　acuta, viii, 11.
　anceps, viii, 4, 8, 9.
　apiculata, xxv, 7.
　Baileyi, xxiv, 24.
　dilatata, viii, 1.
　gracile, lxviii, 25.
　gracilis, viii, 10.
　Icostaurum, cxii, 9.
　inflata, xxv, 11.
　legumen, viii, 5.
　linearis, " 3.
　lineata, cxii, 11.
　monogramma, lxviii, 6, 7.
　Phoenecenteron, viii, 12, 15.
　phyllodes, cxii, 9.
　producta, viii, 7.
　pteroidea, xxiv, 26.
　punctata, viii, 1.
　pusilla, lxviii, 24.
　salina, viii, 6.
　sigma, xxiv, 25.
　staurophaena, xxiv, 24.
　Stodderii, xxv, 12.

STEPHANODISCUS.
　astraea, cxii, 13.
　carconensis, lxvi, 26, 27.
　lineatus, lxii, 4.
　Niagara, lxvi, 28, 29.

STEPHANOGONIA.
　actinoptycus, lxvii, 6.
　Californicum, " 11, 12.
　polygona, " 8, 15, 16.

STEPHANOPYXIS.
　aculata, lxii, 5.
　apiculata, lxii, 16.
　appendiculata, lxii, 12–15.
　Campeachiana, lxvii, 9, 10.

STEPHANOPYXIS.
　corona, lxii, 1, 6; lxvii, 20.
　crassipina, lxii, 3.
　ferox, lxvi, 5, 6, 7.
　limbata, lxvii, 7.
　rudis, " 21.
　spino, lxii, 2.
　spinosissima, lxxxiv, 5.
　superba, " 8, 9.
　turgida, lxvi, 1, 2.
　turris, " 3, 4

STICTODESMIS.
　craticula, lxix, 11.

STICTODISCUS.
　Californicus, lxxv, 5–8.
　Grevillianus, lxxi, 8.
　Hardmanianus, lxxv, 3.
　Kittonianus, " 9, 10.
　simplex, " 4.
　? " 13–15.

STRIATELLA.
　unipunctata, li, 1, 12, 13.

SURIRELLA.
　anceps, liii, 24, 25.
　angusta, lii, 16, 17.
　apiculata, liii, 9, 10.
　arctissima, " 27.
　Baileyi, " 22, 23.
　Beldjeckii, " 4.
　bifrons, liv, 2.
　biseriata, lii, 1, 2.
　campylodiscus, cviii, 13–15.
　cardinalis, liv, 6.
　cocconeis, v, 10.
　crenulata, lvi, 5.
　cruciata, lii, 6, 7.
　crumena, liii, 5, 6.
　delicatissima, liii, 17.
　elegans, i, 1; liv, 1; lv, 3, 10, 11.
　engiypta, liii, 20, 21.
　fastuosa, xxxiv, 21; lii, 10; lv, 5.
　Febegerii, lv, 1.
　fluminensis, lii, 9.
　gemma, lii, 11.
　Geroltii, lvi, 3, 4.
　inducta, liii, 14; lv, 2.
　intermedia, liii, 26.
　laevigata, lv, 12.
　leptoptera, lvi, 2.
　liosoma, v, 6.
　Mexicana, liii, 1–3.

SURIRELLA.
  microcora, liii, 11.
  minuta, " 18, 19; lvi, 8.
  Mississippica, v. 5.
  Molleriana, lii, 14, 15.
  nobilis, cxi, 11.
  Norwegica, lii, 4, 5.
  oblonga, " 12, 13.
  ovata, liii, 15, 16.
  ovalis, liii, 7, 8.
  panduriformis, lvi, 11.
  pulchra, cvii, 7.
  Ratrayi, lxix, 10.
  recedens, lii, 3, 8.
  reflexa, lvi, 1.
  regina, lv, 7.
  robusta, liv, 8.
  Saxonica, lv, 4.
  scutis, liv, 4.
  spiralis, liii, 12, 13.
  splendida, i, 2-6; liv. 3, 7; lv, 8, 9.
  striatula, lv, 6.
  valida, liv, 5.
SYMBOLOPHORA.
  acuta, xxii, 7.
  trinitatus, lxxviii, 14.
SYNDENDRIUM.
  diadema, lxiii, 11-13, 38.
SYNEDRA.
  acus, xli, 2-4.
  acuta, xli, 21-23, 30.
  affinis, xli, 10-12, 33, 34.
  amphyrhynchus, xli, 15.
  biceps, xli, 13, 14.
  capitata, xxxix, 5.
  Chasei, v, 1, 2.
  crotonensis, xli, 19, 20.
  crystallina, xxxix, 2, 3.
  Danica, xl, 28-30; xli, 5, 6.
  fulgens, xxxix, 4, 9.
  Gallionii, " 10, 11.
  Goulardii, xxv, 18, 19.
  investiens, xli, 27, 29.
  invertens, xxxix, 15.
  lanceolata, xl, 21; xli, 9.
  longiceps, xl, 31.
  longissima, xxxix, 7; xl, 32.
  obtusa, xli, 7.
  onyrrhynchus, xxxix, 12.
  pulchella, xli, 16-18, 26.
  radians, xxxix, 17.

SYNEDRA.
  robusta, xxxix, 1.
  rumpens, " 13, 14.
  spathulifera, xli, 1.
  superba, xli, 24, 25, 32.
  ulna, xxxix, 8; xli, 31.
  undulata, xxxix, 6.
  valens, v, 18.
  vaucheriæ, xxxix, 16.
  vitrea, xli, 8.
SYRINGIDIUM.
  Americana, lxiv, 38.
  simplex, " 37.
SYSTEPHANEA.
  Raena, lxiii, 7.
TABELLARIA.
  fenestrata, 1, 1-6.
  flocculosa, 1, 7-11.
  nodosa, lxviii, 22.
  robusta, " 22.
TERPSINOE.
  Americana, lxiii, 6.
  intermedia, lxxviii, 1, 2.
  magna, lxi, 7.
  minima, lxi, 12.
  musica, " 6, 13-15.
  tetragramma, lxi, 8, 9, 10, 11.
TESSELA.
  interrupta, xlviii, 14.
TETRACYCLUS.
  emarginatus, 1, 12, 15, 16.
  lacustris, 1, 13, 14, 17-20.
TOXONIDIA.
  Gregoriana, lxix, 12.
TRICERATIUM.
  aculeatum, xxv, 21.
  alternans, c, 16, 18; cvi, 2.
  amblyceros, lxxvii, 3.
  Americanum, cv, 7; cxi, 18.
  amoenum, ci, 14.
  amphitetras, civ, 2.
  antillarum, ci, 1.
  arcticum, cv, 5, 8; cvi, 7.
  biquadratum, civ, 10, 17.
  Brownianum, ci, 10.
  bullosum, c, 3.
  Californicum, xcix, 7; cv, 13.
  Campeachianum, cx, 2.
  cinnamomium, lxxv, 11; cv, 18.
  condecorum, cii, 6.
  consimile, xcix, 10; cvi, 6.

TRICERATIUM.
  dubium, cii. 13.
  elegans, ci, 6, 7.
  Ehrenbergii. civ, 13, 16.
  favus, xcix, 1, 2.
  fimbriatum, xcix, 34.
  Fischeri, cii, 8.
  fraglich? cvi, 9, 11.
  grande, civ, 1.
  Grun, xcix, 5, 6.
  Harrisonianum, c, 13, 14.
  Heilprinianum, lxxviii, 8, 9.
  heteroporum, c, 6.
  indentatum, lxxviii, 7.
  inelegans, xcix, 8, 9, 12; ciii, 4.
  interpunctatum, c, 8.
  irregulare, c, 8.
  jucatense, cii. 4.
  Kainii, lxxviii, 10; civ, 8; cv, 16.
  Kittonianum, ci, 2.
  lithodesmium, civ, 18, 19.
  Marylandicum, ci, 8.
  membranaceum, cv, 6.
  Montereyii, cii, 1; cvii, 6, 9.
  muricatum, cii, 7.
  obliquum, cvi, 10.
  obscurum, cii, 10.
  obtusum, civ., 7; cviii, 3.
  ornatum, ci, 15.
  parallelum, c, 11, 12, 15.
  pentacrinus, lxix, 6, 7; ci, 16.
  pileus, cviii, 2.
  pileolus, cviii, 1.
  punctatum, ci, 9, 12.

TRICERATIUM.
  quadrangulare, cvi, 8.
  receptum, ci, 13.
  regina, cv, 17.
  reticulum, civ, 5, 6; cviii, 16.
  robustum, ci, 11.
  scitulum, cvi, 1.
  sculptum, " 3, 5; cxi, 4.
  secernendum, c, 17.
  semicirculare, cv, 2, 3, 4.
  Shadboltii, lxiv, 10.
  solenoceros, ci, 3.
  spinosum, xxv, 22; lxxvii, 1, 9, 10; c, 11, 25.
  striolatum, civ, 11, 12, 14.
  subcornutum, ci, 4, 5.
  subrotundatum, cii, 3; cxii, 7.
  tabellarium, c, 1.
  tesselatum, cii. 9.
  tridactylum, cv, 1.
  tripartitum, c, v.
  trisculum, c, 9, 10.
  uncinatum, cii, 12.
  undulatum, civ, 3, 9; cv. 14.
  validum, cii, 11.
  variabile, lxiii, 3, 5.
  venulosum, c, 2, 4. 7.

TROCHOSIRA.
  spinosa, lxvi, 13, 14.

XANTHIOPSIS.
  umbonatus, lxxxiv, 10.

XANTHIOPYXIS.
  cingulata, lxii, 27.
  umbonatus, lxxiv, 16.

PLATE I.

Figures magnified 500 diameters.

Fig. 1. SURIRELLA ELEGANS, showing the single nucleus and the radial distribution of the endochrome.

" 2. SURIRELLA SPLENDIDA, showing the two nuclei and the germinal (?) dot. Also the distribution of the endochrome in the canaliculi and the arrangement of the lobes of the endochrome with regard to the germinal (?) dot.

Figs. 3, 4. The sporangium of the Surirella Splendida after the complete fusion of the endochrome; the valves of the parent frustules adhering.

" 5, 6. SURIRELLA SPLENDIDA, early stage of conjugation; the fusion of the two endochromes resulting in the enlargement of the internal contents, and the separation of the valves of the frustules which adhere to the mucous mass of the sporangium.

Fig. 7. EPITHEMIA VENTRICOSA, Thallus containing matured frustules and spores the same as are found in the diatom itself.

" 8. Thallus containing several diatoms in early and later stages of development.

" 9. Two diatoms (Navicula) conjugating and surrounded by a gelatinous covering; the valves of frustule separated and the endochrome contracted into spherical masses.

Figs. 10-12. CYMBELLA, phases of growth of the sporangium in its mucous envelope.

" 13. Process of division and multiplication of diatoms; a Navicula; F.F., the silicious epiderm; G.G., the sides sliding one over the other; A., the nucleus; B.B.B., endochrome surrounding the spore or primordial cell; D. D., central cavities.

" 14. A., the spores dividing and the end of the epiderm (F.F.) sliding apart.

" 15. A.A., spore fully divided and the frustule separated by forming dividing membranes; thus there are two frustules approaching complete development.

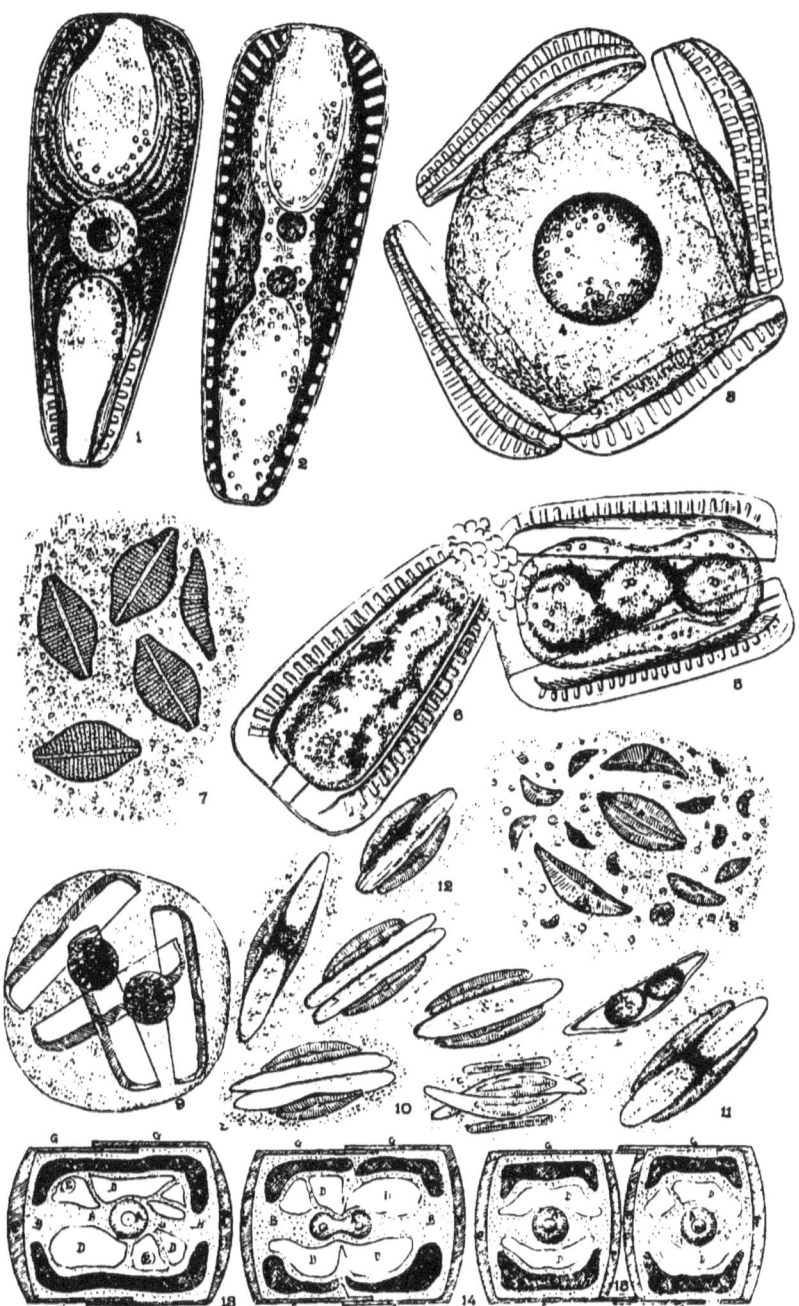

PLATE II.

Figures magnified 500 diameters.

Fig. 1. AMPHIPRORA (Amphipleura) Lindheimeri, var. Grun. Oregon, Texas, etc., T. R. M. S. 1877, p. 179.
" 2. PLEUROSIGMA TRANSVERSALE, W. S.; S. B. D. 11, p. 96, Sm. Sp. T. 415.
" 3. " INTERMEDIUM, W. S.; S. B. D. 1, p. 64, 21, f. 200, Sm. Sp. 405.
Figs. 4, 5. AMPHIPRORA PULCHRA, Bail. front and side view; Bail. M. O. p. 38—2 f, 16; V. H. 22, f. 1, 2.
Fig. 6. " LEPIDOPTERA, Greg. T. M. S. 1857, p. 76, 1, f. 39; V. II., 22, f. 2, 3; Sm. Sp. T. 23.
Figs. 7, 8. " HYALINA, Eulenst. Probably a var. of A. Paludosa, W. S.; also very near *Amphora intermedia*, Lewis; for Grev.'s fig. v. V. II., 22, f. 17.
Figs. 9-11. " QUADRIFASCIATA, Bail. M. O. p. 38, 2, f. 2-4 does not possess much value, the bands of colored endochrome mean nothing.
Figs. 12, 13. " ORNATA, Bail. M. O. p. 38, 2, f. 15, 23; H. L. S., this according to Bail.'s fig.; for larger form v. Pl. 48, f. 21.
Figs. 14, 15. " CONSTRICTA, Ehrb., marine and brackish water, E. Amer. 2-6, f. 28; S. B. D. Vol. I, p. 44, 15, f. 126; Prit. p. 9, 23, 12, f. 1; Sm. Sp. T., 21.
Fig. 16. " CONSPICUA, Grev., (a pulchra var.) T. M. S., p. 86, (1861), Pl. 10, 16; V. II., 22 b. f. 3, Hudson River, N. Y.
Figs. 17, 18. " MAXIMA, Greg., D. C., p. 35, 4, f. 61; Lens, Vol. II, No. 11. p. 89; V. II., 22, f. 4.
Fig. 19. " DECUSSATA, Grun., V. II., 22, f. 13.
Figs. 20, 21. " ALATA, Ehrb., three views, K. B., p. 107, 3, f. 60; S. B. D., Vol. I, p. 44, 15, f. 124; V. II., 22, f. 11, 12.
Figs. 22, 23. " PALUDOSA, W. S.; S. B. D., p. 44, 31, f. 269; V. II., 22, f. 10.

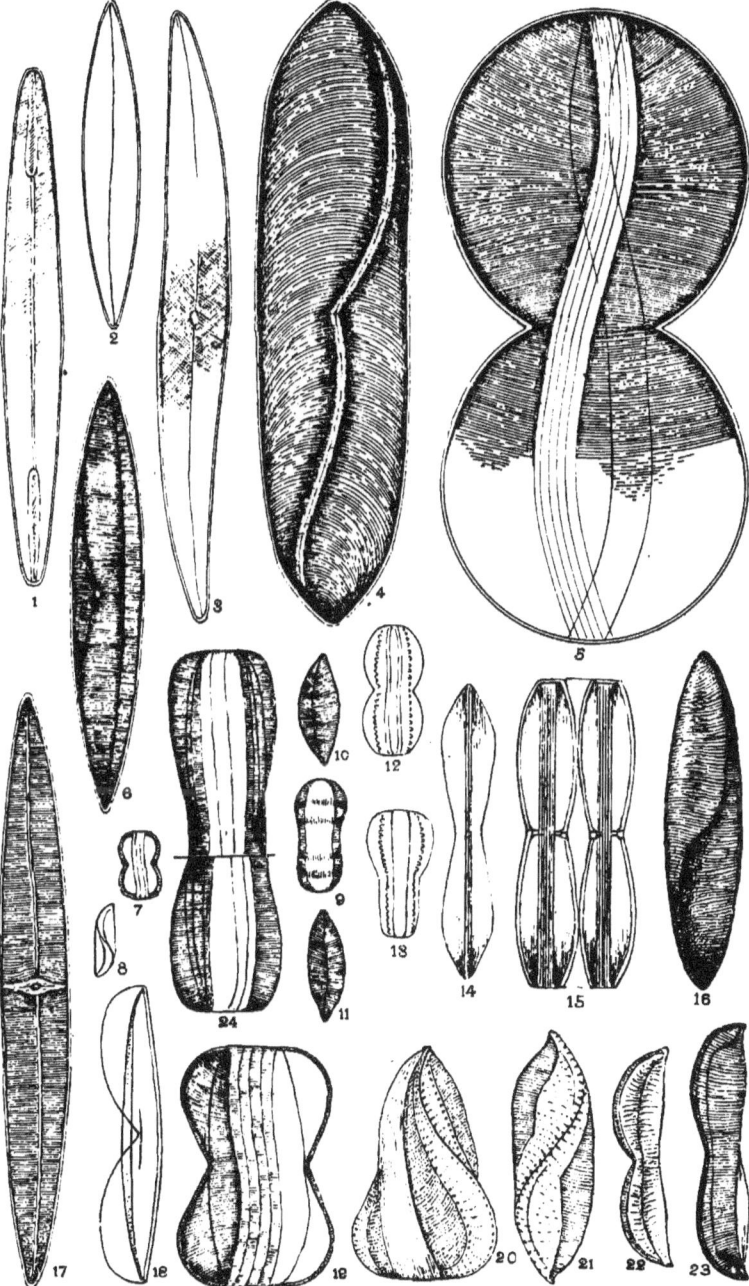

PLATE III.

Figures magnified 500 diameters.

Fig. 1-3.   AMPHORA ROBUSTA, Grev., D. C., p. 44, 5, f. 79; Lens, p. 8, No. 14, 3, f. 2; Schm. At., 27, f. 38.
"   4.   "   EXCISA, Greg., H. L. S., in Lens, p. 75, No. 26, 2, f. 4; Schm. At., 39, f. 3.
"   5.   "   SCHMIDTII, Grun., Schm. At., 28, f. 2, 3, Camp. Bay.
"   6.   "   GRANULATA, Greg. (the back view) Schm. At., 27, f. 66; Gulf of Mexico.
Figs. 7, 8.   "   SPECTABILIS, Greg., Schm. At., 40, f. 18-23.
Fig. 9.   "   PELLUCIDA, Greg., H. L. S., Lens, p. 78, No. 36, 2, f. 15; Schm. At., 27, f. 11, 36; 39, 65.
Figs. 10, 11.   "   PROTEUS, Greg., D. C., Trans. Roy. Soc., Edinb., xxi, Pl. 4, 1857; Schm. At., 27, f. 63; f. 1-7; Salt Lake.
"   12, 13.   "   VENETA, Kütz., K. B., p. 108, 3, f. 25; H. L. S.; Lens, p. 84, No. 58, 67; Schm. At., 27, f. 16.
Fig. 14.   "   LIBYCA, Ehrb., Schm. At., 26, f. 102-105; K. B., p. 107, 29, f. 28; Sm. Sp. T., No. 34.
"   15.   "   EXSECTA, Grun., Schm. At., 27, f. 54, 55; 39, f. 4; Campeachy Bay.
"   16.   "   OBTECTA, L. W. B.; B. J. N. H., p. 348, f. A. B., H. L. S., Lens, p. 77, No. 34, 2, f. 12, a c.
"   17.   "   PELLUCIDA, Greg., G. D. C., Pl. iv, f. 73. H. L. S., Lens, p. 78, No. 36, 2, f. 15; Schm. At., 27, f. 11, 36, 37, 65.
Figs. 18, 19.   "   TURGIDA, Greg., Schm. At., 25, f. 22, etc.; H. L. S., Lens, p. 85, No. 66, 3, f. 27.
"   20, 21.   "   EGREGIA, Ehrb., Abh., 1872, Pl. 2, f. 20; Schm. At., 28, f. 13-15; comp. Pl. 2, f. 1.
Fig. 22.   "   LAEVIS, Greg., H. L. S., Lens, p. 70, No. 1, 1, f. 13; Schm. At., 26, f. 8.
"   23.   "   RIMOSA, Ehrb., Mik., 5, 1, f. 27; 13, 2, f. 17; H. L. S., Lens, p. 81, No. 49, 3, f. 12.
Figs. 24, 25.   "   MEXICANA, Schm. At., 27, f. 47-49.
Fig. 26.   "   MICANS, Schm. At., 27, f. 18, Campeachy Bay.
"   27.   "   TRUNCATA, Greg., H. L. S., Lens, p. 76, No. 29, Schm. At., 28, f. 4, Campeachy Bay.
"   28.   "   SULCATA, Breb., (Obtecta?) Schm. At., 26, f. 46, 47; 27, f. 12, 13, H. L. S., Lens, p. 75, No. 29, 2, f. 11.
Figs. 29, 30.   "   OBTUSA, Greg., T. M. S., 1857, p. 72, 1, f. 34; H. L. S., Lens, p. 70, No. 5, 1, f. 5, Schm. At., 40, f. 4-7.

Plate III.

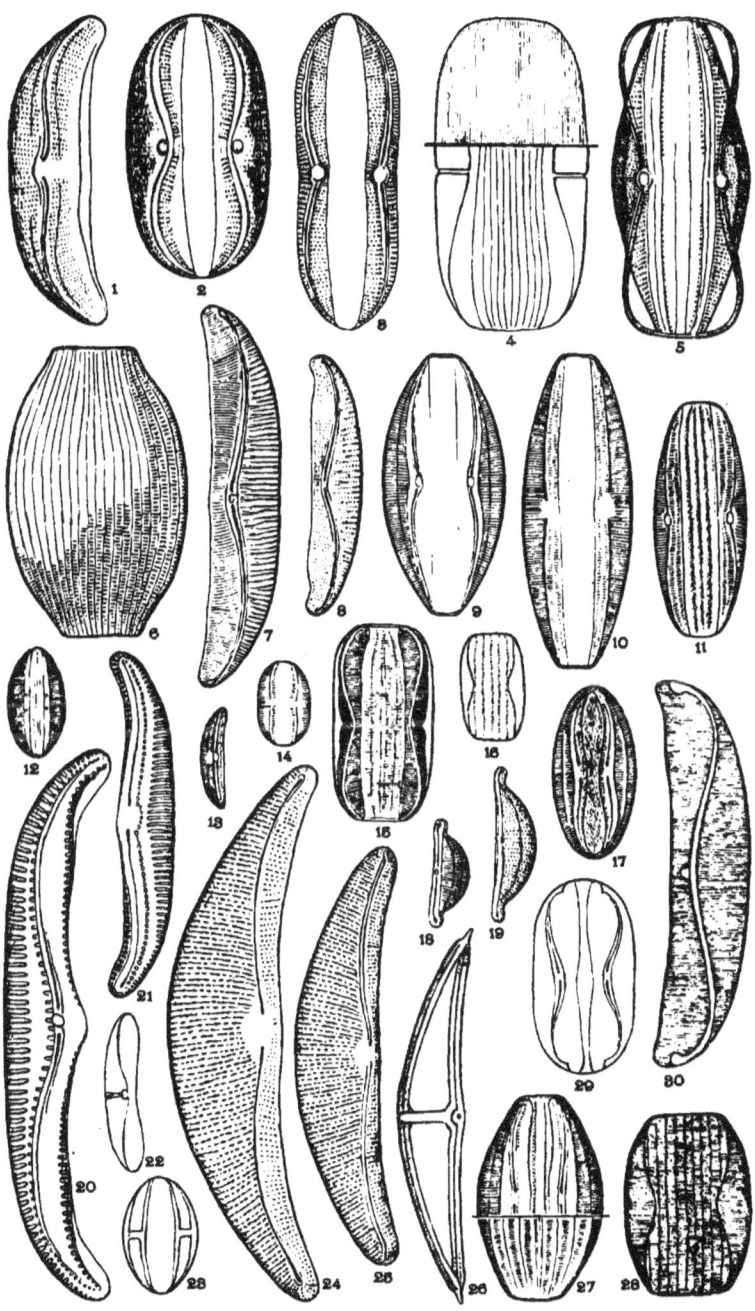

PLATE IV.

Figures magnified 500 diameters.

———

Fig. 1. AMPHORA EGREGIA, Ehrb., 1872, Pl. 2, f. 20; Schm. At., 28, f. 13-15.
" 2. " FARCIMEN, Grun., Schm. At., 27, f. 57.
" 3-6. " LAEVISSIME, Greg., H. L. S., in Lens, p. 73, No. 81, 1, f. 14.
Figs. 7, 8. " BIGIBBA, Grun., Schm. At., 25, f. 66, 67, etc.
Fig. 9. " RICHARDTIANA, Grun., H. L. S., Lens, p. 84, No. 62; Schm. At., 39, f. 33-35. Doubtfully an Amphora.
" 10. " CYMBIFERA, Greg., H. L. S., Lens, p. 85, No. 63, 3, f. 26; Schm. At., 27, f. 17-19; 26, f. 33; 39, f. 18.
" 11. " COFFEÆFORMIS, K. B., p. 108, 5, f. 37, H. L. S., Lens, p. 82, No. 53, 3, f. 17, Schm. At., 26, f. 56-58.
Figs. 12, 13. " INFLATA, Grun., Schm. At., 25, f. 29, 30.
Fig. 14. " LIBYCA, Ehrb., comp. Pl. 2, f. 14.
" 15. " AREOLATA, Grun., Schm. At., 39, f. 28.
" 16. " LANCEOLATA, Cleve., M. J., 1874, p. 256, 8, f. 3; Schm. At., 25, f, 6.
Figs. 17, 18. " CYMBIFERA, vide 10, above.
" 19, 20. " COFFEÆFORMIS, vide 11, above.
Fig. 21. " CLEVIA, A., Schm. At., 25, f. 46-48.
" 22. " EULENSTEINII, Grun., Schm. At., 25, f. 1-3, v. nearly allied to A. lanceolata.
Figs. 23, 24. " DELPHINEA, L. W. B.; B. J. N. H. VII, Pl. 1, f, 1. Lens, p. 73, No. 22, 1, f. 18; Schm. At. 40, f. 25-27.
" 25, 26. " DUBIA, Greg., H. L. S., Lens, 80, No. 42, Schm. At., 27, f. 20-26.
Fig. 27. " CINGULATA, Cleve., Schm. At., 26, f. 17; I. M., 1879, p. 29, 2, f. 15.
" 28. " GREVILLIANA, Greg., T. M. S., 1857, p. 73, 1, f. 35. Lens, p. 77, No. 32, 2, f. 9. Campeachiana? Grun.
" 29-32. " OVALIS, K. B., p. 107, 5, f. 35, 39; S. B. D., p. 19, 2, f. 26, H. L. S., Lens, 80, No. 45, 2, f. 17, Schm. At., 26, f. 106-111.
Figs. 33, 34. " FORMOSA, Cleve., Schm. At., 27, f. 58, 28, f. 6, 34, 39, f. 2.
Fig. 35. " GRANULATA, Greg., front view? Schm. At., 27, f. 67; small form, A. gigantea, Grun., Gulf of Mexico.
Figs. 36, 37. " CRASSA, Greg., H. L. S., Lens, 76, No. 29, 2, f. 5, T. M. S., 1857, p. 72, 1, f. 35.
" 38, 39. " GRÜNDLERII, Grun., Schm. At., 28, f. 24-27; 39, f. 25.

Plate IV.

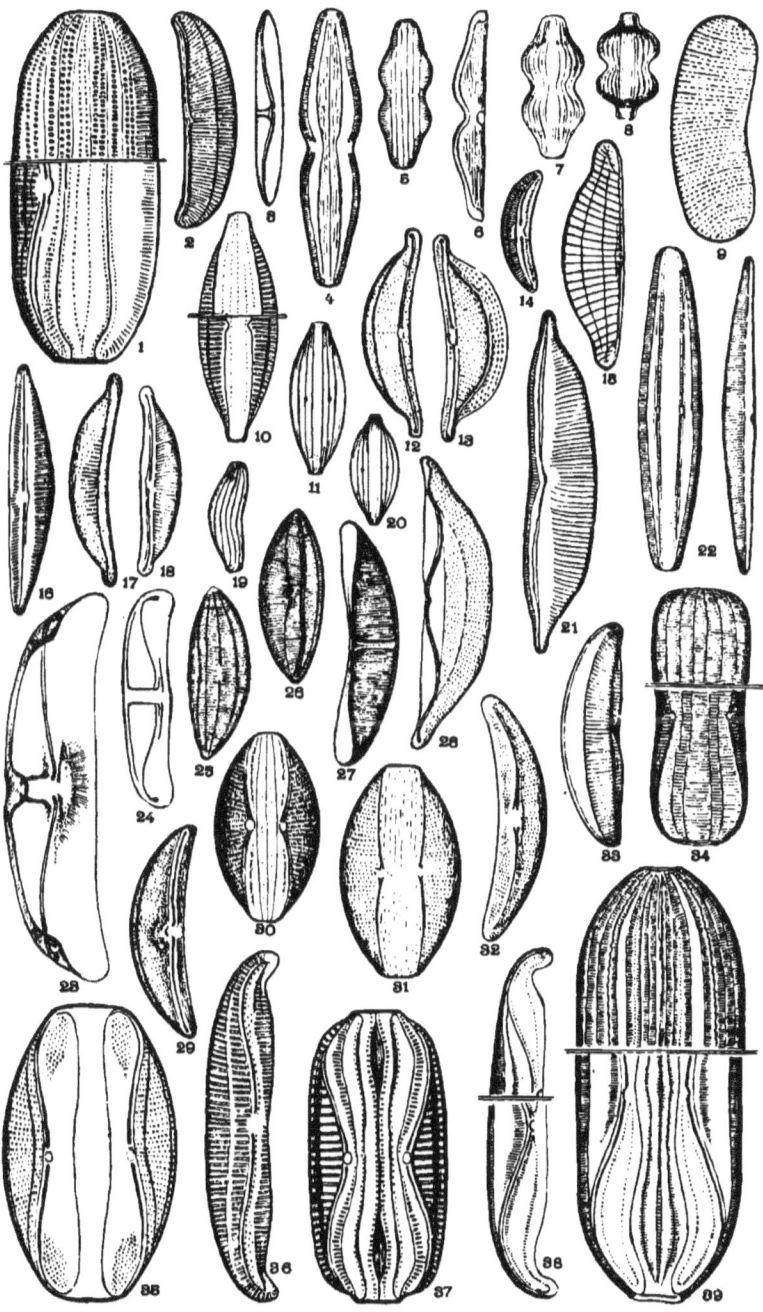

PLATE V.

Figures magnified 500 diameters.

Figs. 1, 2. SYNEDRA CHASEI, Thomas; W. and C., 1, p. 4, 2, f. 3. A var. of S. *longissima*, S. B. D., p. 72, 12, f. 95.
Fig. 3. AMPHIPRORA ELEGANS, Greg., T. M. S., 1857, p. 10, 1, f. 30; H. L. S., Lens, p. 73, No. 19, 1, f. 17.
" 4. AMPHERA GIGAS, Ehrb. Mik., 6, 2, f. 13; H. L. S., Leus, p. 73, *sub* No. 22; p. 80, *sub* No. 45.
" 5. SURIRELLA MISSISSIPPICA, Ehrb. Mik., 35, A. 8, f. 5; Mississippi.
" 6. " LIVSOMA, Ehrb. Mik., 33, 14, f. 25, Connecticut, probably one of the plates of Rhabdonema.
" 7. COCCONEIS GEMMATA, Esrb. Mik., 37, 3, 2, Oregon.
Figs. 8, 9. COCCONEIS RHOMBIFERA, Bail. and Harv., p. 175, 9, f. 3, 4. California.
Fig. 10. SURIRELLA COCCONEIS, Ehrb. Mik., 35, A., 8, 3; a dubious form, not a Surirella, nor a Cocconeis.
" 11. COCCONEIS, REGALIS, Grev., M. J., 1859, p. 156, 7, f. 1. California guano.
" 12. COCCONEIS SULCATA, Bail. and Harv., p. 175, 9, f. 5.
" 13-15. AMPHIPRORA VITREA, S. B. D., Vol. I, p. 44, 31, f. 270; V. H., 22, f. 29, Mobile, Ala.
" 16. DIMEREGRAMMA NOVÆ CÆSAREÆ, (frustules attached) Kain and Schultze, Bot. Bull., Mar., 1889, Vol. XVI, No. 3, artesian well, Atlantic City, N. J.
" 17. FRAGILARIA FREMONTII, Ehrb. Abh., 1870, 56, 2, 1, f. 5, of doubtful value, probably a hoop of some diatom, Salt Lake.
" 18. SYNEDRA VALENS, Ehrb., Prit., 782, 12, 44.
" 19. FRAGILARIA VENTER, Ehrb. Abh, 1869, 1, a Mexico. Probably merely a hoop of Odontidium.
" 20. FRAGILARIA NEWBERYI, Ehrb., 1870, 56, 3, 1, 12, Nevada. A doubtful form, may be a Biblarium.
" 21. FRAGILARIA INFLEXA, Ehrb., not a good specimen or species; probably a hoop of Eunotia.
" 22. DIMEREGRAMMA NOVÆ CÆSAREÆ, K. and S., (frustules detached), Typ. form to right, var. Obtusa to left, Bot. Bull. Vol. XVI, No. 3, 1889, artesian well, Atlantic City, N. J.
" 23. DIMEREGRAMMA FULVUM, Ralfs., V. H., 36, f. 7, 8.
" 24. MASTOGLOEA APICULATA, S. B. D., II, p. 65, 62, f. 3, 87.
" 25. ENTOMOGASTER WOODWARDII, Ehrb. Abh., 1870, var. California. Probably a form of Amphora.

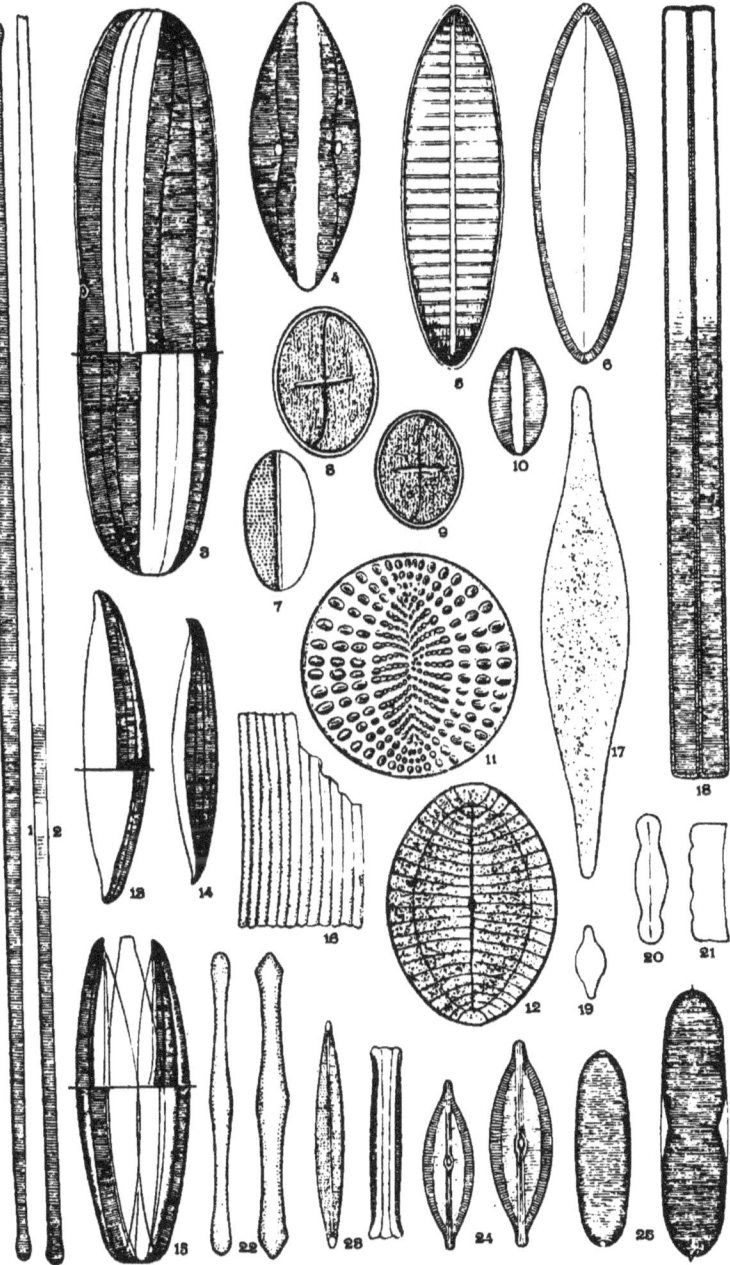

PLATE VI.

Figures magnified 500 diameters.

Figs. 1, 2.   COCCONEMA LANCEOLATUM, Ehrb. Mik., 39, 3, f. 17, etc.   S. B. D., I, p. 75, 23, f. 219; Schm. At., 10, f. 8–10.
Fig. 3.   Granules figured, not coarse enough.
"   4.   COCCONEMA MEXICANUM, Ehrb. Mik., 33, 7, f. 6; Abh., 1869, 2, 1, f. 1; Schm. At., 10, f. 32, 33; 71, f. 82.
"   5.   COCCONEMA JANISCHII, Schm. At., 10, f. 34, 85; 71, f. 81.   Nearly related to C. cistula.
Figs. 6, 7.   CYMBELLA CISTULA, Hop., Mik., 38, A., 20, f. 5; 37, 3, f. 3; S. B. D., I, p. 76, 23, f. 221. Schm. At., 10, f. 1–4.
"   8, 9.   COCCONEMA PARVUM, W. S., S. B. D., I, p. 76, 28, f. 222; 24, f. 222; Schm. At., 10, f. 14, 15.
Fig. 10.   CYMBELLA CYMBIFORMIS, Breb. Al. Fal., p. 49, Pl. 7; V. II., 2, f. 11.
"   11.   COCCONEMA LANCEOLATUM, vide f. 1–3, above.
"   12.   CYMBELLA CISTULA, vide f. 6, 7.   State of development.
"   13.   CYMBELLA VENTRICOSA, Ag., Consp., p. 9; K. B., p. 80, 6, f. 16; Schm. At., 9, f. 32; 72, f. 11.
"   14.   COCCONEMA LANCEOLATUM, W. S., in vegetative state, S. B. D., p. 75, 23, f. 219.

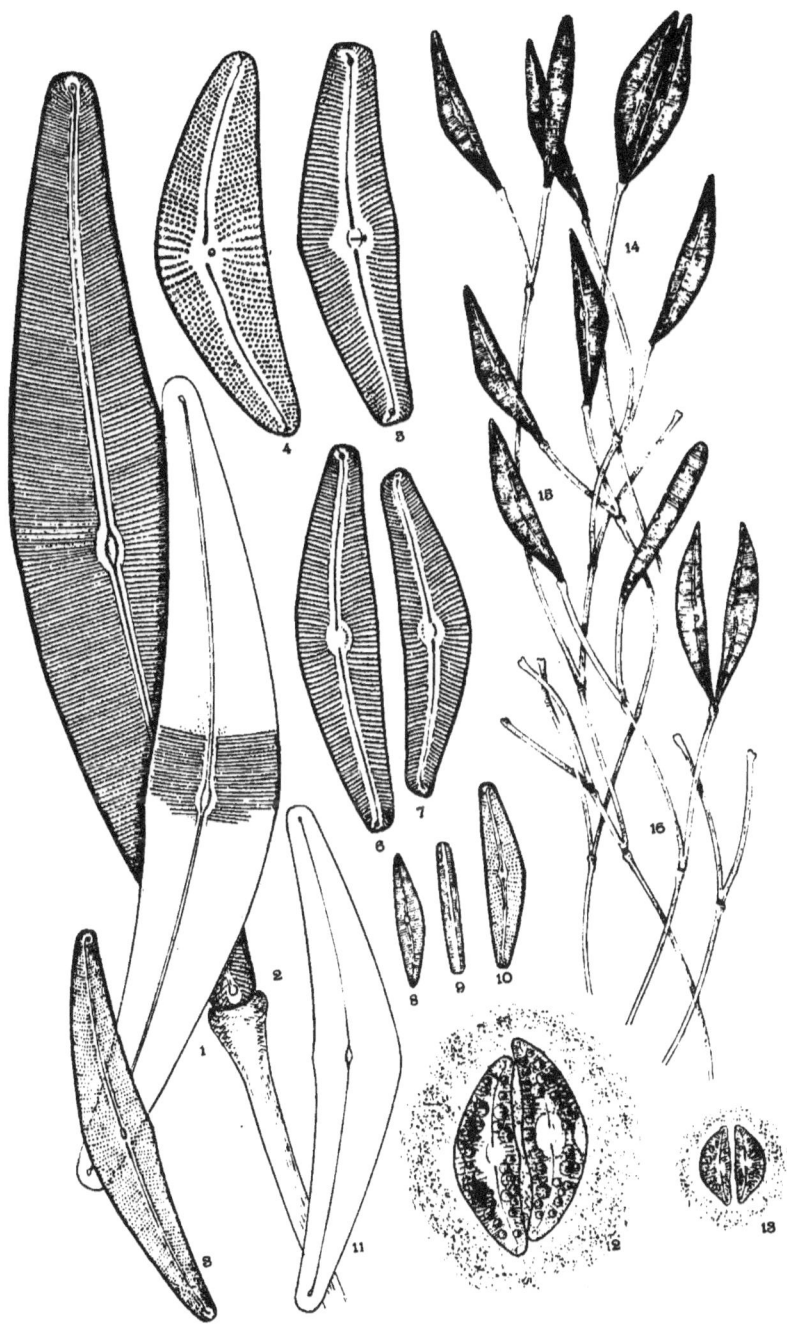

PLATE VII.

Figures magnified 500 diameters.

Figs 1, 2. CYMBELLA HETEROPLEURA, Kg., Schm. At., 9, f. 3–5; Cherry-field, Me.
Fig. 3. " EHRENBERGII, Kg., K. B., p. 79, 6, f. 11; Schm. At., 9, f. 6–9, 16–18; S. B. D., I, p. 17, 2, f. 21.
Figs. 4, 5. " GASTROIDES, Kg., K. B., p. 79, 6, f. 4; Schm. At., 9, f. 1, 2, 72, f. 12–14. V. H., 2, f. 8; f. 4, v. near C. lanceolata.
Fig. 6. " HELVETICA, Kg., K. B., p. 79, 6, f. 13; S. B. D., I, p. 18, 2, f. 24; Schm. At., 10, f. 18, 21; V. H., 2, f. 15.
" 7. " NAVICULIFORMIS, Auers., Heib. Consp., p. 108, 1, f. 3, V. H., 2, f. 5.
" 8. " LEPTOCEROS, Kg., K. B., p. 80, 6, f. 14. Rab. S. D., p. 22, 7, f. 14, V. H., 3, f. 24.
Figs. 9, 10. " EXCISA, Kg., K. B., p. 80, 6, f. 17. Schm. At., 71, f. 35, 36. More than usually rostrate.
Fig. 11. " KAMTSCHATICA, Grun., Moll., T., p. 3, 2, 10, Schm. At., 10, f. 31.
Figs. 12, 13. " GASTROIDES, Kg., conjugating and developing in gelatinous envelope.
Fig. 14. " NAVICULIFORMIS, Auers., same as figure 7.
" 15. " AMPHICEPHALA, Naeg., Schm. At., 9, f. 62, 64, 66; 71, f. 52; V. H., 2, f. 6.
" 16. " CUSPIDATA, Kg., K. B., p. 79, 3, f. 40, S. B. D., I, p. 18, 2, f. 22; Schm. At., 9, f. 50, 53, 55.
Figs. 17, 18. " STOMATOPHORA, Grun., Schm. At., 10, f. 28–30.
Fig. 19. " PARVA, V. H., 2, f. 14. (Cocconema parvum.)
" 20. " CUSPIDATA, var. from Canada, for reference vide f. 16.
" 21. " EHRENBERGII, Kg., a smaller form than f. 3, above. Nearly allied to cuspidata.
" 22. " MACULATA, Kg., K. B., p. 79, 67, f. 2, S. B. D., I, p. 18, 2, f. 23; V. H., 2, f. 16, 17.
" 23. " GASTROIDES, Kg., comp. f. 4, 5, and 12, 18.
" 24. " AMERICANA, A. S., Schm. At., 9, f. 15, 20. Probably var. of Ehrenbergii.
" 25. " EHRENBERGII, forma minor, Kg., comp. f. 3, 21, above.
" 26. " GRACILIS, Kg., K. B., p. 79, 6, f. 9; Rab. S. D., p. 22, 7, f. 12; H. L. S., Sp. T., No. 119.
" 27. " TURGIDULA, A. S., Schm. At., 9, f. 23–26.
" 28. " CYMBIFORMIS, Breb., Alg., Fal., p. 49, Pl. 7; V. H., 2, f. 11.
" 29. " CYSTULA, Brun., Al., p. 58, 3, f. 18, V. H., p. 64, 2, f. 12, 13.
" 30. " CURTA, A. S., Schm. At., 9, f. 47.
Figs. 31, 32. " AFFINIS, Kg., K. B., p. 80, 6, f. 15; S. B. D., I, p. 18, 30, f. 250; Schm. At., 9, f. 20, 38; 72, f. 28, 29.
" 33, 34. " ANGLICA, Lag., Schm. At., 9, f. 63; V. H., 2, f. 4.
Fig. 35. " CISTULA, Brun., forma minor, V. H., Pl. 11, f. 13.

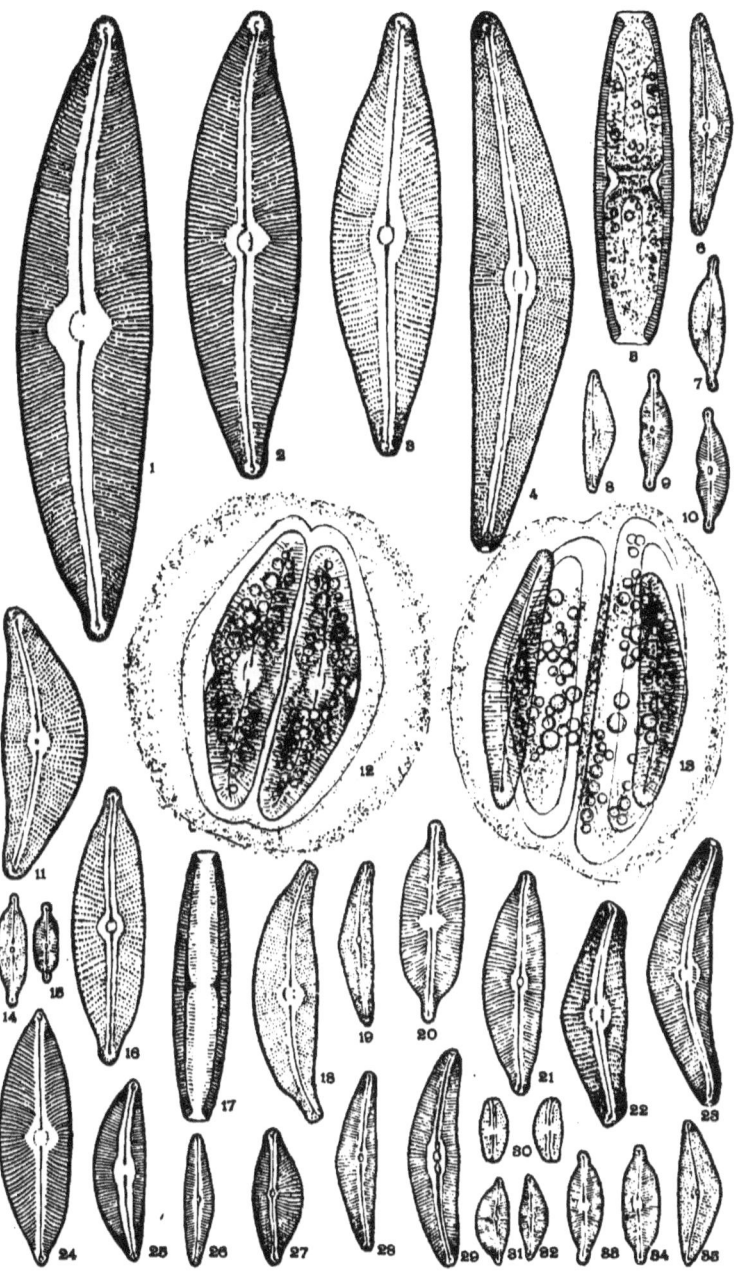

PLATE VIII.

Figures magnified 500 diameters.

Fig. 1. STAURONEIS PUNCTATA, Kg., K. B., p. 106, 21, f. 9; S. B. D., I, p. 61, 19, f. 189; H. L. S., Sm. Sp. T., No. 498.
" 2. " DILATATA, W. S., S. B. D., I, p. 60, 19, f. 191; Rab., E. D., p. 248.
" 3. " LINEARIS, Ehrb., S. B. D., I, p. 60, 19, f. 193.
" 4. " ANCEPS, Ehrb., K. B., p. 105, 29, f. 4; Rab. S. D., p. 48, 9, f. 14; S. B. D., I, p. 60, 19, f. 190; V. H., 4, f. 4, 5, 6.
" 5. " LEGUMEN, Ehrb. Mik., 39, 3, f. 104; K. B., p. 107, 29, f. 11; M. J., 1856, Pl. 1, f. 9; V. H., 4, f. 11.
" 6. " SALINA, W. S., S. B. D., I, p. 60, 19, f. 188; V. H., 10, f. 16.
" 7. " PRODUCTA, Grun., V. H., 4, f. 12; probably a variety of anceps.
Figs. 8, 9. " ANCEPS, Ehrb., var. Amphicephala, vide references f. 4; V. H., 4, f. 4, 6, 7.
Fig. 10. " GRACILIS, Ehrb. Mik., 16, 1, f. 4; 17, 2, f. 15; 17, 1, f. 5; S. B. D., I, p. 59, 19, f. 186; Abh., 1870, 2, f. 41.
" 11. " ACUTA, W. S., S. B. D., I, p. 59, 19, f. 187, V. H., 4, f. 3.
Figs. 12-15. " PHOENICENTERON, Ehrb. Mik., 39, 3, f. 105; Abh., 1862, p. 64, 1, f. 6; S. B. D., I, p. 59, 19, f. 185; V. H., 4, f. 2. Fig. 14, var. Baileyi.
Fig. 16. ENCYONEMA GRACILE, Rab. S. D., p. 24, 10, f. 1; Schm. At., 10, f. 36, 37, 39, 40; V. H., 3, f. 20, 22.
" 17. " The same under lower power, in gelatinous tube.
Figs. 18, 19. " VENTRICOSA, Kg., Schm. At., 10, f. 59; 71, f. 13; V. H., 3, f. 15-17.
Fig. 20. " CAESPITOSUM, Kg., Rab. S. D., p. 24, 7, f. 5; S. B. D., II, p. 68, 55, f. 346; Schm. At., 10, f. 57, 58, 60, 62.
" 21. " AUERWALDII, Rab. S. D., p. 24, 7. f. 2. Probably only a variety of Fig. 20.
" 22. " TRIANGULATUM, Kg., Schm. At., 10, f. 54; 71, f. 10. Cymbella gibba of Bailey.
" 23. " PARADOXUM, Kg., K. B., p. 82, 22, f. 1, Rab., S. D., p. 24, 7, f. 3; Schm. At., 10, f. 67, 69.
" 24. " PROSTRATRUM, Ralfs., Rab. S. D., p. 24, 7, f. 1; S. B. D., II, p. 68, 54, f. 345; Schm. At., 10, f. 64, 66. Evidently varieties of Fig. 23.

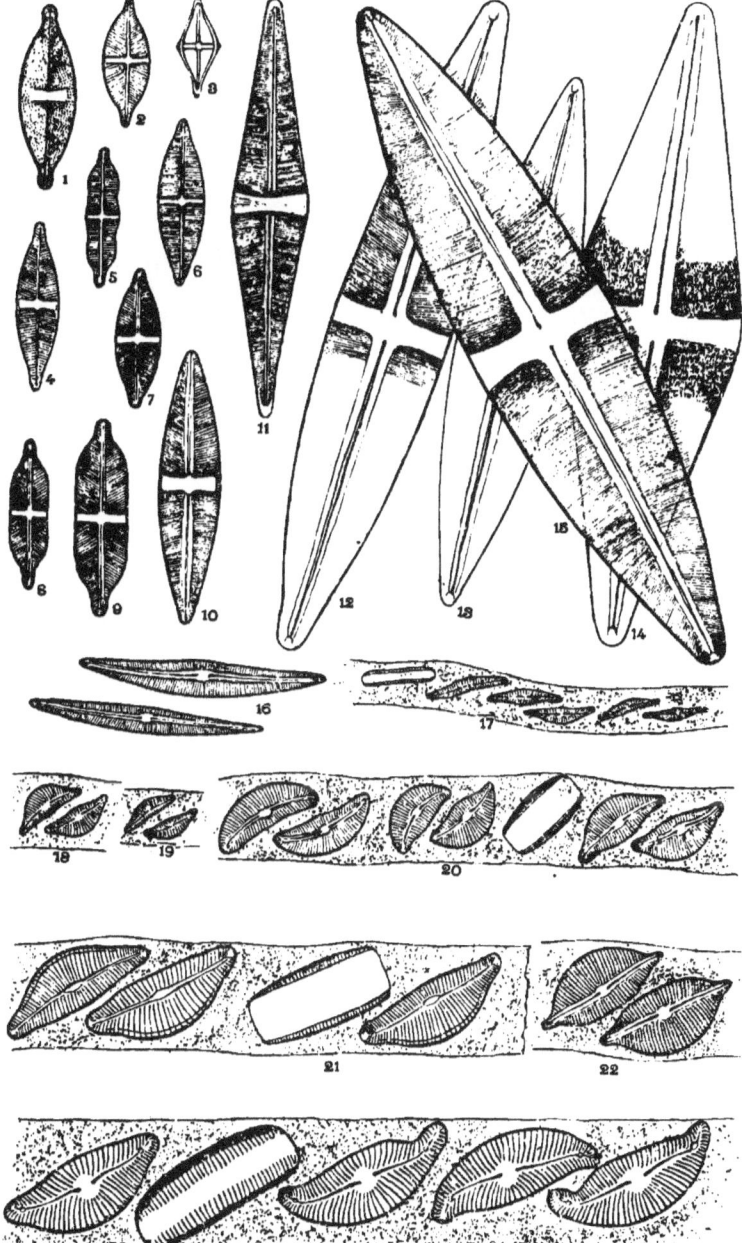

PLATE IX.

Figures magnified 500 diameters.

Fig. 1. AMPHORA COFFEAEFORMIS, Kg., K. B., p. 108, 5, f. 37; H. L. S., Lens, p. 82. No. 53, 3, f. 17; Schm. At., 26, f. 56–58.
" 2. " HYALINA, Kg., p. 108, 30, f. 18; H. L. S., Lens, p. 74, No. 24, 2, f. 2; S. B. D., I, p. 19, 2, f. 28; Schm. At., 26, f. 52–55.
" 3. " COSTATA, W. S., S. B. D., I, p. 20, 30, f. 253; H. L. S., Lens, p. 83. No. 55, 3, f. 28.
" 4. " LAEVISSIMA, Greg., D. C., p. 41, 4, f. 72; H. L. S., Lens, p. 73, No. 81, 1, f. 14; V. H., 1, f. 15.
" 5. " LINEOLATA, Ehrb., Mik., 13, 1, f. 19; Rab., S. D., p. 36, 9, f. 9, 10; H. L. S., Lens, p. 74, No. 23, 1, f. 22.
" 6. " OBTECTA. L. W. B., B. J. N. H., p. 348. f. A. B.; H. L. S., Lens, p. 77, No. 34, 2, f. 12.
" 7. " FUSCA, A. S., Schm. At., 27, f. 68.
" 8. " SALINA, W. S., S. B. D., I, p. 19, 30, f. 251; H. L. S., Lens, p. 84, No. 59, 3, f. 29; Schm. At., 26, f. 81.
" 9. " BINODES, var. interrupta, Grun., Schm. At., 25, f. 65.
" 10. " FLEXUOSA, Grev.? Schm. At., 25, f. 82; this is not Grevilli's figure which see Pl. 112, f. 12.
" 11. " ARCUATA, A. S., Schm. At., 26, f. 27–29.
" 12. " NOVA CALEDONICA, Grun., Sc m. At. 26, f. 16, 24, (A. porcellus?)
" 13. " AFFINIS, W. S., S. B. D., I, p. 19, 2, f. 27; K. B., p. 95, 18, f. 65; Kg.'s form not W. S.; near A. proboscidia, Greg., and A. commutata, Grun.
" 14. " RECTANGULARIS, Greg., T. M. S., 1857, p. 70, 5, 29; H. L. S., Lens, p. 82, No. 52, 3, f. 13.
Figs. 15, 16. " LINEATA, Greg., H. L. S., Lens, Pl. 3, f. 21, Schm. At., 26, 59.
Fig. 17. " LINEATA, Greg., small form, Schm. At., 26, f. 82, 26, H. L. S., Lens, p. 8, 2, No. 54, 3, f. 21; Schm. At., 26, f. 59, 82, 86; 27, f. 15.
" 18. " GIBBA, A. S., Schm. At., 39, f. 32.
" 19. " PROTEUS, Greg., D. C., p. 46, 5, f. 81; H. L. S., Lens, p. 79, No. 41, 3, f. 1; Schm. At., 27, 63.
" 20. " GIGANTEA, Grun., Schm. At., 27, f. 46, 67.
" 21. " CONTRACTA, Grun., Schm. At., 25, f. 54, 55, 57, 62. Probably a var. of Amp. Janischii, A. S., North American?
" 22. NAVICULA ELEGANS, W. S., S. B. D., I, p. 49, 16, f. 137; Donk., B. D., p. 23, 4, f. 1. Striae should be more radial and somewhat flexuose. H. L. S.
Figs. 23, 24. " BOREALIS, Kg., K. B., p. 96, 28, f. 68, 72; Schm. At., 45, f. 15, 21; V. H., 6, f. 34.
" 25, 26. " CYNTHIA, A. S., Schm. At., 8, f. 41; O'M., L. D., p. 375, 33, f. 10.

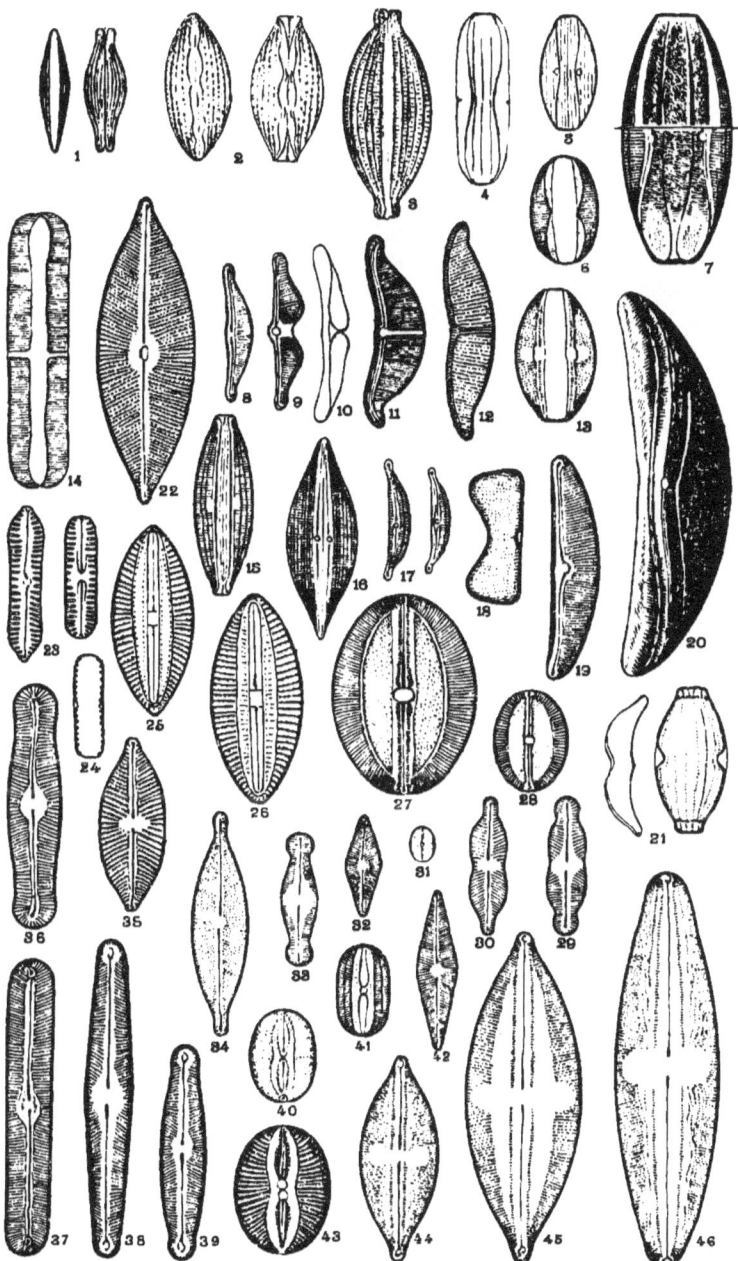

Fig. 27.   NAVICULA CIRCUMSECTA, Grun., Cl., 1880, p. 12, (*N. polysticta* Schm. At., 3, f. 27.)
"   28.   "   POLYSTICTA, Grev., Calif., p. 28, 4, f. 12· Schm. At., 3, f. 26.
Figs. 29, 30.   "   TERMIS, A. S., Schm. At., 45, f. 67, 71; V. H., 6, f. 12 13; probably a var. of N. mesolepta.
"   31.   "   FELLICULOSA, Hilse., V. H., 14, f. 32; *Frustulia pelliculosa*, Grun., 1860, 5, f. 18.
"   32.   "   LANCEOLATA, Kg., K. B., p. 94, 28, f. 38; 30, f. 48, Schm. At., 47, f. 49; V. H., 8, f. 16, 17; Kg.'s form not W. S.'s.
"   33.   "   GLORICEPS, Ralfs., V. H., A., f. 13. Figure perhaps rather large for an average size.
"   34.   "   RHYNCOCEPHALA, Kg., K. B., p. 152, 30, f. 35; S. B. D., I, p. 47, 16, f. 132. Schm. At., 47, f. 28-32.
"   35.   "   PLACENTULA, Ehrb., K. B., p. 94, 28, f. 57, Lewis, W., M. D., 2, f. 7. A somewhat dubious form; Ehrb., Kg. and Lewis do not quite agree.
"   36.   "   RECTANGULATA, Greg., D. C., p. 7. 1, f. 7; Donk., B. D., p. 66, 10, f. 5; V. H., A. S., f. 7.
"   37.   "   ELONGATA, Grun., Schm. At., 50, f. 27-29, merely a form of N. viridis.
Figs. 38, 39.   "   DECURRENS, Kg. K. B., p. 99, E. Mik., 11, f. 28; Schm. At., 45, f. 29, 30.
Fig. 40.   "   NUMMULARIA, Grev., Cal., p. 29, 4, f. 6; Schm. At., 70, f. 37, 38.
"   41.   "   PAPULA, Grun., Schm. At., 7, f. 45-48.
"   42.   "   LANCEOLATA, Kg. Same as Fig. 32. Kg.'s form has divergent striae; W. S. has striae parallel.
"   43.   " . NUMMULARIA. Grev., same as Fig. 40, larger form.
Figs. 44, 45.   "   SCULPTA, Ehrb., Mik., 10, 1, f. 5, a. b. Schm. At., 49, f. 46-48; V. H., 12, f. 1.
Fig. 46.   "   BOHEMICA, Ehrb., Mik., 10, 1, f. 4; Schm. At., 49, f. 43-45.

PLATE X.

Figures magnified 500 diameters.

Fig. 1.	NAVICULA	ANGLICA, Ralfs., Donk., B. D., p. 35, 5, f. 11; O'M., I. D., p. 414, 24, f. 24; V. H., 8, f. 29, 30.
" 2.	"	GASTRUM, Kg., K. B., p. 94, 28, f. 56; Lewis, W., M.D., p. 11, 2, f. 17; Donk., B. D., 22, 3, f. 10; B. I. N. H., p. 335, f. 34.
" 3.	"	PARVA, Ehrb. (Stauroptera parva, Ehrb.), V. H., 6, f. 6.
" 4.	"	DISCREPANS, A. S., Schm. At., 8, f. 8.
" 5.	"	PARCA, A. S., Schm. At., 8, f. 20–22.
Figs. 6, 16.	"	PALPEBRALIS, Breb., S. B. D., 1, p. 50, 31, f. 273; V. H., 11, f. 9.
" 7, 17.	"	BREVIS, Greg., var. Greg.'s typical form has produced ends; V. H., 11, f. 18, 19; Donk. B. D., p. 19, 3, f. 4.
" 8, 9.	"	QUINQUENODIS, Grun. (a var. of mutica), 1860, p. 522, 1, f. 33; 1863, p. 149, 13, f. 9; V. H., 10, f. 21.
Fig. 10.	"	EUGENIA, A. S., Schm. At., 8, f. 44, 45; O'M. I. D., p. 395, 33, f. 17.
" 11.	"	LITTORALIS, var. Donk. B. D., p. 5, 1, f. 2; Schm. At., f. 28, 4.
" 12.	"	LITTORALIS, Schm. At., 7, f. 12, V. H.; B. S., f. 22.
" 13.	"	CRYPTOCEPHALA, Kg., K. B., p. 95, 3, f. 20, 26; Rab. S. D., 6, f. 71; V. H., 8, f. 15.
" 14.	"	COCCONEIFORMIS, Greg., M. J., 1856, p. 6, 1, f. 22, S. B. D., II, p. 92.
" 15.	"	SERIANS, punctulata, Breb., K. B., p. 92, 30, f. 23; 28, f. 43; Rab., S. D., p. 38, 6, f. 51; S. B. D., I, p. 47, 16, f. 130.
Figs. 18, 19.	"	PUSILLA, W. S.; S. B. D., I, p. 52, 17, f. 145; V. H., 11, f. 17.
Fig. 20.	"	VIRIDULA, var. minor, Kg., K. B., p. 91, 30, f. 47; 4, f. 10, 15; Schm. At., 47, f. 48, 53, 56.
" 21.	"	RHYNCOCEPHALA, Kg., K. B., p. 52, 30, f. 35; S. B. D., I, p. 47, 16, f. 132; Schm. At., 47, f. 28, 29.
" 22.	"	GRACILIS (var. Silesiaca), Ehrb., Mik., 39, 3, f. 85; 16, 3, f. 29; 16, 1, f. 14, a. b., etc.
" 23.	"	GRACILIS, K. B., p. 91, 3, f. 48; 30, f. 57; Rab., S. D., p. 38, 6, f. 64.
" 24.	"	SEJUNCTA, A. S., Schm. At., 57, f. 50; 70, f. 55, 56.
" 25.	"	INFLEXA, Ralfs., Schm. At., 46, f. 69–72.
" 26.	"	BREBISSONII, Kg. (Pinn. stauroneiformis, W. S.), K. B., p. 93, 3, f. 49; 30, f. 39; Schm. At., 44, f. 16.
" 27.	"	DUBIA, Ehrb., Mik., 39, 2, f. 82; K. B. 96, 28, f. 61; Schm. At., 49, f. 11, 24.
" 28.	"	TABELLARIA, leptogongyla, Grun., Ehrb., K. B., p. 96, 28, f. 79, 80; Schm. At., 43, 5.
Figs. 29, 30.	"	BACILLUM, and var. Ehrb., Mik., 38a, 20, f. 3, 29, 3, f. 81; 27, 3, f. 8. Sill. Jour., May, 1851, f. 46, K. B., p. 96, 28, f. 69; B. I. N. H., p. 335, f. 20.

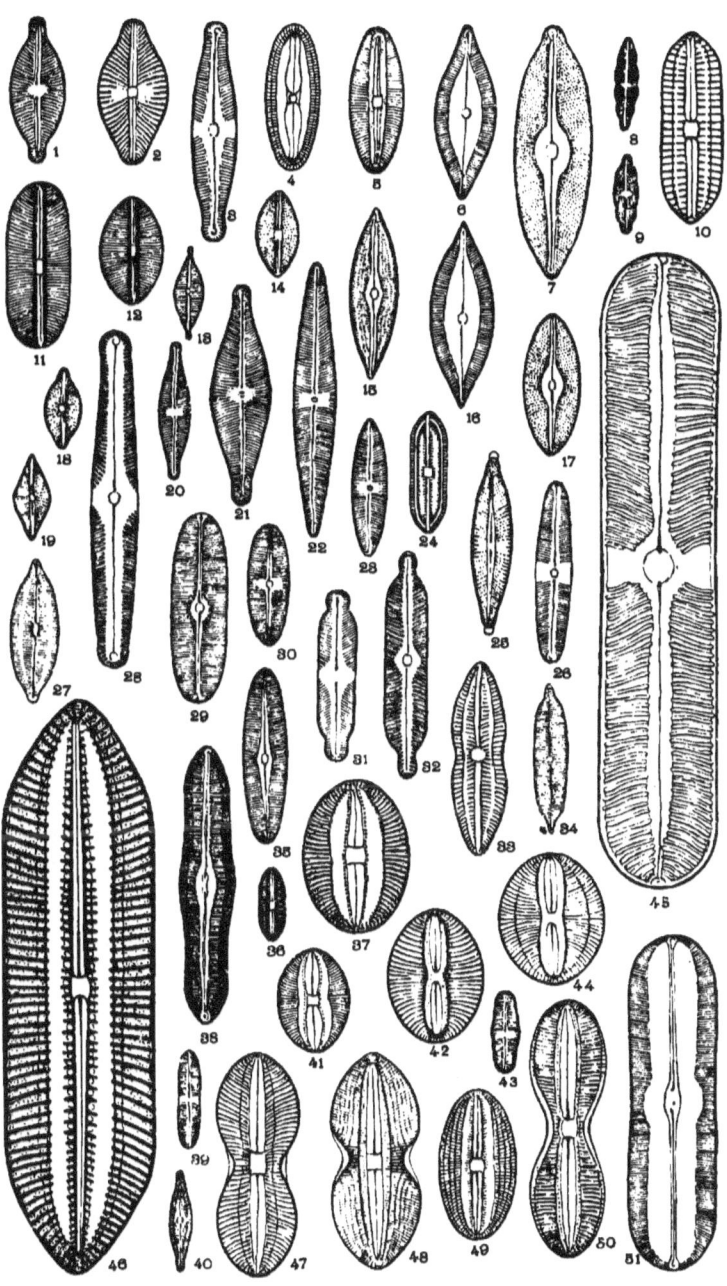

Fig. 31.	NAVICULA	INTERRUPTA, Bail., K. B., p. 100, 29, f. 93; Schm. At., 12, f. 10, 11; Sill. Jour., 1842, 2, f. 18.
" 32.	"	INTERRUPTA, W. S., var. bicapitata, Lagerst., V. II., 6, f. 9.
" 33.	"	CRABRO, Ehrb., var. M. J., 1847, p. 8, 3, f. 11, Donk., B. D., 46, 7, f. 1; Schm. At., 69, f. 1.
" 34.	"	ANGUSTA, Grun., 1860, p. 528, 3. f. 19, V. II., 7, f. 17.
" 35.	"	GIBBERULA, var. subinflata, Kg., K. B., p. 101, 3, f. 50, Rab., S. D., f. p. 45, 6, f. 30.
" 36.	"	OCULATA, Breb., I. G. C., 1870, p. 37, 1, f. 5, V. II., 9, f. 10.
" 37.	"	SUBORBICULARIS, Ralfs., Prit., p. 898, Donk., B. D., p. 8, 1, f. 9; Schm. At., 1, f. 3–5. A var. of N. Smithii.
" 38.	"	GIBBERULA, Kg., O'M., I. D. P., 368. This the type form, Fig. 35 a variety.
" 39.	"	LACUNARUM, Grun., V. II., 12, f. 31.
" 40.	"	EXILIS, Kg., K. B., p. 95, 4, f. 6, Rab., S. D., p. 39, 6, f. 84.
" 41.	"	NUMMULARIA, var. of Suborbicularis, Grev., Cal., p. 29, p. 4, f. 6; Schm. At., 70, f. 37, 38.
Figs. 42, 44.	"	NUMMULARIA (typical form) Schm. At., 70, f. 37, 30.
Fig. 43.	"	SAUGERI, Desmez., V. II., 11, f. 8a, (16b, var.)
" 45.	"	CARDINALIS, Ehrb., Schm. At., 44, f. 1, 2, O'M. I. D., p. 341, 30, f. 2; V. II., a Fig. 5, Prit. p. 896; 12, f. 72.
" 46.	"	GEMMATA, var. Spectabile, Grun., Schm. At., 8, f. 38.
" 47.	"	BOMBUS, var. Schm. At., 69, f. 28, 29; Donk., B. D., p. 50, 1, f. 7.
" 48.	"	BOMBUS, var. Küntzingii, Schm. At., 13, f. 22, 24.
" 49.	"	NOTABILIS, Grev., T. M. S., 1863, p. 18, 1, f. 9, Schm. At., 8, f. 46-52.
" 50.	"	BOMBUS, var. interrupta, Bail., K. B., p. 100, 29, f. 93; Schm. At., 12, f. 10, 11; Sill. J., 1842, 2, f. 18.
" 51.	"	AMERICANA, Ehrb., Mik., 2, f. 16; 12, f. 16; O'M. I. D., p. 351, 30, f. 30; V. II., 12, f. 37.

PLATE XI.

Figures magnified 500 diameters.

Fig. 1.	NAVICULA	SILLIMANORUM, Ehrb., Mik., 2, 2, f. 13; Lewis, W., M. D., p. 11, 2, f. 8; W. and C., 1, p. 6, 2, f. 2, Crane Pond, Mass., etc.
" 2.	"	PERMAGNA, var. Esox, K., Edwards, M. J., 1860, p. 129; Lewis, N. and R., p. 12, 2, f. 11; H. L. S. T., No. 308, White Mt.
" 3.	"	ANGELORUM, var. excavata, Cleve., N. L. K. D., p. 8, 2, f. 20, S. Monica.
" 4.	"	POWELLII, Lewis, N. and R., Sp., p. 7, 2, f. 2. The typical form is somewhat incurved; White Mountain.
" 5.	"	GREENLANDICA, Cleve., N. L. K. D., p. 7, 1, f. 13, Davis Strait.
" 6.	"	CRUCIATA, Cleve., N. L. K. D., p. 6, 1, f. 11.
" 7.	"	FLORIDIANA, Cleve., N. L. K. D., p. 6, 1, f. 10, near N. fluminensis, Grun.
" 8.	"	FEBEGERII, Cleve., N. L. K. D., p. 9, 2, f. 21. Perhaps a var. of N. Anglica, Ralfs.
" 9.	"	PENSACOLA, Cleve., N. L. K. D., p. 14, 3, f. 39.
" 10.	"	MARGINULATA, Cleve., N. L. K. D., 11, 3, f. 29.
" 11.	"	CLUTHENSIS, var. minesta, Greg., D. C., p. 6, 1, f. 2, Prit., p. 909, 7, f. 73.
Figs. 12, 13.	"	GEMINA, Ehrb., Ber., 1840, p. 19; Schm. At., 13, f. 4-9. May be a var. of Bombus, Richmond, Va.
Fig. 14.	"	INTERPOSITA, Lewis, W., M.D., p. 18, 2, f. 19; Sm. Sp. T. No. 258. May be a var. of N. firma.
" 15.	"	OREGONICA, Kg., K. S. A., p. 71; Prit., p. 907: Abh., 1870, 2, 1, f. 10, 11.
" 16.	"	MARGINATA, Lewis, (Mastogloia) Lewis, N. and R., Sp., p. 26, 2, f. 1; Delaware River.
" 17.	"	POLYONCA, Breb. Rab., S. D., p. 41; Lewis, N. and R., Sp., 2, f. 7; V. H., A., f. 14.
Figs. 18, 19.	"	DIRHYNCHUS, Ehrb., Mik., 35, A., 14, f. 3, Rab., S. D., p. 40, 6, f. 48, Donk., B. D., p. 29, 5, f. 3.

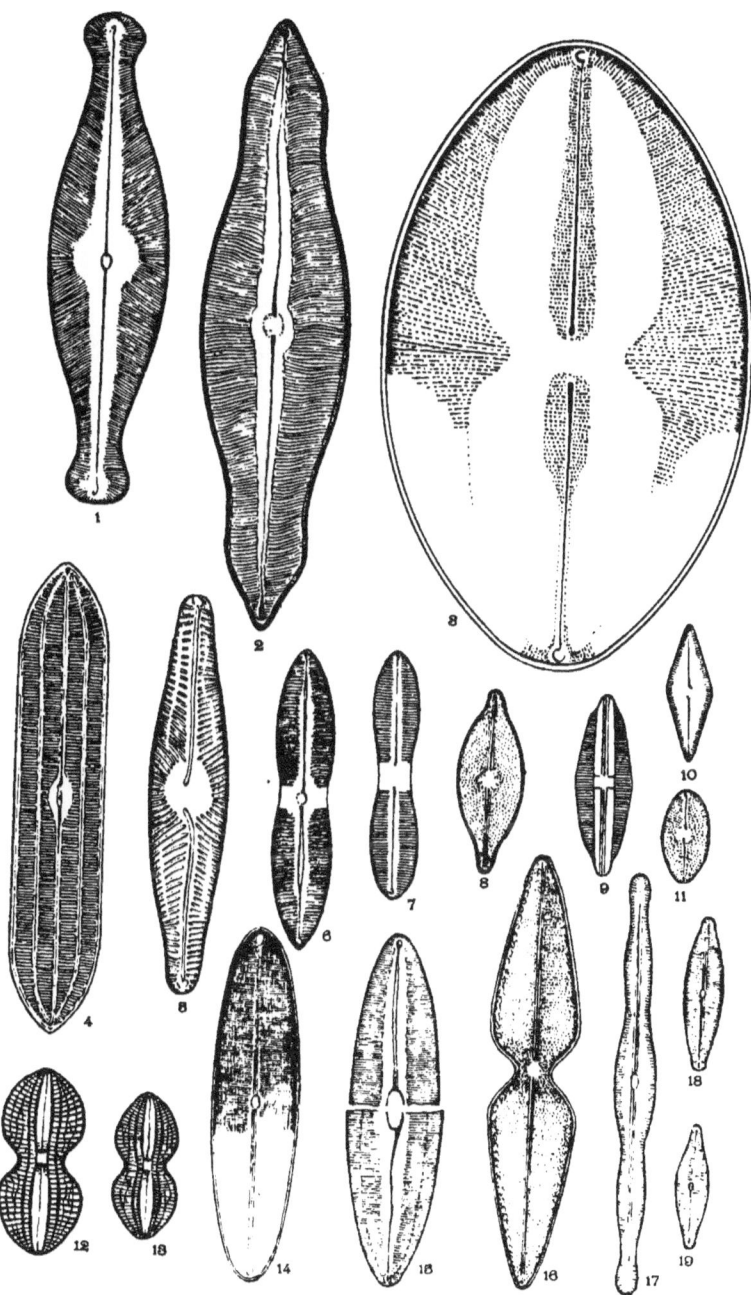

PLATE XII.

Figures magnified 500 diameters.

Fig. 1. NAVICULA TABELLARIA, Kg., K. B., 98, 28, f. 79, 80; Prit., p 896, 12, f. 21; Schm. At., 43, f. 4.
" 2. " SUBNUDA, var. A. S., Schm. At., 12, f. 34.
" 3. " DIPLOSTICHA, Grun., Schm. At., 13, f. 25-30. Gulf of Mexico, etc.
" 4. " INTERRUPTA, Bail., var. Microstaurum, O'M. I. D., p. 354, 30, f. 36.
" 5. " SPLENDIDA, Greg., var. T. M. S., 1856, p. 44, 5, f. 14; Schm. At., 12, f. 31-35; 13, f. 31-34; 69, f. 22, Gulf of Mexico.
" 6. " ENTOMON, Kg., K. B., p. 100, 28, f. 74; Donk., B. D., p. 49, 7, f. 5; Schm. At., 13, f. 43-46, Camp. Bay.
" 7. " MUSCAEFORMIS, Grun., Schm. At., 13, f. 42, 47; var. of didyma, Campeachy Bay.
" 8. " GEMINA, Ehrb., Ber., 1840, p. 19; Schm. At., 13, f. 4; Campeachy Bay.
" 9. " SALVA, A. S., Schm. At., 46, f. 23, Camp. Bay.
" 10. " VACILLANS, A. S., Schm. At., 8, f. 61; 8, f. 42, 43, 52, V. H., 9, f. 9; Camp. Bay.
" 11. " SMITHII, Breb., S. B. D., II, p. 92, Schm. At., 7, 19. Frequent marine.
" 12. " DIDYMA, var. Kg., K. B., p. 100, 4. f. 7; S. B. D., I, p. 53, 17, f. 154; Schm. At., 13, f. 1-3; Lens, Vol. II, p. 235, 4, f. 8.
" 13. " BIOCULATA, Grun., Schm. At., 70, f. 9, 10. Camp. Bay.
" 14. " DIRRHOMBUS, A. S., Schm. At., 11, f. 21, 22; 69, f. 9, Gulf of Mexico, etc.
" 15. " CAMPYLODISCUS, Grun., Schm. At., 8, f. 9, 10, 12, 70, f. 64, 65. Camp. Bay.
" 16. " CUSPIDATA, Kg., K. B., p. 94, 3, f. 24; 37; Rab., S. D., p. 37, 5, f. 16; S. B. D., 16, f. 131.
" 17. " BOMBOIDES, A. S., var. Schm. At., 13, f. 36-40; V. H. B., f. 19. Camp. Bay.
" 18. " CARDINALIS, Ehrb., var. Not as large as Fig. 45, Pl. 9, Schm. At., 44, f. 1, 2.
Figs. 19, 21. " SCHMIDTIANA, Grun., Schm. At., 48, f. 19, 20. Camp. Bay.
" 20, 22. " PEREGRINA, Kg., K. B., p. 97, 28, f. 52; Schm. At., 47, f. 57-60; O'M. I. D., p. 408, 34, f. 6; V. H., 7, f. 2, Utah, etc.
" 23, 24. " LONGA, Ralfs., Schm. At., 47, f. 6, 8, 10; Donk., B. D., p. 54, 8, f. 3. Camp. Bay.
Fig. 25. " FUSCA, Ralfs., Schm. At., 7, f. 2-4, 7-9; 8, f. 33-37; V. H. B., f. 2, marine.
" 26. " SUSPECTA, A. S., Schm. At., 11, f. 12, 13, 25-27. Gulf of Mexico, etc.
" 27. " SCOLIOPLEURA, A. S., Schm. At., 46, f. 63.
" 28. " LACRIMANS, A. S., Schm. At., 12, f. 59, 60, 61. Campeachy Bay, Gulf of Mexico.

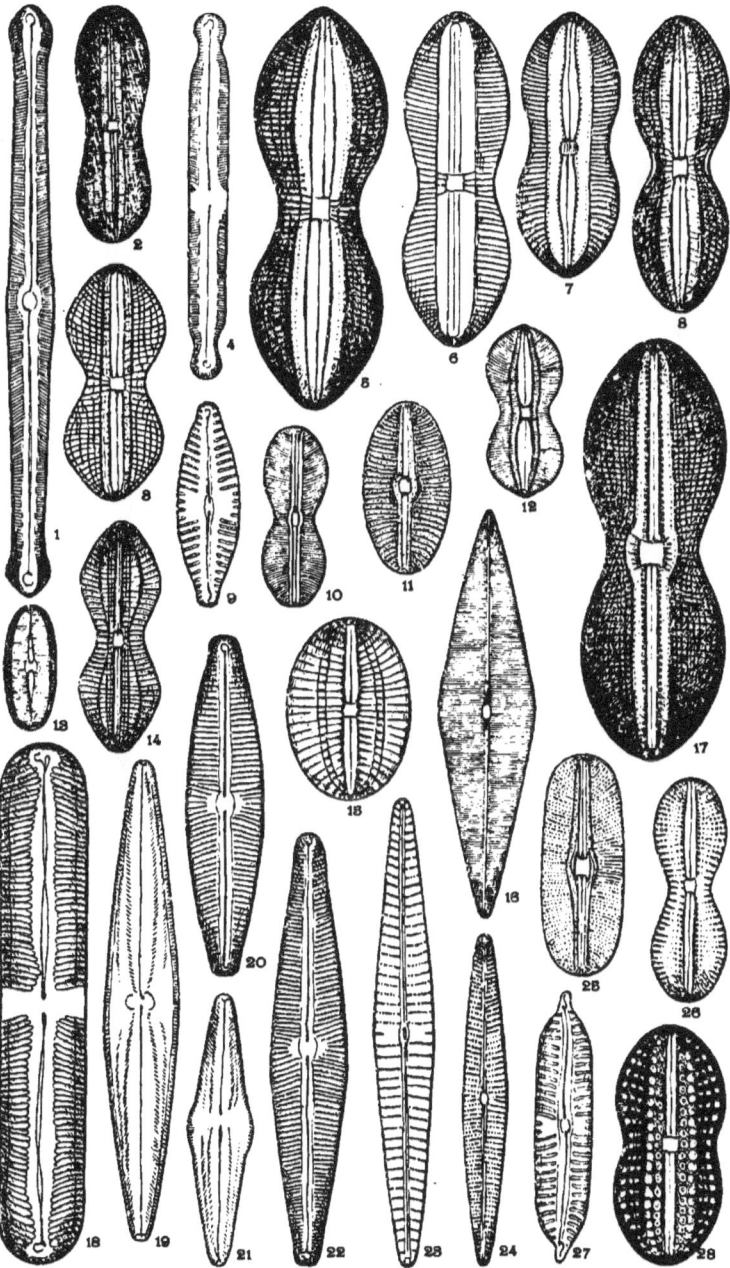

PLATE XIII.

Figures magnified 500 diameters.

Fig. 1.   NAVICULA VIRIDIS, Kg., K. B., p. 97, 4, f. 18; 30, f. 12; Prit. p. 907, 9, f. 133-136.
Figs. 2-4.   "   NOBILIS, Ehrb., Mik., 11, f. 24a,b., Donk., B. D., p. 68, 11, f. 1; vide Fig. 6, below.
Fig. 3.   "   MAJOR, Kg., K. B., p. 97, 4, f. 19, 20; Prit., p. 896, 7, f. 65; 12, f. 15, 31; 16, f. 1-6; Schm. At., 42, f. 8-10, Sm. Sp. T., No. 294.
"   5.   "   GIGAS, Kg., Schm. At., 42, f. 1, 4; K. B., p. 98.
"   6.   "   NOBILIS, Ehrb. (K. B.), p. 98, 4. f. 24; Mik., 15, A., f. 13; 15 B. f. 7; Schm. At., 43, f. 13.
Figs. 7, 9.   "   VIRIDIS, Kg., same as Fig. 1, above. One of our most common and widely distributed fresh water diatoms; very variable in size.
"   9, 10.   "   VIRIDIS, front views of frustules as often seen in pairs, or sometimes four or more united.

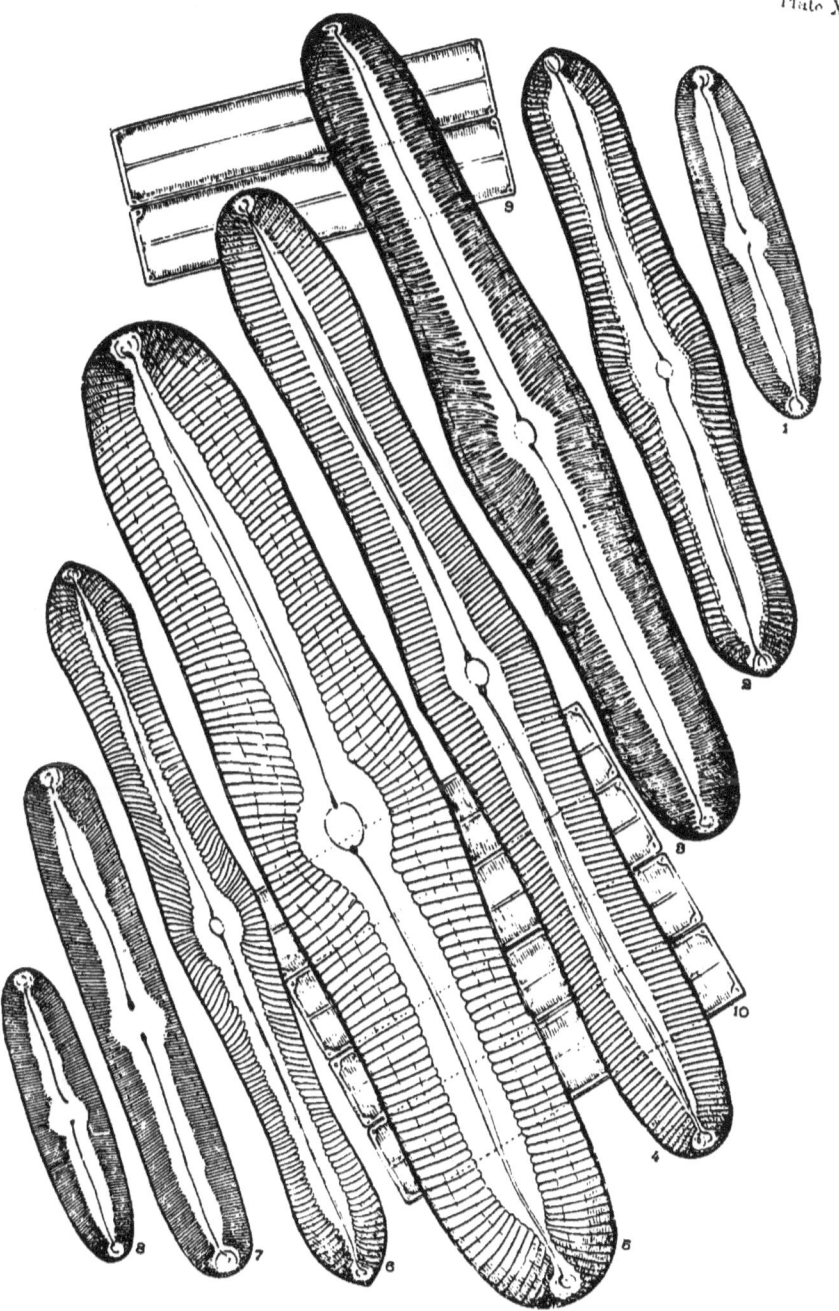

Plate XIII

PLATE XIV.

Figures magnified 500 diameters.

Figs. 1, 2.	NAVICULA	AFFINIS, Ehrb. Mik., 39, 3, f. 79, etc.; K. B., p. 79, 28, f. 65; 30, f. 45, 46; Rab., S. D., p. 40, 6, f. 58; S. B. D., 1, p. 50, 16, f. 143.
" 3, 4.	"	AMPHIRHYNCUS, Ehrb., K. B., p. 95, 4, f. 13; 2, f. 11; S. B. D., I, p. 51, 16, f. 142; Prit., p. 901, 12, f. 6.
Fig. 5.	"	SERRATULA, Grun., Schm. At., 7, f. 42, 43. Campeachy Bay.
" 6.	"	EUDOXIA, O'M. I. D., p. 397, 33, f. 19; Schm. At., 8, f. 39, 40; 70, f. 71, Monterey.
" 7.	"	MARINA, Ralfs., Donk., B. D., p. 19, 3, f. 5, Schm. At., 6, f. 9, Roxbury, Mass.
" 8.	"	NITESCENS, Ralfs., Prit., p. 898; Donk., B. D., p. 8, 1, f. 7; Schm. At., 7, f. 37-41, smaller form.
Figs. 9, 10.	"	PENNATA, N. distens, var. A. S., Schm. At., 48, f. 41-43.
Fig. 11.	"	FISCHERI, A. S., (Maculata, Edwards), Schm. At., 6, f. 38, Neuse River.
" 12.	"	SMITHII, Breb., S. B. D., II, p. 92; Schm. At., 7, f. 19, Campeachy Bay.
Figs. 13, 14.	"	APICULATA, Breb., Lewis, W., M.D., 2, f. 7; Schm. At., 46, f. 56, 58, a smaller form.
Fig. 15.	"	EXCAVATA, Grev., T. M. S., 1866, p. 130, 12, f. 15; Schm. At., 3, f. 22-25.
" 16.	"	PELAGI, A. S., Schm. At., 7, f. 25, 26; Campeachy Bay.
" 17.	"	CALIFORNICA, Grev., Cal., p. 29, 4, f. 5, Schm. At., 3, f. 15, 16.
Figs. 18, 19.	"	ELLIPTICA, Kg., K. B., p. 98, 30, f. 55; Schm. At., 7, f. 27-32; 54, 55; V. H., 10, f. 10.
Fig. 20.	"	Undetermined, A. S., Schm. At., 6, f. 29, from Gulf of Mexico, near small form of N. Brasiliensis, Fig. 21.
" 21.	"	BRASILIENSIS, Grun., 1863, p. 152, 14, f. 10; Schm. At., 6, f. 19, 20, Campeachy Bay.
" 22.	"	NITESCENS, Ralfs., larger form, comp. Fig. 8 above.
" 23.	"	MARINA, allied to N. maculata, comp. Fig. 7 above.
" 24.	"	CALIFORNICA, var. Camp. Bay, Grev., Schm. At., 3, f. 24.
Figs. 25, 26.	"	IMPRESSA, Grun., Schm. At., 6, f. 17, 18, Camp. Bay.
Fig. 27.	"	EXCAVATA, Grev., comp. Fig. 15 above, Camp. bank.
Figs. 28, 29, 30.	"	HENNEDYI, var. W. S., S. B. D., II, p. 93; T. M. S., 1856, p. 40, 5, f. 3; Prit., p. 898, 7, f. 69; Schm. At., 3, f. 3-5; 3, f. 17, 18; Camp. Bay, etc.

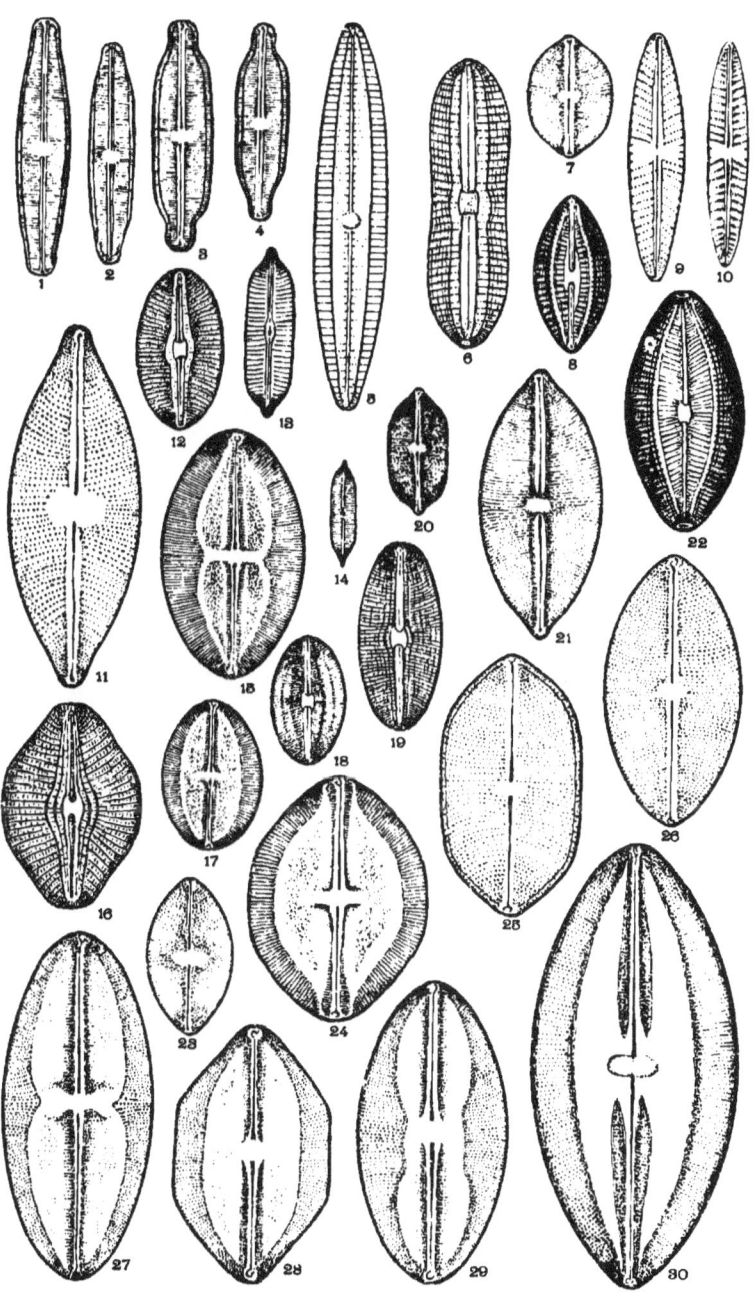

PLATE XV.

Figures magnified 500 diameters.

Fig. 1. NAVICULA LACRIMANS, A. S., Schm. At., 12, f. 59, 60, 61, Gulf of Mexico.
" 2. " MULTICOSTATA, Grun., 1860, p. 524, J, f. 13; Schm. At., 11, f. 14-20, Camp. Bay.
" 3. " COARCTATA, A. S., Schm. At., 11, f. 30-32; 69, f. 11, Camp. Bay.
" 4. " EXEMPTA, A. S., Schm. At., 11, f. 28, 29; 69, f. 13, 40, Camp. Bay.
" 5. " PANDURA, Breb., D. C., p. 15, f. 4. T. M. S., 1856, p. 43, 5, f. 11, Schm. At., 11, f. 1, 2, 4, 8, 9 (*N. crabro* var.), Gulf of Mexico.
" 6. " PRISCA, A. S., Schm. At., 12, f. 67-70, Richmond, Va.
" 7. " CRABRO, Kg., S. B. D., II, p. 94; M. J., 1857, p. 8, 3, f. 11; Schm. At., 69, f. 1. Santa Monica.
" 8. " GIEBELII, A. S., Schm. At., 12, f. 73, Camp. Bay.
" 9. " CRABROXIFORMIS, Grun., Schm. At., 11, f. 24; Gulf of Mexico.
" 10. " PUELLA, A. S., Schm. At., 12, f. 13-15; 69, f. 25, 15, California.
" 11. " COARCTATA, A. S., comp., Fig. 3 above.
" 12. " CONFECTA, A. S., Schm. At., 12, f. 46, Camp. Bay.
" 13. " PRISCA, A. S., same as Fig. 6, above.
" 14. " SPLENDIDA, var. Greg., T. M. S., 1856, p. 44, 5, f. 14, Schm. At., 12, f. 31-35; 13, f. 31-34; 69, f. 22, Gulf of Mexico, etc.
" 15. " DEMTA, A. S., Schm. At., 69, f. 34, Santa Monica.
Figs. 16, 18. " PUELLA, A. S., same species as Fig. 10, above.
Fig. 17. " DONKINII, A. S., Schm. At., 12, f. 63, 64, Camp. bank
" 19. " CHERSONENSIS, Grun., Schm. At., 12, f. 40; 69, f. 21.
" 20. " ORNATA, A. S., Schm. At., 69, f. 5, Monterey.

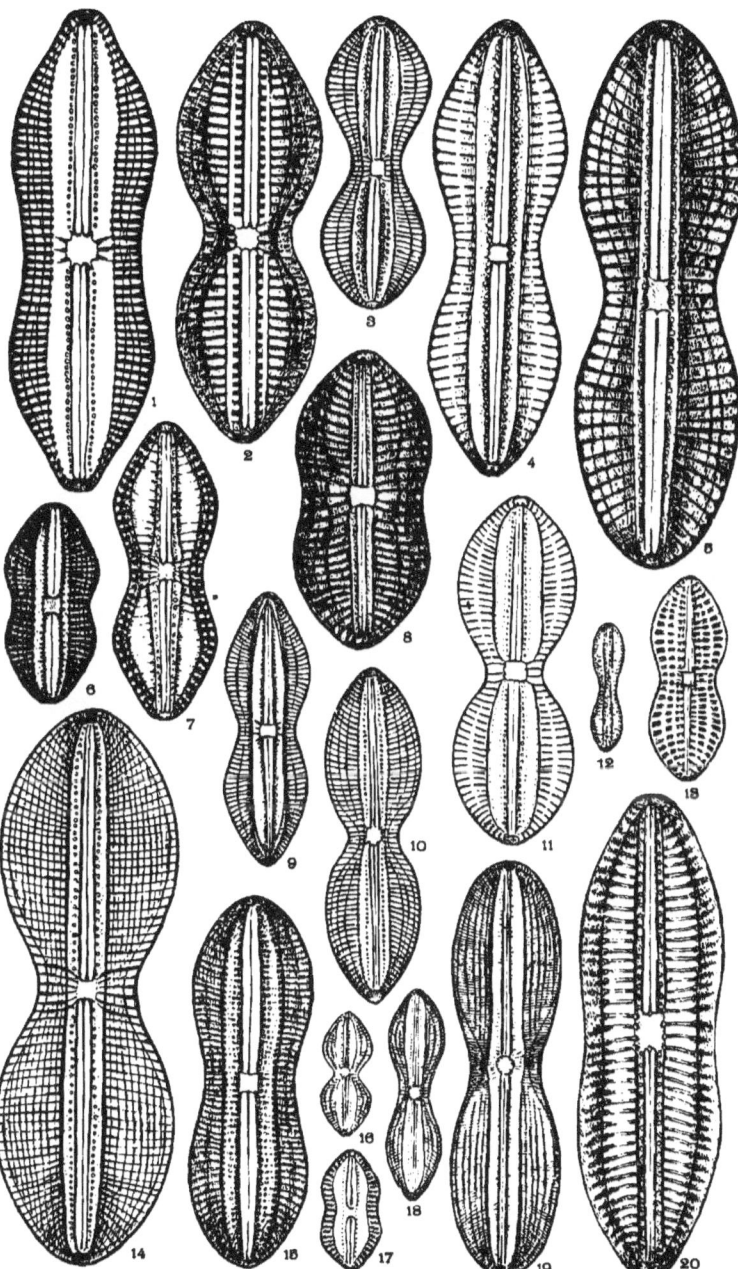

PLATE XVI.

Figures magnified 500 diameters.

Fig. 1. NAVICULA CALIFORNICA, var. Grev., Cal., p. 29, 4, f. 5, Schm. At., 3, f. 15, 16, Camp. bank.
" 2. " IRRORATA, Grev., Cal., p. 27, 4, f. 1, Schm. At., 2, f. 19, 22, 23, Gulf of Mexico.
" 3. " SPECTABILIS, Greg., D. C., p. 98, 1, f. 10; Donk., B. D., p. 12, 2, f. 5; Schm. At., 3, f. 20, 21, 29.
" 4. " LYRA, var. recta, Ehrb. Amer., 1, 1, f. 9; K. B., p. 94, 28, f. 55; Rab., S. D., 5; f. 15; Schm. At., 2, f. 18, 2, f. 24, 25, 32; 3, f. 11, 12, J. M. S., 1878, p. 509; 44. f. 1, Gulf of Mexico.
" 5. " EXSUL, A. S., Schm. At., 2, f. 13, Camp. bank.
" 6. " LYRA, Typ., Camp. bank, Schm. At., 2, f. 16, comp. Figs. 4, 9, 14. Frequent.
" 7. " COUPERI, Ralfs., Prit., p. 898, Schm. At., 2, f. 12, Camp. Bay.
" 8. " ROBERTSIANA, Grev., S. P. D., Pl. 3, f. 9; Schm. At., 2, f. 7.
" 9. " LYRA, var. dilatata, A. S. Schm. At., 2, f. 26, comp. Figs. 4, 6, 14, Gulf of Mexico.
" 10. " DISTENTA, A. S., Schmm., 1867, p. 58, 2, f. 53; Schm. At., 2, f. 14, Camp. bank.
" 11. " DIFFUSA, A. S., Schm. At., 2, f. 28, Mexico.
" 12. " PYGMAEA, Kg., S. B. D., II, p. 91, Schm. At., 70, f. 6, 7, Camp. Bay.
" 13. " FORCIPATA, Grev., M. J., 1859, p. 83, 6, f. 10; Schm. At., 70, f. 17; V. H., 10, f. 3, N. Smithii var.?
" 14. " LYRA, Ehrb., subtypical, Schm. At., 2, f. 32, Neuse River, comp. Figs. 4. 6, 9, 14.
" 15. " CARINIFERA, Grun., Schm. At., 2, f. 1, 2; 70, f. 42, Camp. Bay.
" 16. " ACUMINATA, Ralfs., Prit., p. 909; O'M. 1. D., p. 354, 30, f, 41.
" 17. " IRRORATA, Grev., Cal., p. 27, 4, f. 1, Schm. At., 2, f. 19, 22, 23, comp. Fig. 2, above.
" 18. " APPROXIMATA, Grev., Cal., p. 28, 4, f. 4, var. Schm. At., 2, f. 20, 21, Camp. Bay.
" 19. " APIS, Kg., K. B., p. 100, 28, f. 76, Donk., B. D., p. 48, 7, f. 3; Schm. At., 12, f. 16-25; Schm. At., 69, 41, 43, 44.
" 20. " FORCIPATA, Grev., comp. Fig. 13, above.
" 21. " DIFFLUENS, A. S., Schm. At., 2, f. 15, Camp. bank.
" 22. " SCIOTILLANS, A. S., Schm. At., 70, f. 61, Camp. bank.
" 23. " INFLATA, Kg., K. B., p. 99, 3, f. 36; Rab. S. D., 5, f. 10; S. B. D., II, p. 50, 17, f. 158.
" 24. " CARIBAEA, A., Schm. At., 70, f. 48; Cleve., 1878, p. 5.
" 25. " WEISFLOGII, A. S., Schm. At., 12, f. 26-32, var. 12, f. 30, Gulf of Mexico.
" 26. " LYRA, var. signata, A. S., Schm. At., 2, f. 4, Gulf of Mexico.
" 27. " CLUTHENSIS, Greg., D. C., p. 6, 1, f. 2, Prit., 7, f. 13, Gulf of Mexico.

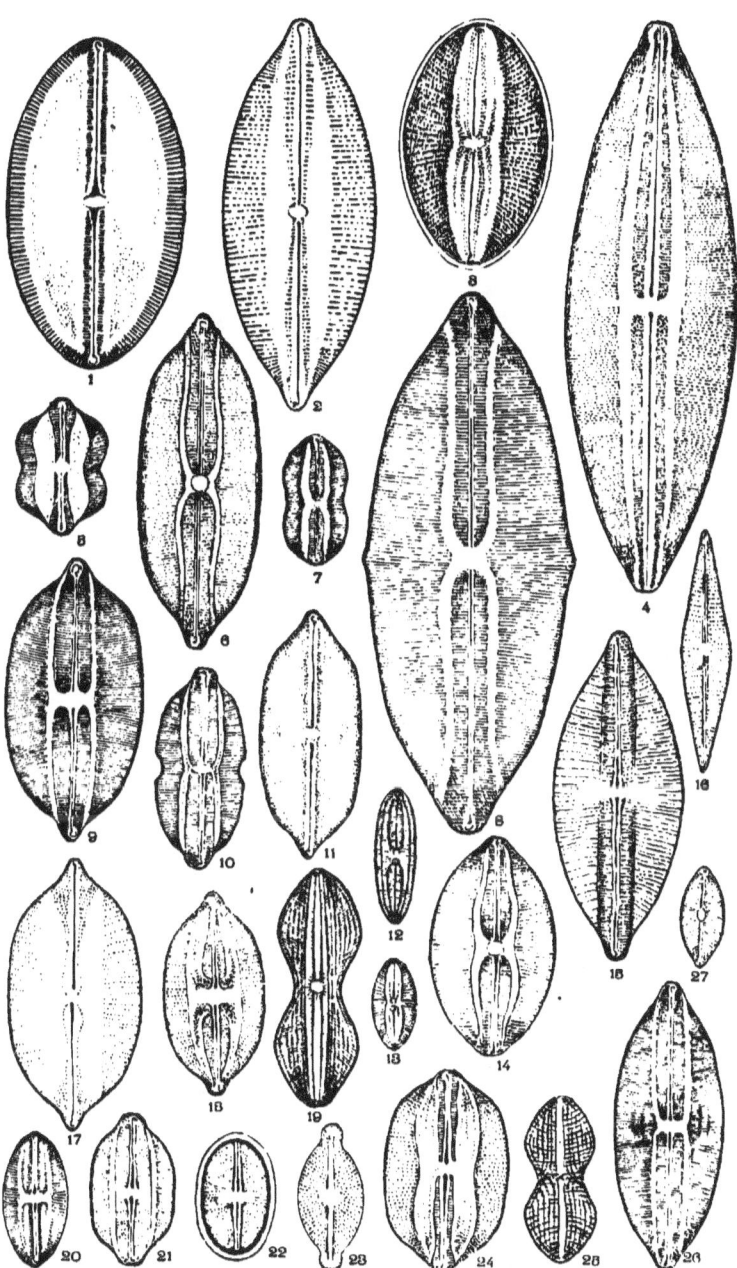

PLATE XVII.

Figures magnified 500 diameters.

Fig. 1.	NAVICULA MACULATA, Edwards, M. J., 1860, p. 129; Sm. S. T., No. 293.
" 2.	"	COUPERII, Ralfs., Prit., 808, Schm. At., 2, f. 12; (var. of Lyra), Camp. Bay, etc.
Figs. 3, 4.	"	PERMAGNA, Edwards, T. M. S., 1866, p. 127, 12, f. 18-21; Lewis, N. and R., p. 12, 2, f. 11, Florida, etc.
Fig. 5.	"	DARIANA, A. S., Schm. At., 42, f. 24, 25, Neuse River.
" 6.	"	SOLARIS, Greg., T. M. S., 1856, p. 43, 5, f. 10; Schm. At., 46, f. 16, N. Providence.
" 7.	"	OSCITANS, A. S., Schm. At., 6, f. 41, Monterey.
" 8.	"	FUTILIS, A. S., Schm. At., 13, f. 17, 18, 69, f. 36; California. Near a var. of didyma.
Figs. 9, 10.	"	RHOMBOIDES, Ehrb., K. B., 94, 28, f. 45; 30, f. 44; S. B. D., I, p. 46, 16, f. 129; Lewis, W., M. D., p. 10, 2, f. 10, 11; H. L. S., M. M. J., 1876, p. 279.
Fig. 11.	"	INDICA, Grev., T. M. S., 1862, p. 95, 9, f. 13; Schm. At., (var. 1, 3, f. 7).
" 12.	"	HUMEROSA, Breb., S. B. D., II, p. 93; Donk., B. D., p. 18, 3, f. 3; Schm. At., 6, f. 3-5.
" 13.	"	SPHAEROPHORA, Kg., K. B., 95, 4, f. 17, Rab., S. D., 40, 6, f. 65; S. B. D., I, p. 52, 17, f. 148.
" 14.	"	GRANULATA, Breb., T. M. S., 1858, p. 17, 3, f. 19, Schm. At., 6, f. 15, 16.
" 15.	"	HUMEROSA, Breb.
Figs. 16, 17.	"	RETUSA, Breb., S. B. D., II, p. 92; Schm. At., 46, f. 45, 46, Creswell.
Fig. 18.	"	COMMUTATA, Grun., Schm. At., 45, f. 22, 23, 35, (= N. viridis, var.)
Figs. 19, 20.	"	HEMIPTERA, Kg., K. B., 30, f. 11; Schm. At., 43, f. 26-30, O'M., I. D., p. 349, 30, f. 22.
Fig. 21.	"	DICEPHALA, Ehrb. Mik., 16, 1, f. 17; 17, 2, f. 7; S. B. D., I, p. 53, 17, f. 157; Schm. At., 72, f. 29-33; 44, f. 33, 34.
" 22.	"	DUBIA, Ehrb. Mik., 39, 2, f. 82; K. B., 96, 28, f. 61; Schm. At., 49, f. 11, 24.
" 23.	"	BINOIDES, Ehrb. Ber., 1840, p. 18, K. B., p. 100, 3, f. 35, Rab., S. D., p. 41, 5, f. 5; S. B. D., I, p. 53, 17, f. 150.
Figs. 24, 25.	"	LATISSIMA, Greg., T. M. S., 1856, p. 40, 5, f. 4, Prit., p. 903, 7, f. 70, Schm. At., 6, f. 7, var. N. humerosa.

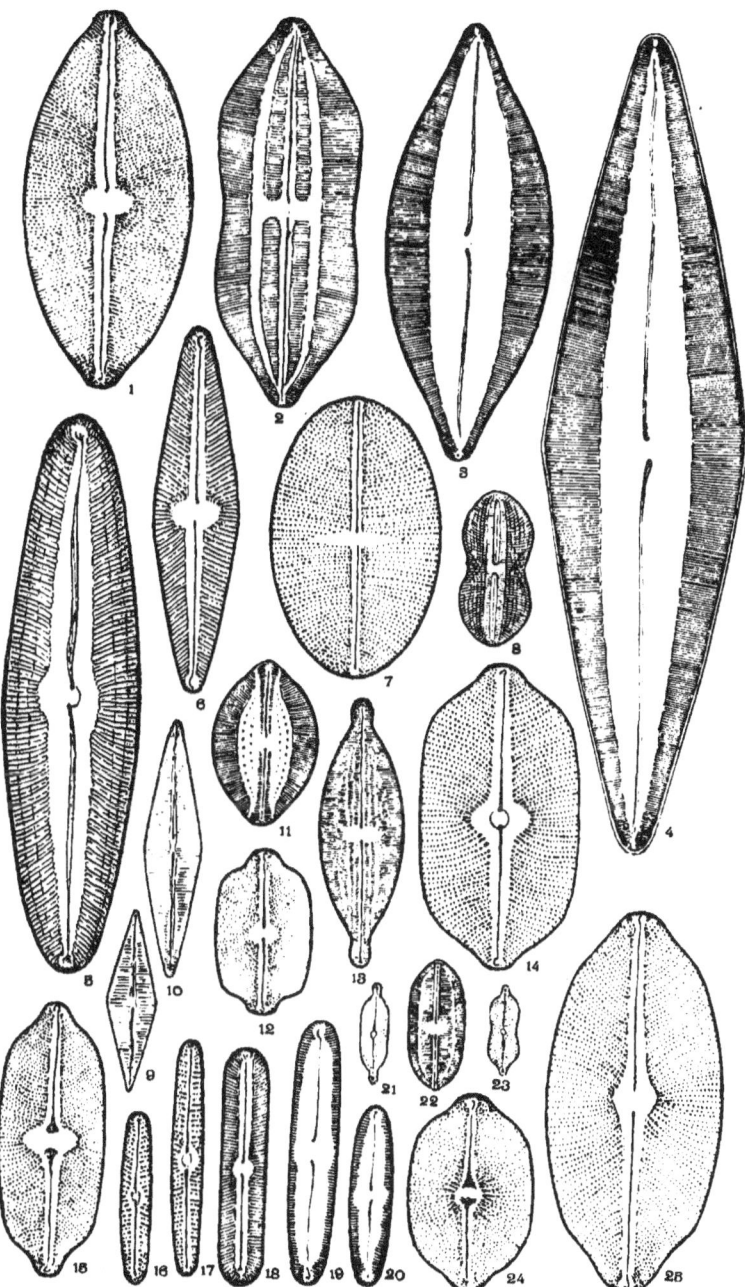

PLATE XVIII.

Figures magnified 500 diameters.

Fig. 1. NAVICULA COLUMNARIS, Ehrb. Mik., 14, f. 23; 7, 1, f. 4; Schm. At., 49, f. 1. New York, Mass.

" 2. " EXCENTRICA, Grun., 1860, p. 545, 1, f. 1; Schm. At., 50, f. 6, 7, marine.

" 3. " TUMESCENS, Grun., Schm. At., 49, f. 10; (*N. firma* var.) Maine.

" 4. " IRIDIS, Ehrb., K. B., 28, f. 42; Donk., B. D., p. 30, 5, f. 6; Schm. At., 49, f. 2, (larger var. of *N. firma?*) New York.

" 5. " DILATATA, Ehrb. Mik., 3, f. 25, Rab., S. D., p. 37, comp. Fig. 7, typical form.

" 6. " COLUMNARIS, *vide* Fig. 1, above.

" 7. " DILATATA, var. Ehrb. Mik., 16, 3, f. 25; 8, 3, a., f. 9, Schm. At., 49, 6, 9; Maine, Mass.

" 8. " DILATATA, Schm. At., 49, 9. Monmouth, Maine.

" 9. " BLEISCHII, Jan. and Rab., p. 9, 2, f. 10, Schm. At., 50, f. 22, 25, nearly related to *N. excentrica*.

" 10. " AMPHIGAMPHUS, Ehrb. Mik., 6, 1, f. 20; 7, 1, f. 16; K. B., p. 93, 28, f. 40; Rab., S. D., p. 38, 6, f. 47; Schm. At., 49, f. 31-34. Frequent, usually represented by smaller forms.

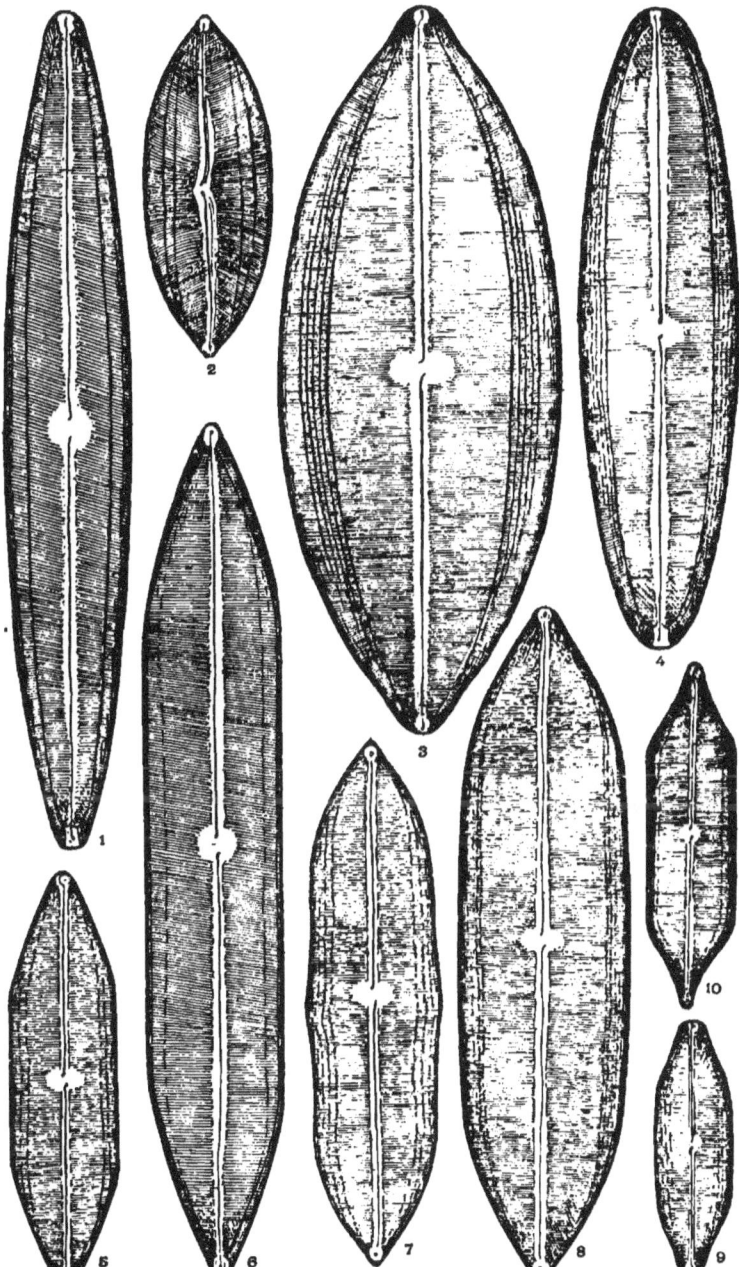

PLATE XIX.

Figures magnified 500 diameters.

Fig. 1. NAVICULA PROBABILIS, A. S., Schm. At., 50, f. 46, Camp. Bay.
" 2. " FORMOSA, Greg., T. M. S., 1856, p. 42, 5, f. 6, Schm. At., 50, f. 8-14. Only a var. of *N. permagna*, H. L. S., Salt Lake; Neuse River, etc.
" 3. " PRODUCTA, a var., W. S., S. B. D., I, p. 51, 17, f. 144, Prit., p. 902, 7, f. 66; Schm. At., 49, f. 37-39, 50, f. 47. Marine.
" 4. " HEXAPLA, A. S., Schm. At., 50, f. 50, Oregon.
" 5. " AMPHIGOMPHUS, Ehrb. Mik., 6, 1, f. 20; 7, 1, f. 6, Rab. S. D., p. 38, 6, f. 47; Schm. At., 49, f. 31-34. Maine, etc.
" 6. " LINEARIS, Grun., 1860, p. 546, 1, f. 2, Schm. At., 50, f. 38; O'M. I. D., p. 371, 31, f. 39. Gulf of Mexico.
Figs. 7, 8. " MORMONORUM, Grun., Schm. At., 44, f. 24-26. Utah.
Fig. 9. " FIRMA, (var. subundulata, Kg., K. B., p. 92, 21, f. 10, S. B. D., 16, f. 138; Schm. At., 49, f. 3, 14; var. Subundulata, f. 16, Hudson River.
" 10. " FIRMA, typical form.
" 11. " FORMOSA, var. Same species, different form, as Fig. 2, above. Utah, etc.
" 12. " PRODUCTA, var., Schm. At., 50, f. 49; says *fralich*. Marine.
" 13. " AMPHIRHYNCHUS, Ehrb., K. B., p. 95, 4. f. 13; 2, f. 11, S, B. D., I, p. 51, 16, f. 142; Prit., p. 901, 12, f. 6. Fresh-water.
" 14. " NODOSA, Ehrb. Mik., 17, 2, f. 12, 13, K. B., 28, f. 82; Rab., S. D., 41, 6, f. 86; Schm. At., 45, 56-58. Fresh-water.
" 15. " PRODUCTA, var. nearer the typical form, Fig. 12, compare refs., Fig. 3. Marine.
Figs. 16-19. " BREBISSONII, Kg., K. B., 3, f. 49; 30, f. 39, Rab., S. D., 6, f. 54; Schm. At., 44, f. 16-19; Sm. Sp. T., No. 249, O'M. I. D., p. 350, 30, f. 24. Fresh-water.
" 20, 21. " DIVERGENS, Ralfs., Prit., p. 896, Schm. At., 44, f. 6, 14, 15, 42. Fresh-water.
Fig. 22. " VACILLANS, A. S., Schm. At., 8, f. 61, V. II., 9, f. 9. Marine.
" 23. " AMPHISBAENA, Borg. Mik., 10, 1, f. 7; 7, 2, f. 5, etc., S. B. D., I, p. 51, 17, f. 147; Donk. B. D., p. 36, 5, f. 13. Small form, fresh-water.

Plate XIX.

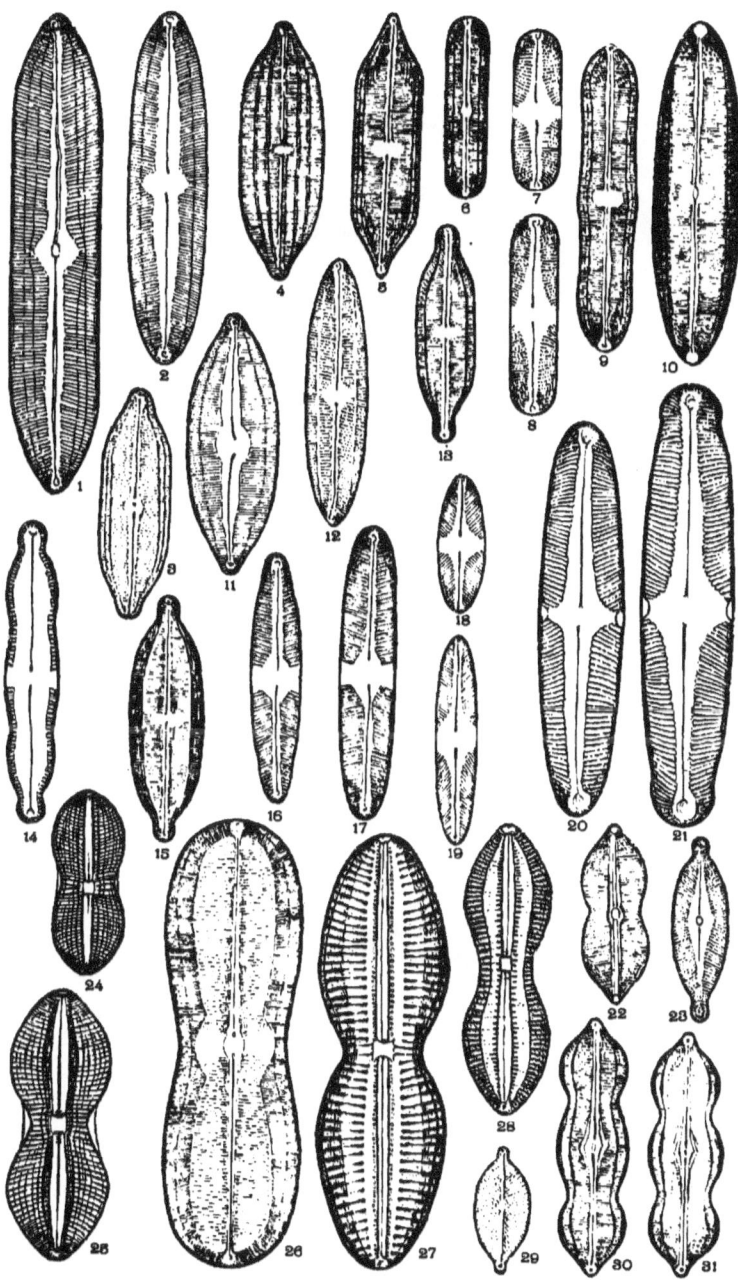

Figs. 24, 25.	"	DIDYMA, Kg., K. B., 100, 4, f. 17; S. B. D., I, p. 53, 17, f. 154; Prit., p. 893, 7, f. 61; Schm. At., 13, f. 1-3. Marine.
Fig. 26.	"	MOESTA, A. S., Schm. At., 69, f. 18, 19. Fresh-water.
" 27.	"	SEPARABILIS, A. S., Schm. At., 11, f. 3, 5, 6, 7, 10, 11. Marine.
' 28.	"	PRAESTES, A. S., Schm. At., 12, f. 57, 58. Marine.
' 29.	"	INFLATA, Kg. K. B., 3, f. 66; S. B. D., II, p. 50, 17, f. 158, Sm. Sp. T., No. 284. Fresh-water.
Figs. 30, 31.	"	HITCHCOCKII, Ehrb. Mik., 53, f. 11; 33, 12, f. 24; Prit., p. 89, 7, f. 62; Schm. At., 49, f. 35, 36. Rather unusually large form; Pennsylvania waters. Ordinarily a smaller diatom than *N. amphisbaena*, Fig. 23.

PLATE XX.

Figures magnified 500 diameters.

Fig. 1.   NAVICULA PRETEXTA, Ehrb., Greg., D. C., p. 9, 1, f. 11; Schm. At., 3, f. 30, 34. Marine.
Figs. 2, (18).   "   SINGULARIS, A. S., Schm. At., 43, f. 30, a variety of acrosphaeria, Kg. Fresh-water.
Fig. 3.   "   MESOLEPTA, Ehrb., var. undulata, Grun., K. B., p. 101, 28, f. 73; Rab., S. D., 41, 6, f. 72; Schm. At., 45, f. 52, 53, 70. Fresh-water.
"   4.   "   ASPERA, (Stauroneis), Kg., A. B., 29, f. 12, Prit., 9, 14; V. H., 10, f. 13; Sm. Sp. T., No. 407.
"   5.   "   ACUMINATA? W. S., fig. probably represents merely a var. of N. viridis. Fresh-water.
"   6.   "   HEMIPTERA, Kg., var. macilenta, Grun., K. B., p. 99, 30, f. 11; Grun., 1860, p. 519, 2, f. 20; Schm. At., 43, f. 26, 28. Fresh-water.
"   7.   "   PRETEXTA, Ehrb., var. of Fig. 1. Marine.
"   8.   "   SMITHII, Breb., S. B. D., II, p. 92; Schm. At., 7, f. 9, O'M., I. D., p. 382, 32, f. 8. Fresh-water.
Figs. 9, 10, 11.   "   GIBBA, Kg., 28, f. 70; Schm. At., 45, f. 45-51, V. H., A., f. 12. Fresh-water.
"   12, 17.   "   TABELLARIA, (var. macilenta), Kg. K. B., 98, 28, f. 7, 9; Prit., 896, 12, f. 21; Schm. At., 43, f. 4. Fresh-water.
Fig. 13.   "   MESOLEPTA, Ehrb. Mik., 17, 2, f. 11; K. B., 101, 28, f. 73. Fresh-water.
Figs. 14, (16).   "   MESOLEPTA, var. Stauroneiformis, Lewis, (N. mesotyla), Ehrb. Mik., 10, 1, f. 10; Rab., S. D., p. 41, 5, f. 6; Schm. At., 45, f. 52, 55. Fresh-water.
Fig. 15.   "   MESOLEPTA, Ehrb., comp. Figs. 3, 13, 14, 16.
"   18.   "   SINGULARIS, vide Fig. 2.
Figs. 19, 20.   "   OBTUSA, Ehrb. Mik., 2, 3, f. 7; 13, 2, f. 9, 20, 1, f. 51, etc., Abh., 1870, 2, 1, f. 37. Fresh-water.
Fig. 21.   "   SEMEN, Ehrb. Mik., 38, A., 20, f. 2; 39, 3, f. 88, 89; Abh., 1871, 1, D., f. 2, 3; Sill. J., May, 1851, f. 48, 49, Schm. At., 72, f. 1. Fresh-water.
"   22.   "   ELGINENSIS, Ralf., Prit., p. 902, also near dicephala, Ehrb. Mik., 16, 1, f. 17; Abh., 1872, 1, E, f. 4; K. B., 28, f. 60, 82. Fresh-water.

Plate XX

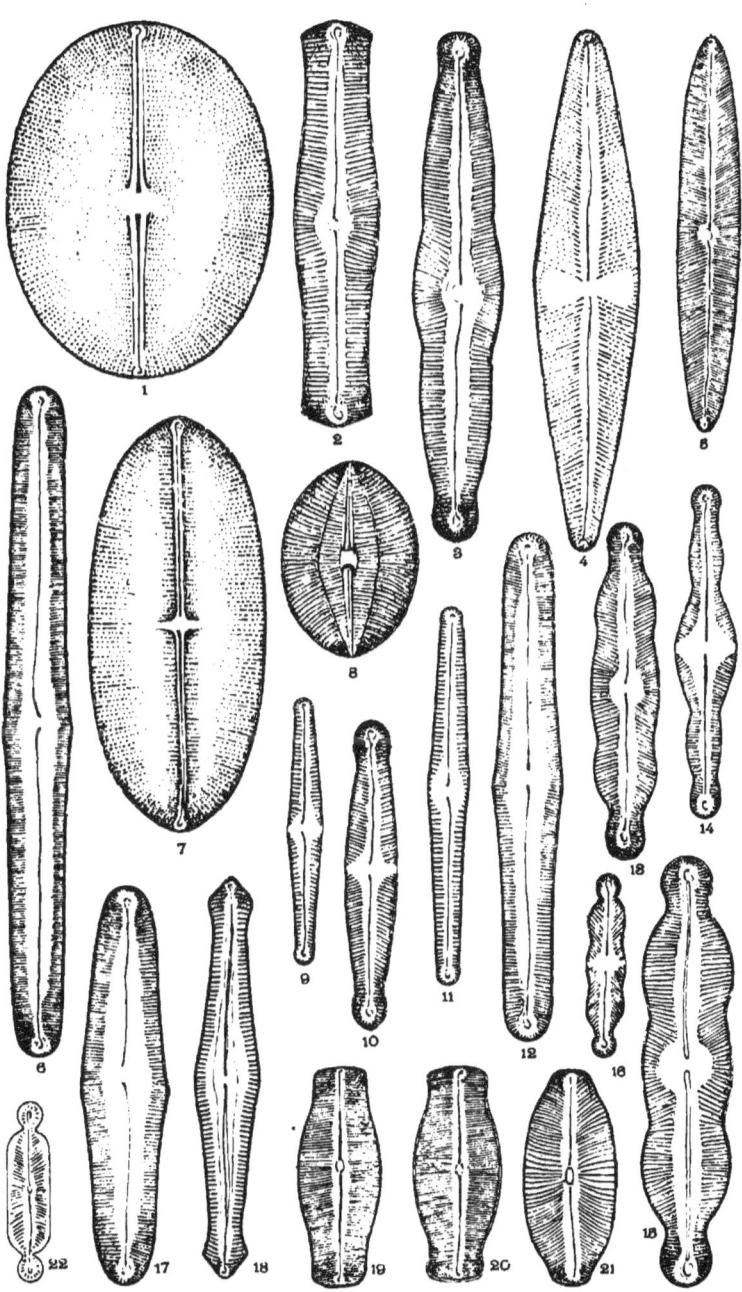

PLATE XXI.

Figures magnified 500 diameters.

Figs. 1, 2, 3. NAVICULA DACTYLUS, Kg., K. B., p. 98, 28, f. 59. Schm. At., 42, f. 6; V. H., 5, f. 1. Fresh-water.
Fig. 4. " DACTYLUS, Kg., var. subgigas, A. S. Fresh-water.
" 5. " TABELLARIA, Ehrb., (var. transversa, n. var.), K. B., p. 98, 28, f. 79, 80; Prit., 896, 12, f. 21, Schm. At., 43, f. 4. Fresh-water, New York.
Figs. 6, 7. " RADIOSA, Kg., K. B., 4, f. 23; Schm. At., 47, f. 50-52; Sm. Sp. T., No. 311. Fresh-water.
" 8, 9. " ACROSPHAERIA, Kg. K. B., 5, f. A., Schm. At., 43, f. 16, 21, 22. Fresh-water.
Fig. 10. " OBLONGA, Kg. (var., curta, n. var.), K. B., 4, f. 21; Schm. At., 47, f. 63-68; Sm. Sp. T., No. 300. Marine.
" 11. " ACROSPHAERIA, forma major. Fresh-water.

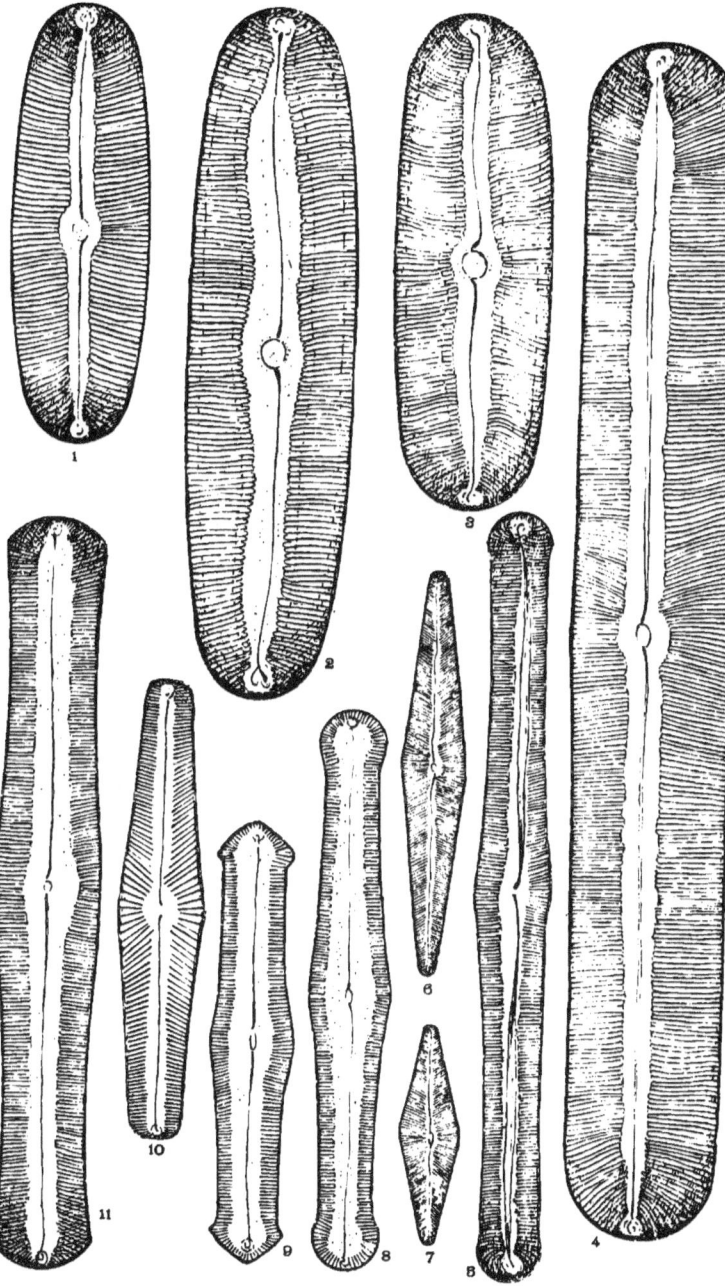

PLATE XXII.

Figures magnified 500 diameters.

Fig. 1. BIDDULPHIA TUOMEYI, var. Roper, T. M. S., 1859, p. 8, 1, f. 1, 2; Prit., p. 848, 6, f. 10; V. H., 98, f. 23. Marine.
Figs. 2, 3, 4. PYXIDICULA GLOBATA. This fig. copied from Pritchard has no value; it is no diatom.
Fig. 5. BIBLARIUM (HYLOBIBLIUM), eccentricum, Ehrb. Mik., 33, 12, f. 31. Marine.
" 6. " (STYLOBIBLIUM) divisum, Ehrb. Mik., 33, 12, f. 30.
" 7. SYMBOLOPHORA ACUTA, Ehrb. Mik., 33, f. 21, Mic. Dicty, 43, f. 54. Of no value; probably only a fragment of a Triceratium.
Figs. 8, 9. GOMPHONEMA OREGONICUM, Ehrb. Mik., 37, 2, f. 12, 13, G. Herculaneum, Ehrb. a large var. of G. Capitatum, Ehrb. Fresh-water.
Fig. 10. NAVICULA TUSCULA, Ehrb. Mik., 6, 1, f. 13, a. b., V. H., 10, f. 14. Fresh-water.
" 11. " SUBINFLATA, Grun., M. S. S., Cleve, Vega., p. 470, f. 50, var. of *limosa*, Grun., V. H., 12, f. 20. Fresh-water.
" 12. " UNDULATA, var. of *limosa*, Grun., V. H., 12, f. 22. Fresh-water.
" 13. " GIBBERULA, var. *limosa*, Grun., V. H., 12, f. 19. Fresh-water.
" 14. " SCHUMANNIANA, (*trochus?* Ehrb.) Grun., V. H., 11, f. 21. Fresh-water.
Figs. 15, 16. " TENELLA, Breb., Prit., p. 904, Schm. At., 47, f. 45, 46, — *N. radiosa*, var. Fresh-water.
Fig. 17. NITZSCHIA CURSORIA, (Bacilaria Cursoria), Prit., 784, 4, f. 20, Grun., Cleve, 1880, p. 89; V. H., 62, f. 19. Fresh-water.
" 18. NAVICULA DEWITTIANA, K. and S., Bull. Tor. Bot. Club, August, 1889, Atlantic City, N. J. Fossil.
" 19. GOMPHONEMA CAPITATUM, Ehrb. Mik., 16, 3, f. 37; K. B., 16, f. 2; Rab. S. D., 8, f. 15; S. B. D., 28, f. 237. Fresh-water.
" 20. SCHIZONEMA VULGARE, Thwaites, A. N. H., 1848, 12, f. 41, V. H., 17, f. 6. Marine.
" 21. NAVICULA RHOMBOIDES, Ehrb., K. B., 28, f. 45; 30, f. 44; Rab., S. D., 5, f. 15; S. B. D., 16, f. 129. Fresh-water.
" 22. PLEUROSIGMA DELICATULUM, W. S., S. B. D., 21, f. 302, Prit., p. 918; Sm. Sp. T., No. 398. Fresh-water.
Figs. 23, 24. EPITHEMIA MARINA, Donk., T. M. S., 1868, p. 29, 4. f. 14.
Fig. 25. AMPHORA PLICATA, (*A. lineolata*, Ehrb.) Greg., T. M. S., 1857, p. 70, 1, f. 3; H. L. S., Lens, p. 75, No. 25, 2, f. 3; Sch m. At., 26, f. 50.
Figs. 26, 27. PODOSPHENIA (LICMOPHORA), papeana, Grun., 1863, p. 138, 14, f. 11, Fig. 27, var. elongata.
Fig. 28. COLLECTONEMA VULGARE, Thwaites, A. N. H., 2d Ser., Vol. I, 12, f. H., S. B. D., II, p. 70, 56, f. 35, Sm. Sp. T. Fresh-water.
" 29. BIDDULPHIA LUNATA, Ehrb. Mik., 18, f. 53. Very doubtfully a Biddulphia.

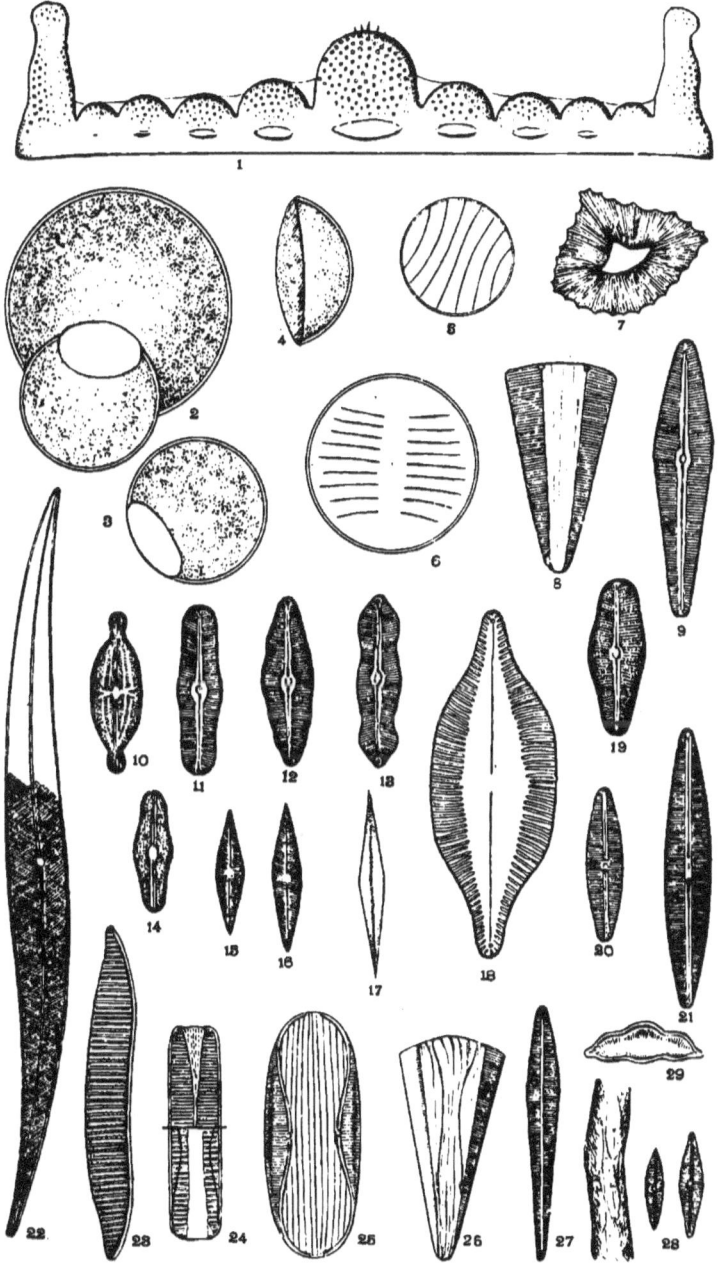

Plate XXII.

PLATE XXIII.

Figures magnified 500 diameters.

Fig. 1.   NAVICULA PRESTIOPHORA, Jan., Schm. At., 70, f. 72, (= *N. Crabro* var.) Marine.
" 2. " MARGINATA, O'M. not Lewis's *Marginata* which is a *Mastogolia*. This may be *N. incurvata*, Greg.
" 3. " BOMBUS, Kg., Greg. D. C., p. 12, 1, f. 12; Schm. At., 69, f. 28, 29; V. H. B., f. 22.
" 4. " SIMULANS, O'M., figure, p. 373, 31, f. 47. Marine. O'Mara's figure too angular and Donkin's figure acuminate. H. L. S.
Figs. 5, (11). " LIBER, W. S., S. B. D., I. p. 48, 16, f. 133; Donk., B. D., p. 62, 9, f. 5; Schm. At., 50, f. 16-18. Freshwater.
Fig. 6. " ISOCEPHALA, Ehrb., E. Amer., p. 133; K. B., p. 101; O'M. I. D., 30, f. 31.
" 7. " VENETA, Kg., K. B., 30, f. 76; Rab. S. D., 6, f. 83; Sm. Sp. T., No. 328; V. H., 8, f. 3, 14, f. 34.
" 8. " DUPLICATA, Ehrb., Mik., 21, f. 35; K. B., 28, f. 78; Rab. S. D., 5, f. 4.
" 9. " DISPHENIA, Kg., K. B., p. 93, 28, f. 54; Prit., 908.
" 10. " FUSIDUM, Ehrb. Mik., 7, 1, f. 16, 5, f. 4; K. B., p. 96.
" 11. " LIBER, W. S. *vide* above, Fig. 5.
" 12. " SILLIMANORUM, Ehrb. Mik., 2, f. 13; Lewis, W., M. D., p. 11, 2, f. 8; W. and C., 1, p. 6, 2, f. 8.
" 13. " JOHNSONII, O'M. I. D., p. 373, 31, f. 46; V. H. B., t. 28; Sm. Sp. T., No. 286. A very delicate diatom.
" 14. " ESOX, Kg., K. B., p. 94, 28, f. 53; Prit. 896, 12, f. 43; O'M. I. D., 31, f. 33. An exaggerated form of *N. permagna*?
Figs. 15, 16. " TRUNCATA, Kg. These figures from Rab. S. D., 6, f. 67. Entirely distinct in Donk.'s form, which see Plate at end.
Fig. 17. " REINHARDTI, Grun., V. H., 7, f. 5, 6; Sm. Sp. T., No. 311. Var. of *N. varians*, Greg.
" 18. " UNDOSA, Ehrb., Mik., 39, 3, f. 90; Abh., 1871, 1, f. 4; K. B., 28, f. 83; Rab. S. D., 6, f. 82.
Figs. 19, 20. " PUPULA, Kg., K. B., 30, f. 40; Rab. S. D., 6, f. 82; V. H., 13, f. 15, 16.
Fig. 21. " VELOX, Kg., K. B., 3, f. 66; Rab. S. D., 5, f. 12.
" 22. GOMPHONEMA, AMERICANUM, (*capitatum*), Ehrb. Mik., 16, 13, f. 33; 17, 1, f. 53; 5, 1, f. 38; 9, 1, f. 35.
" 23. EUNOTIA, HEPTODON, Ehrb. Mik., 16, 2, f. 26; 17, 1, f. 34.

Plate XXIII.

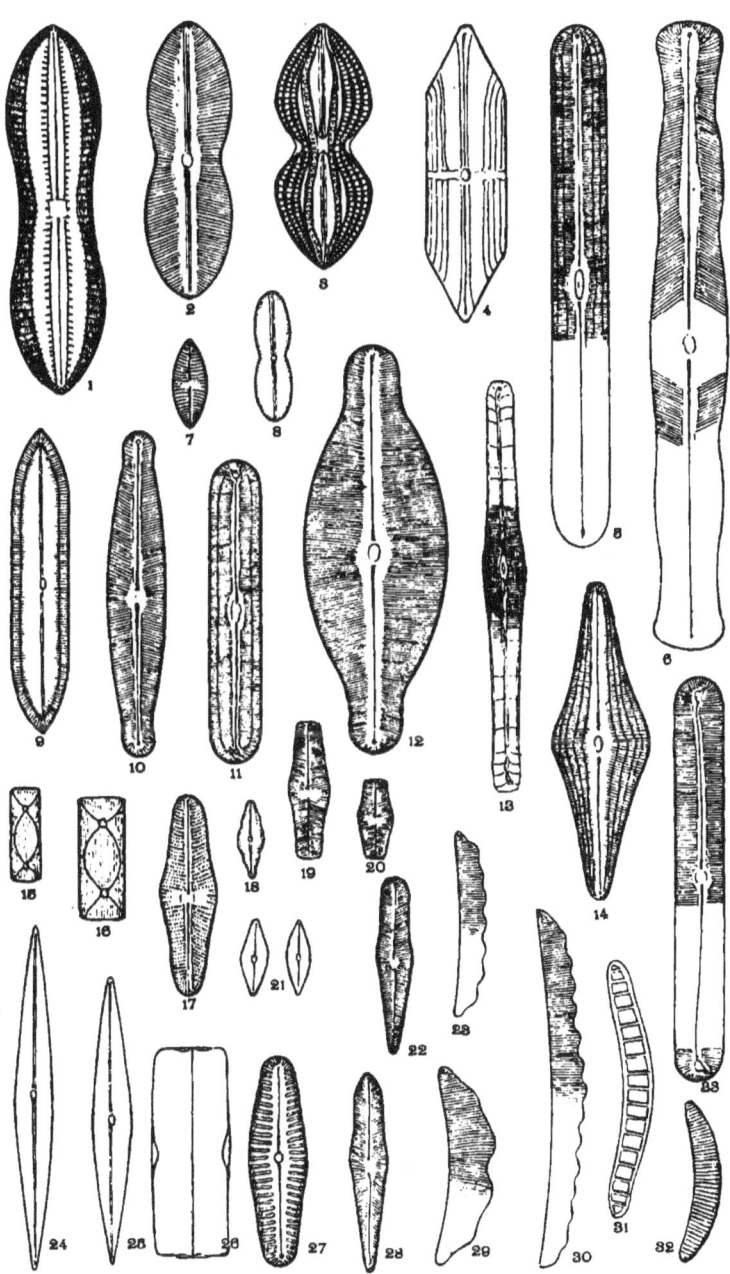

Figs. 24, 25. NAVICULA AMPHIONYS, Ehrb. Mik., 39, 3, f. 80; 37, 3, f. 5, 6, Sill. Jour., May, 1851, p. 47 · K. B. 28, f. 37; Rab. S. D., 6, f. 63. Very like *N. fusiformis*, Grun., V. H., 14, f. 33.
" " 26, 27. " LATA, Breb., V. H., 6, f. 1, 2; Donk., B. D., 55, 18, 167.
Fig. 28. GOMPHONEMA, AMERICANUM, Ehrb. (= *G. acuminatum*), Mik., 16, 3, f. 33.
" 29. EUNOTIA, SELLA, Ehrb. Mik., 33, 12, f. 17; K. B., 29, 50; Rab. S. D., p. 17, 1, f. 28.
" 30. " TREDENARIU, Ehrb., 33, 10, f. 9.
Figs. 31 32. EPITHEMIA, ZEBRA, Kg., forms of; K. B., 5, 12, f. 6; Rab. S. D., 1, f. 8; S. B. D., 12, 1, f. 4; V. H., 31, f. 9, 11, 14. Same as *E. Zebrina*, Ehrb. Mik., 12, f. 25, 26.

PLATE XXIV.

Figures magnified 500 diameters.

Fig. 1. NAVICULA DIGITUS, Ehrb. Mik., 5, 3, f. 2, Ralfs., Prit., p. 908.
Hardly admits of separation from *N. dactylus*.
" 2. " LEWISIANA, Grev., T. M. S., 1863, p. 115, 1, f. 7.
Marine, N. America?
Figs. 3, 4. " BAYLEYANA, Grun., Schm. At., 6, f. 26, 27. Marine.
Fig. 5. " SCHULTZII, Kain., Bull. Tor. Bot. Club, March, 1889, p. 75, Pl. 89, f. 2. Fossil Atlantic City. Nearly allied to N. maculata, H. L. S.
Figs. 6, 7. " COSTATA, Kg., K. B., p. 93, 3, f. 56, Prit., 906.
Fig. 8. " DIOMPHALA, Ehrb., K. B., 28, f. 63, Prit., p. 900.
" 9. " ELEGANS, S. B. D., I, p. 49, 16, f. 137, Sm. Sp. T., No. 270. Striae wavy.
" 10. " SIGMA, Ehrb. Mik., 14, f. 21; 18, f. 61, Sill. J., May, 1842, 2, f. 24 *Pleurosigma delicatulum* of later observers.
" 11. " SIGMA, Ehrb. Pleurosigma angulatum, S. B. D., 21, f. 205; V. H., 18, f. 24.
" 12. " FORMICA, Ehrb. Mik., 17, 2, f. 10; 4, 3, f. 8. Fresh-water.
" 13. " POLYONCA, a var., Breb., Lewis, N. and R. Sp., 2, f. 7, V. H., A., f. 14. Fresh-water.
" 14. " LEPTOSTIGMA, Ehrb., Mik., 33, 12, f. 21, E. Ber., 1845 Fossil, U. S.
" 15. " PUMILA, Grun., var. of N. veneta, Kg., V. H., 8, 1. 6, 7; 14, f. 35.
" 16. " NODULOSA, Kg., K. B., 3, f. 57; 28, f. 71, Sm. Sp. T., No. 297, var. of *N. mesolepta*.
Figs. 17, 18. " RHOMBICA, Greg., M. J., 1855, p. 40, 4, f. 16, Prit., 903, 7, f. 71. Striae indistinct, radiating. F. W.
Fig. 19. " CRUCIFORMIS, Donk., M. J., 1861, p. 10, 1, f. 7, V. H., A., f. 8.
" 20. " CARASSIUS, Ehrb., K. B., 28, f. 67, Rab., S. D., 6, f. 57, Lewis, W., M.D., 2, f. 21.
Figs. 21, 22. " TERMES, (nodulosa, *vide* Fig. 16), A. S., Schm. At., 45, f. 67, 71, V. H., 6, f. 12, 13.
Fig. 23. " TUMIDULA, Rab., S. D., 4, f. 9, Prit., p. 895. Fresh-water.
" 24. STAURONEIS STAUROPHAENA, Ehrb. Mik., 6, 1, f. 18; 2, 3, f. 11; 5, 3, f. 16, K. B., p. 105; Prit., p. 913.
" 25. " SIGMA, Ehrb. Ber., 1844, Mik., 18, f. 63, evidently a form of Pleurosigma acuminatum, Grun., Sm. Sp. T., No. 393, V. H., 21, f. 12.
" 26. " PTEROIDEA, Ehrb. Mik., 14, f. 5; 3, 3, f. 7, Prit., p. 913. Fresh-water.
" 27. " BAILEYI, Ehrb. Mik., 6, 1, f. 17; 35, A., 5, f. 9; 5, 1, f. 13, Rab., S. D., f. 48, Prit., 913. Fresh-water.

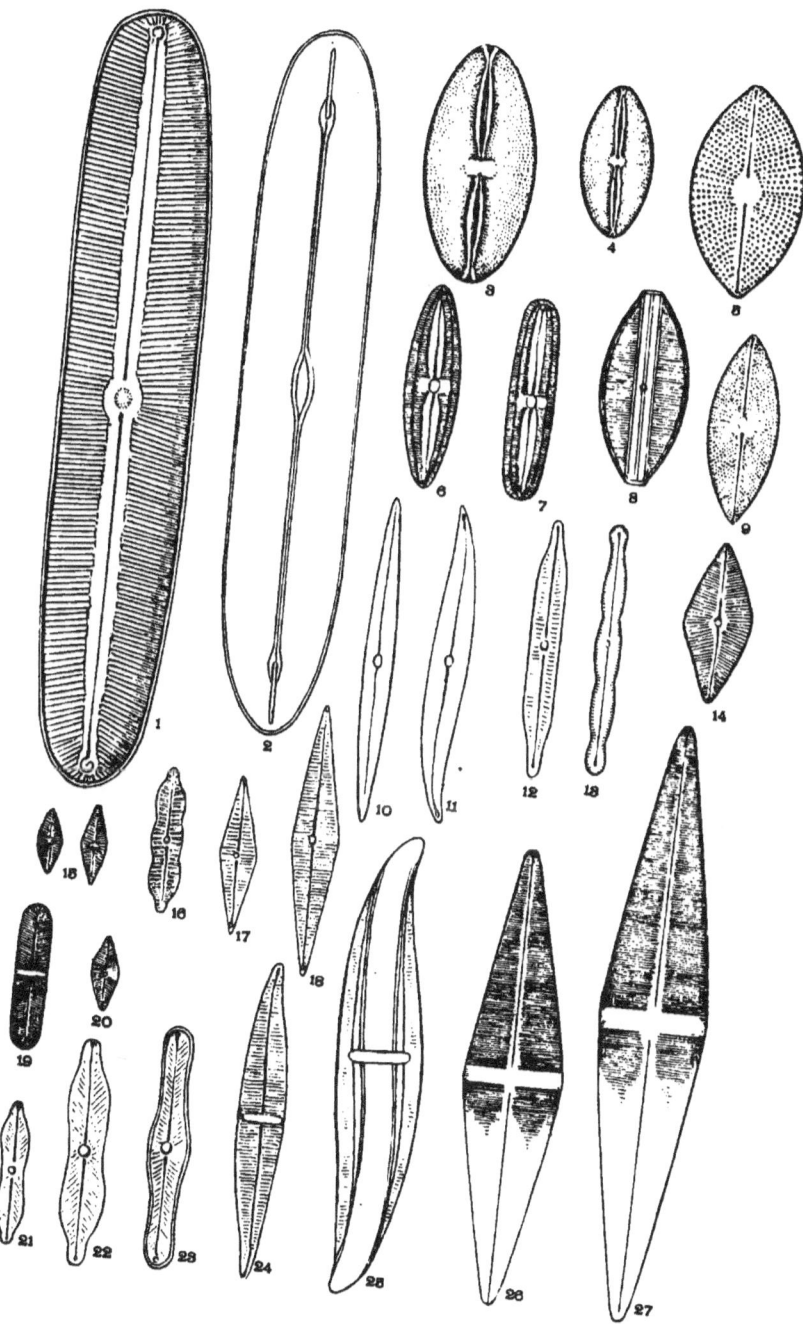

PLATE XXV.

Figures magnified 500 diameters.

Fig. 1. NAVICULA MESOGONGYLA, Kg., K. B., p. 99, Prit., p. 895, a var. of N. nobilis, Abh., 1870, 2, 1, f. 16.
" 2. " INCOMPERTA, Lewis, W., M.D., p. 18, 2, f. 20. Oregon, White Mountain.
" 3. " DECURRENS, Ehrb. Mik., 11, f. 28, Schm. At., 45, f. 29, 30.
" 4. " SERIANS, var. apiculata, Breb., K. B., 30, f. 23, Lewis, W., M.D., 8, 2, f. 5. White Mountain.
" 5. " SERIANS, var. cruciforme, Lewis, W., M.D., 8, 2, f. 5. White Mountain.
" 6. " DIRHYNCHUS, Ehrb. Mik., 35, A., 14, f. 3, K. B., 28, f. 48, Rab., S. D., 6, f. 48.
" 7. STAURONEIS APICULATA, Grev., Cal., p. 30, 4, f. 8, Prit., 913, striae very fine. Cal.
Figs. 8, 9. NAVICULA BOECKII, Heib., Cons., p. 85, Sm. Sp. T., No. 252. Fresh-water, N. J.
Fig. 10. " STAURONEIFORMIS, Lewis, W., M.D., 2, f. 9. Lake Michigan.
" 11. STAURONEIS INFLATA, Kg., K. B., p. 105, 30, f. 22, Rab., S. D. p. 48, 9, f. 15, Sm. Sp. T., No. 492. Fresh-water.
" 12. " STODDERII, Lewis, W., M.D., p. 13, 2, 6, Proc. Al. N. Sc., Sm. Sp. T., No. 502. Markings on this form are stronger. White Mountain.
Figs. 13, 14, 15. NAVICULA (ALTOIONEIS) Stauntonii, Grun., C. and M., No. 304, from slide, H. L. S.
" 16, 17. NITZSCHIA VIRGATA, Roper., M. J., 1858, p. 23, 3, f. 6, Sm. Sp. T., No. 375, J. R. M. S, 1880, Pl. 13, f. 13.
" 18, 19. SYNEDRA GOULARDII, Breb., Cleve., 1880, p. 107, 6, f. 119, Sm. Sp. T., No. 558. Figures taken from slide.
Fig. 20. GONOTHECIUM ROGERSII, Ehrb., Bail., Sill. J., Mar., 1844, p. 301, Mic., 18, f. 92, 93, M. J., 1856, Pl. 7, f. 43-46, Mic. Dic., 42, f. 39. Marine.
" 21. TRICERATIUM ACULEATUM, Ehrb. Mik., 1856, p. 14. Fig. suggested by the description. Fossil.
" 22. " SPINOSUM, Bril., Sill. J., Vol. XLVI, 1844, Schm. At., 87, f. 7. Fossil, Va.
Figs. 23, 24, 25. HEMIAULUS POLYMORPHUM, Grun., F. Jos., p. 14, 2, f. 42, 49. Marine.
" 26, 27. MASTOGLOIA KINSMANII, Lewis, W., M.D., p. 17, 2, f. 15. White Mountain. Fresh-water.

Plate XXV.

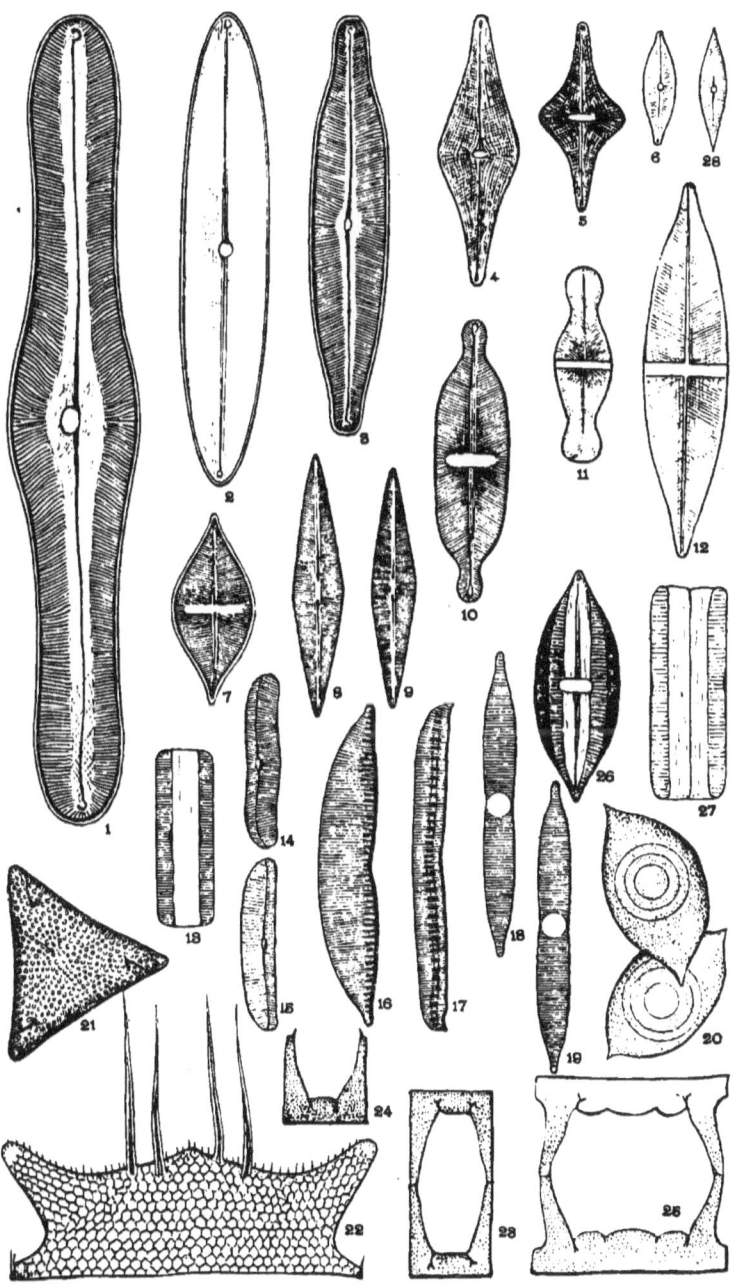

PLATE XXVI.

Figures magnified 500 diameters.

Fig. 1-4, also 10, 11. GOMPHONEMA GEMINATUM, S. and F. views, Ag., Syst., K. B., p. 86; 13, f. 2, Rab., S. D., 8, f. 14, S. B. D., 27, f. 235, Sm. Sp. T., No. 150. Variable in size and form. Fig. 2, var. hybrida.
" 5. GOMPHONEMA MAMMILLA, Ehrb. Mik., 27, 2, f. 10, Abh., 1870, p. 56, V. H., 23, f. 1.
Figs. 6, 7. " CAPITATUM, vars.
" 8, 9. " MAXIMUM, Grun., not separable from G. mammilla, Ehrb., V. H., 23, f. 3. Compare Fig. 5.
" 10, 11. " GEMINATUM, Ag., vide Figs. 1-4.
" 12, 13. " CONSTRICTUM, Ehrb., K. B., 13, f. 1, 2, Rab., S. D., 8, f. 12; S. B. D., I, p. 78, 28. f. 236, Prit., p. 887, 10, f. 187-190.
" 14, 15. " CONSTRICTUM, in vegetative condition under lower power.
Fig. 16. " DICHOTOMUM, Kg., K. B., 85, 8, f. 14, S. B. D., 28, f. 240, Sm. Sp. T., No. 179.
" 17. " DICHOTOMUM, under lower power, in vegetative state.

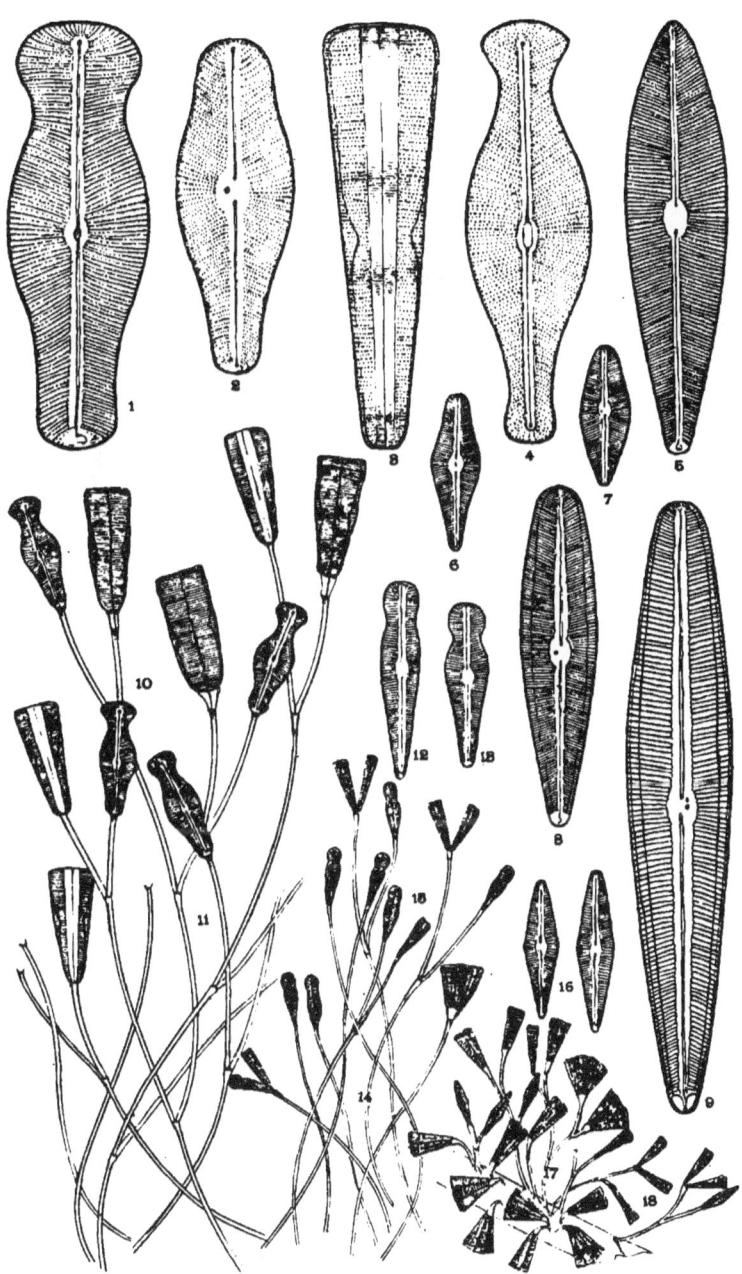

PLATE XXVII.

Figures magnified 500 diameters.

Figs. 1, 2.	GOMPHONEMA	SPHAEROPHORUM, Ehrb. (G. Constrictum, var.), Mik., 37, 11, f. 16; 35, A., 7, f. 14; L. W. Bail. 344, f. 75, V. H., 23, f. 30.
" 3, (11).	"	CRISTATUM, Ralfs., An. H., 1843, p. 463, 18, f. 6, Rab. S. D., 8, f. 19, S. B. D., I, p. 79, 28, f. 239, Sm. Sp. T., No. 178.
Fig. 4.	"	INTRICATUM, Kg., K. B., 9, f. 4, Rab. S. D., 8, f. 27, S. B. D., 29, f. 241.
" 5.	"	VIBRIO, Ehrb. Mik., 39, 3, f. 71; Rab. S. D., 8, f. 9, S. B. D., 28, f. 242, V. H., 24, f. 26.
Figs. 6, 7.	"	GRACILE, Ehrb. Mik., 39, 3, f. 67, etc., Abh., 1869, 1, A. f. 20. Sill. J., May, 1851, f. 13-15. Rab. S. D., 8, f. 26.
Fig. 8.	"	ELEGANS, Grun., V. H., 25, f. 19. Crista, Cal.
" 9.	"	TURGIDUM, Ehrb. A var. of C. Capitatum, V. H., 23, f. 11.
" 10.	"	SARCOPHAGUS, Greg., M. J., 1856, p. 13, 1, f. 42, V. H., 25, f. 2, S. B. D., II, p. 99.
" 11.	"	CRISTATUM, vide Fig. 3.
" 12.	"	MEXICANUM, Grun., V. H., 24, f. 3, G. Commutatum, var.
" 13.	"	TENELLUM, Kg., S. B. D., I, p. 80, 29, f. 243, Rab. S. D., 8, f. 5, Sm. Sp. T., No. 183.
Figs. 14, 15.	"	OLIVACEUM, Ehrb., K. B., 7, f. 13, 15; S. B. D., 29, f. 244, Sm. Sp. T., No. 182, V. H., 25, f. 20.
Fig. 16.	"	SEMIAPERTUM, Grun., V. H., 24, f. 42.
Figs. 17, 18.	RHOICOSPHENIA, CURVATA, var. Marina Rab., M. Dict., 18, f. 19, V. H., 26, f. 4.
" 19, 20, 21.	TYP. CURVATA, Grun., 1868, p. 8, M. Dict., 18, f. 19, V. H., 26, f. 1-3.
" 22, 23, 24.	ACHNANTHES LONGIPES, Ag. Syst., p. 1, K. B., 77, 20, f. 1, S. B. D., II, 35, f. 300, Prit., 873, 7, f. 42, Sm. Sp. T., No. 43, Bail. M. O., 2, f. 1. 22 entire plant; 23 inferior valve; 24 superior valve.
Fig. 25.	"	EXILIS, Kg., S. B. D., II, 29, 37, f. 303, V. H., 27, f. 16-19. Fresh-water.
" 26.	"	" inferior valve, V. H., 27, f. 18.
27.	"	MICROCEPHALUM, Grun., =(Achnanthidium), K. B., 3, f. 13, 19, Rab. S. D., 8, f. 2, S. B. D., II, p. 31, 61, f. 380, Sm. Sp. T., No. 9. The typical form is broader, more inflated.
" 28.	"	HUDSONIS, Grun., valves superior and inferior, V. H., 27, f. 25, 26.
" 29.	"	BREVIPES, Ag., Mik., 6, 2, f. 25; K. B., 20, f. 9, Sill. J., 1842, p. 325, 3, f. 12, Sm. Sp. T., No. 1, S. B. D., 37, f. 301.
" 30.	"	do. inferior valve.
Figs. 31-34.	"	SUBSESSILIS, Kg. Two inferior valves and two plants in situ.

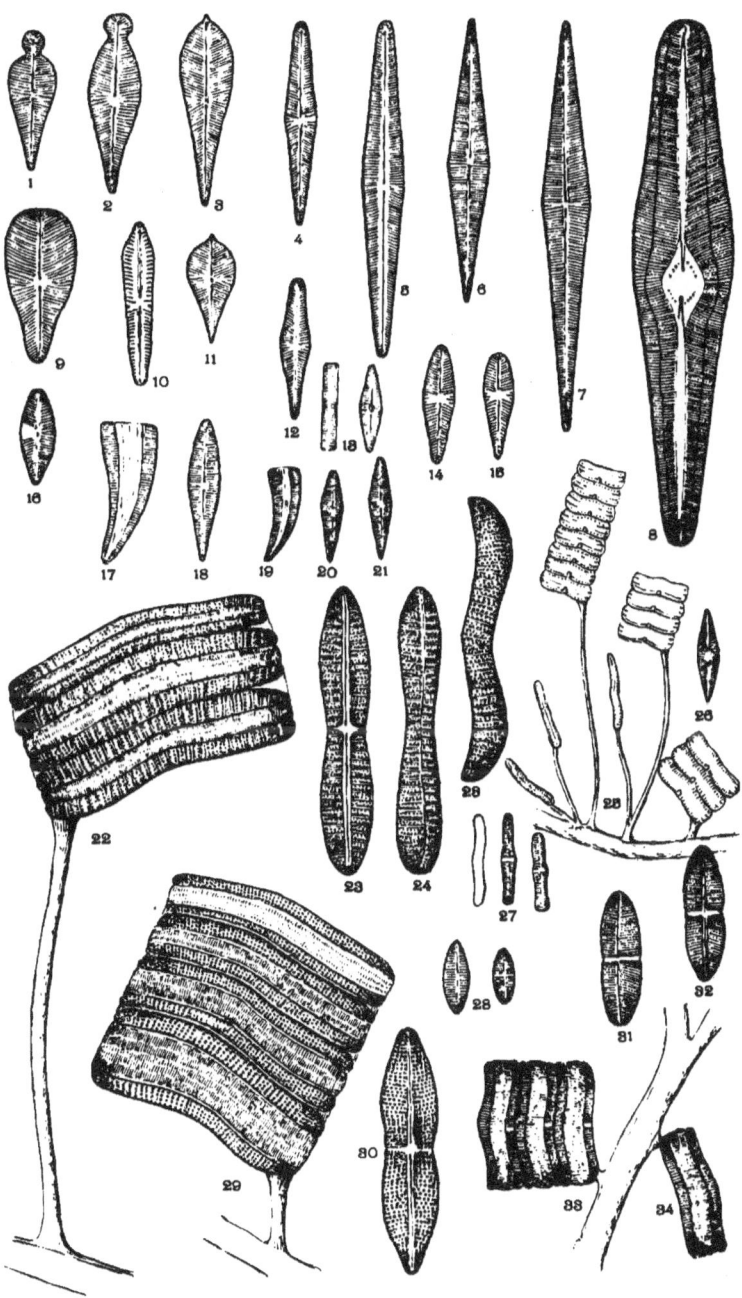

Plate XXVII.

PLATE XXVIII.

Figures magnified 500 diameters.

Figs 1, 2.   SCHIZONEMA RAMOSISSIMA, Ag. Syst., Fig. 1. A frond, natural size; S. B. D., 11, p. 79, 59, f. 369, Sm. Sp. T., No. 478.
"   3, 4.   "   SMITHII, 3, the frustule, 4, frond, natural size, Ag. Syst., p. 10, K. B., 27, f. 5, V. H., 15, f. 4.
"   5, 6.   "   GREVILLEI, 5, frond, nat. size and frustule $\frac{10}{1}$°, Ag., K. B., 5, f. 1; 26, f. 4, S. B. D., 58, f. 364, Prit., p. 928, 8, f. 38.
"   7, 8.   "   IMPLICATUM, 7, frond, nat. size; 8, frustules, S. B. D., 59, f. 367, Sm. Sp. T., No. 471.
"   9, 10.   "   CRUCGIER, frond, nat. size, and frustule $\frac{570}{1}$° S., B. D., 56, f. 354, Prit., 928, V. H., 16, f. 1, 2.
"   11, 12.   "   COMOIDES, (Frustule $\frac{10}{1}$°, and frond, nat. size), Ag., S. B. D., Pl. 57, f. 358, V. H., 16, f. 3.
Fig.   13.   "   AMERICANUM, Grun., V. H., 15, f. 35, (— S. tenue, var.)
"   14.   "   DIVERGENS, S. B. D., Pl. 57, f. 363, Sm. Sp. T., No. 465, V. H., 15, f. 10, probably a var. of S. Smithii.
"   15,(19.) GOMPHONEMA, CONSTRICTUM, var. Minor., Ehrb., K. B., 13, f. 1-3, Rab. S. D., 8, f. 12, S. B. D., II, 98, 28, f. 236.
"   16,(20).   "   CAPITATUM, Ehrb. Mik., 16, 3, f. 37; 17, 1, f. 51, Rab. S. W. D., 8, f. 15, S. B. D., Pl. 28, f. 237, Sm. Sp. T., 177 and 665.
"   17, 18.   "   AUGUR, Ehrb. Mik., 17, 1, f. 55; 9, 1, f. 40, K. B., 29, f. 74, Rab. S. D., 8, f. 19, V. H., 23, f. 29, probably a var. of G. Acuminatum.
Fig.   21.   "   CAPITATUM, Ehrb., comp. Fig. 16 and 20. Larger form.
"   22.   "   ACUMINATUM, var. laticeps, Ehrb.
"   23.   "   "   var. laticeps.
"   24.   "   "   var. trigonocephalum, Ehrb.
"   25.   "   "   var. turris, Ehrb.
"   26.   "   "   var. coronata, Ehrb.
"   27.   "   "   var. intermedia, Grun., Ehrb. Mik., 16, 3, f. 34; 17, 2, f. 37; 16, 2, f. 43, Sill. J., var. July, 1842, p. 323, 3, f. 6, K. B., 13, f. 3, S. B. D., 28, f. 238; M. J., 1854, p. 99, 4, f. 13, Prit., 887, 13, f. 23; Sm. Sp. T., No. 175.

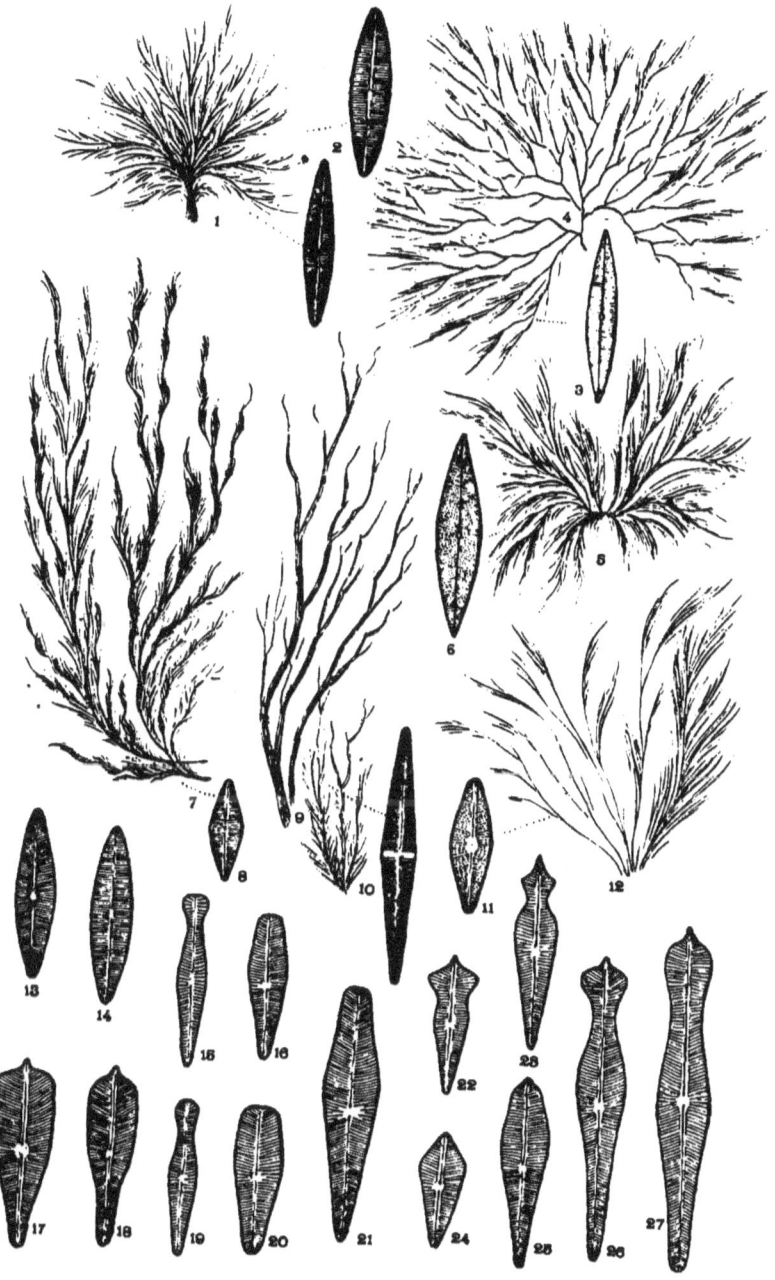

PLATE XXIX.

Figures magnified 500 diameters.

Figs. 1, 2. LICMOPHORA FLABELLATA, W. S., S. B. D., 1, p. 86, Pls. 26 and 32, f. 234, J. R. M. S., 1879, p. 683. Marine.
" 3, 4. RHIPIDOPHORA PARADOXA, fronds and single frustule. Kg., K. B., p. 122, 10, f. 5, S. B. D., 25, f. 231, M. D., 13, f. 19, Sm. Sp. T., No. 699. Marine.
" 5, 6. CYMBOSIRA AGARDHII, K. B., p. 77, 20, f. 3, Prit., p. 875, 14, f. 14, M. D., p. 14, f. 18. Marine.
" 7, 8. CLIMACOSPHAENIA MONILIGERA, Ehrb., F. Amer., 2-6, f. 1, K. B., p. 123, 29, f. 80; Prit., 772, 11, f. 45, 46, M. D., 19, f. 9.
Fig. 9. ACHNANTHES ARENICOLA, Bail., M. O., p. 38, 2, f. 19. Fresh-water.
Figs. 10, 11. " MINUTISSIMA, Ehrb., K. B., 21, f. 2, Rab., S. D., p. 25, 8, f. 2, V. H., 27, f. 35-38, and 41-44. Fresh-water.
Fig. 12. LICMOPHORA JURGENSII, Ag., Cleve., 1880, p. 110, 7, f. 125, V. H., 46, f. 10, 11. Marine.
Figs. 13-15. HOMŒOCLADIA FILIFORMIS, W. S., S. B. D., p. 80, 55, f. 348, Prit., p. 785, 4, f. 25, Sm. Sp. T., No. 197. Fig. 13, frond, under low power, 14, 15, frustules "?". Brackish water.
" 16-18. " CAPITATA, H. L. S., A. Q. J. M., 1878, p. 12, 3, f. 1, Sm. Sp. T., No. 674. Fig. 16, frond, natural size; 17, frustules; 18, section of frond magnified.
Fig. 19. GOMPHONEMA OLOR, Ehrb. Abh., 1870, p. 56, 3, 2, f. 2.

Plate XXIX.

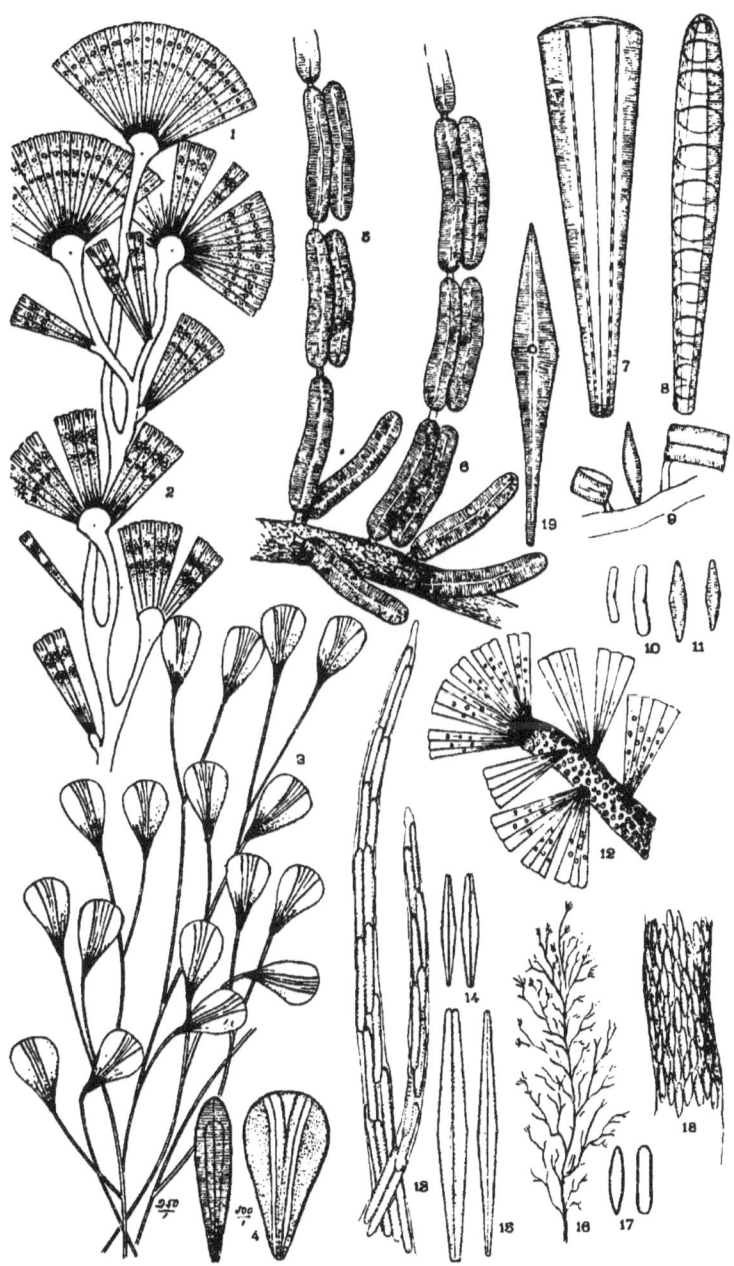

PLATE XXX.

Figures magnified 500 diameters.

Fig. 1.   PLEUROSIGMA OBSCURUM, W. S., S. B. D., I, p. 65, 20, f. 206, Sm. Sp. T., No. 407, Cleve., 1880, p. 49. " Probably the same as P. peisonis, Grun., 1860, p. 562, 4, f. 8," H. L. S.

" 2.   "   ATTENUATUM, W. S., S. B. D., I, p. 68, 22, f. 216, Sm. Sp. T., No. 395, V. H., 21, f. 11.

" 3.   "   RIGIDUM, S. B. D., I, p. 64, 20, f. 198, Sm. Sp. T., No. 410, V. H., 19, f. 3.

" 4.   "   NUBECULA, S. B. D., p. 64, 21, f. 201, Grun., 1860, p. 557.

Figs. 5. (9).   "   ANGULATUM, S. B. D., I, p. 65, 21, f. 205, A. M. H., 1852, p. 7, 1, f. 8, Sm. Sp. T., Nos. 389, 390.

Figs. 6–8.   "   SPENCERII, W. S., S. B. D., I, p. 63, 22, f. 218.

Figs. 10, 11.   "   QUADRATUM, S. B. D., I. p. 65, 20, f. 204, Sm. Sp. T., No. 391. Only a var. of P. angulatum. Fig. 11, found in North America? H. L. S.

Fig. 12.   "   SCIOTENSE, Sullivant, Sill. J., 1859, (March), Prit., p. 917. Probably a var. of P. Spencerii or attenuatum.

" 13.   "   Probably same as *Collectonema eximium*, Breb.

" 14.   "   OBTUSATUM, Sullivant, Sill. J., 1859, Prit., 919.

Figs. 15, 16. MASTOGLOIA EXIGUA, Lewis, N. R. Sp., p. 7, 2, f. 5, V. H., 4, f. 25, 26, Sm. Sp. T., No. 676. Lewis, letters and also annotated copy of his paper describe this diatom as only a variety (immature) of Mast. lanceolata. H. L. S.

" 17, 18.   "   GREVILLEI, W. S., S. B. D., II, 65, 62, f. 389, M. J., 1856, 1, f. 16, V. H., 4, f. 20.

Fig. 19.   "   LANCEOLATA, S. B. D., II, p. 64, 54, f. 340, Sm. Sp. T., No. 214, V. H., 4, f. 15 and 17.

Figs. 20–23.   "   SMITHII, Thw., S. B. D., II, 65, 54, f. 341, Grun., 1860, p. 57, 5, 5, f. 11, V. H., 4, f. 13.

" 24, 25. CERATONEIS ARCUS, Kg., K. B., f. 104, 6, f. 10, Rab., S. D., p. 37, 7, f. 3, Sm. Sp. T., 9, No. 66, V. H , 37, f. 7.

Plate XXX.

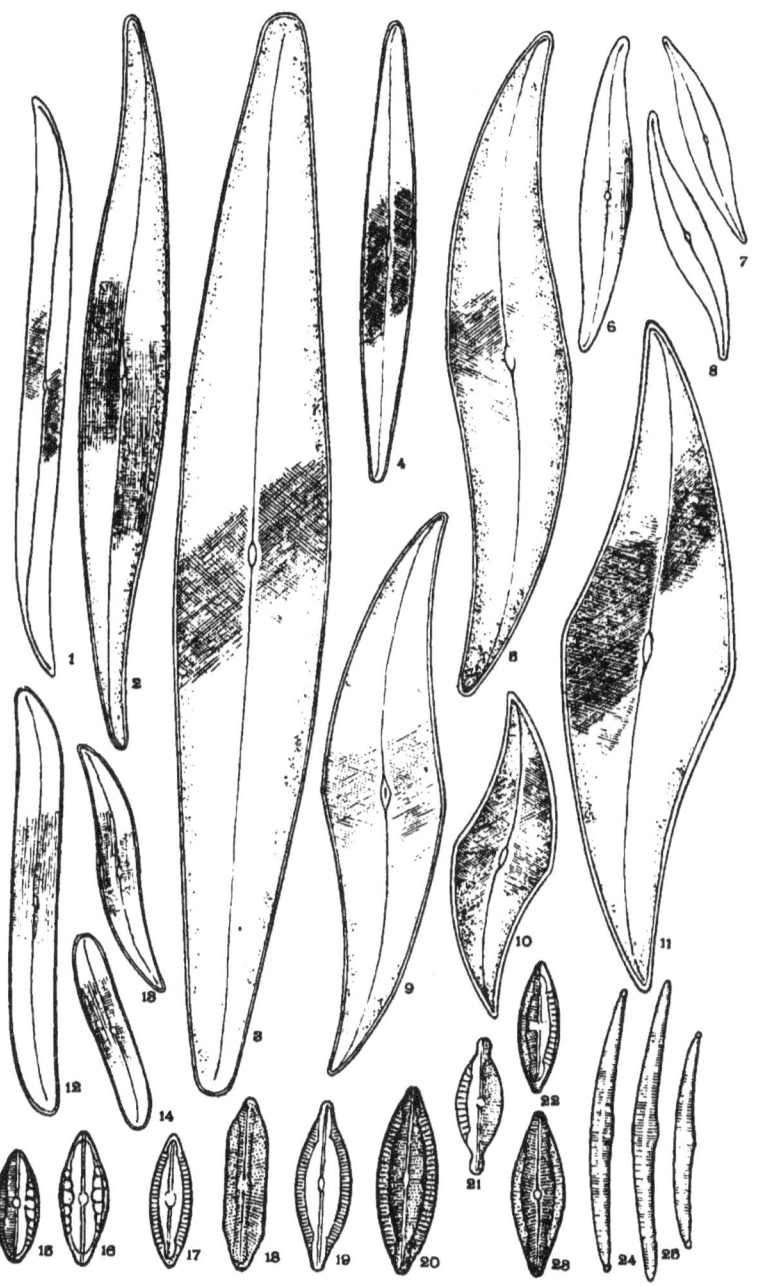

PLATE XXXI.

Figures magnified 500 diameters.

Fig. 1.   AMPHIPLEURA, OREGONICA, Grun., M. M. J., 1877, p. 179.
"   2.   "   LINDHEIMERII, Grun., 1862, p. 469, 11, f. 11, Sm. Sp. T., No. 17.
Figs. 3–5.   "   PELLUCIDA, Kg., K. B., p. 103, 3, f. 52; 30, f. 84, Rab. S. D., 5, f. 5; S. B. D., 15, f. 127; Prit., p. 783, 4, f. 30, Sm. Sp. T., No. 18, 2, front and 2 side views.
Fig. 6.   "   WEISSFLOGII, Janisch, M. M. J., 1877, 195, f. 14.
"   7.   PLEUROSIGMA VIRGINIACUM, A. J. M., Feb., 1877, p. 18; Sm. Sp. T., No. 416.
"   8.   SCOLIOPLEURA LATESTRIATA, Grun., 1878, p. 17, V. H., 17, f. 12; (Navicula convexa, W. S.)
Figs. 9, 10. "   TUMIDA, Breb., (Navicula Jennerii, W. S.), = Nav. Anglica, Ralfs., V. H., 17, f. 11, 13.
Fig. 11. AMPHORA LANCEOLATA, Cleve., 1867, p. 667, 23, f. 2, M. J., 1874, p. 25, 6, 8, f. 3, Schm. At., 25, f. 6, H. L. S., Lens, Pl. 3, f. 34.
"   12. SCOLIOPLEURA ANTILLARUM, Pel., (allowneis antillarum), Cleve. and Grun. Probably only a coarser grained var. of Scol. tumida, Figs. 9, 10.
"   13, (17).   "   JENNERII, Grun., 1860, p. 554. *Navicula, Jennerii*, S. B. D., 16, f. 134. Side and front views.
"   14. AMPHORA, MUCRONATA, H. L. S., A. Q. J. M., p. 17, 3, f. 10, J. M., 1879, p. 134, 6, f. 9, Sm. Sp. T., No. 38.
"   15. BERKELEYA MICANS, Grun., Hedw. vol. 7, p. 5, V. H., 16, f. 11.
"   16. ASTERIONELLA FORMOSA, var. S. B. D., p. 81, T. M. S., 1860, p. 149, 7, f. 8, Prit., p. 779, 4, f. 17, M. Dic., 43, f. 14, Sm. Sp. T., No. 46, V. H., 51, f. 19, 20.

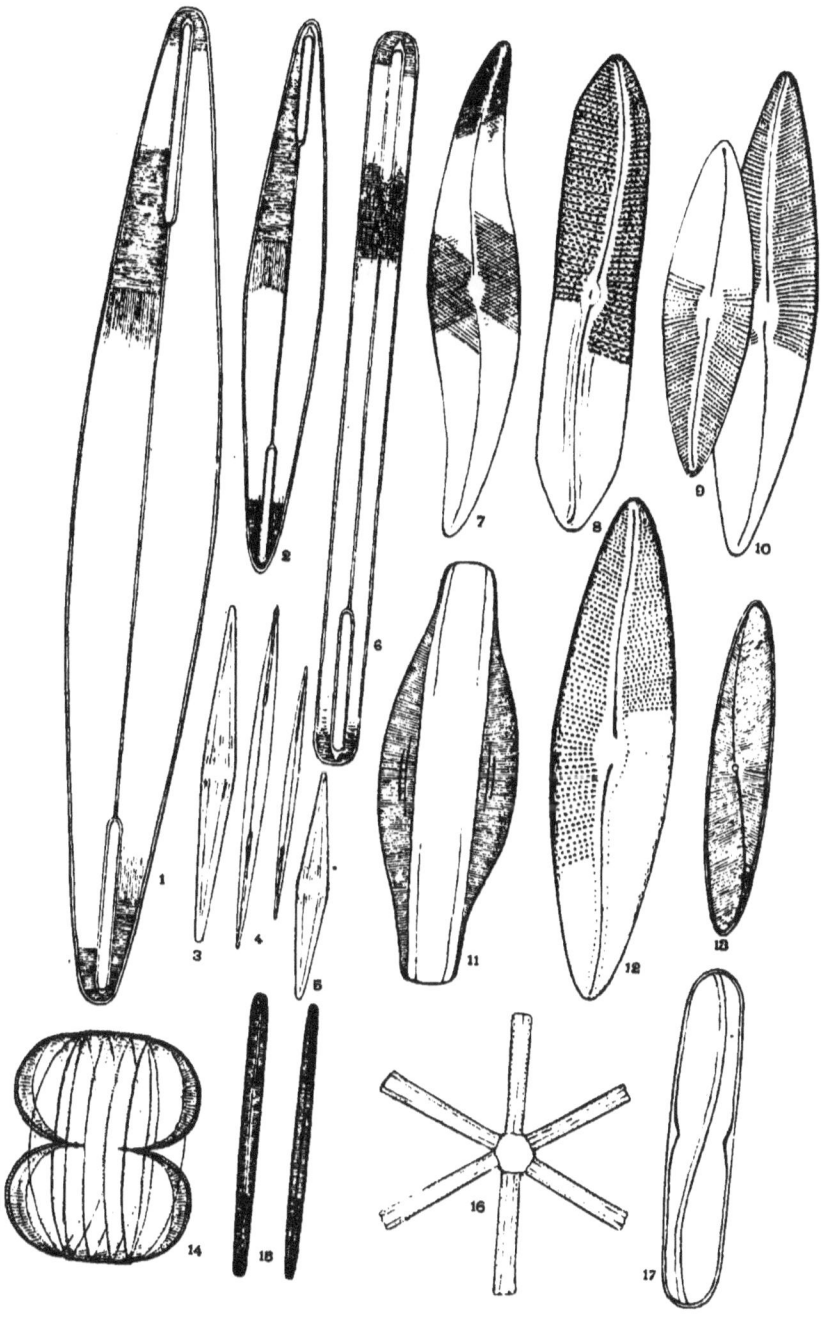

PLATE XXXII.

Figures magnified 500 diameters.

Fig. 1.   PLEUROSIGMA, MACRUM, W. S., S. B. D., I, p. 67, 31, f. 276, (N. Amer.? H. L. S., possibly a local var. of P. Fascicola.)
" 2.   "   ELONGATUM, W. S., S. B. D., I, p. 64, 20, f. 199. Sm. Sp. T. No. 400.
" 3.   "   PULCHRUM, Grun., 1860, p. 556, 4, f. 2, Sm. Sp. T., No. 695.
" 4.   "   DECORUM, W. S., S. B. D., I, p. 63, 21, f. 196, V. II., 19, f. 1, Sm. Sp. T., No. 694. Striation scarcely coarse enough, compared with other figures of plate.
Figs. 5, 6.   "   BALTICUM, S. B. D., 22, f. 207, Prit., 917, 8, f. 33; 9, f. 144, Grun., 1860, p. 558; 1878, p. 18, 3, f. 9, Sm. Sp. T., No. 396.
Fig. 7.   "   HIPPOCAMPUS, W. S., S. B. D., 22, f. 215, Prit., 919, 9, f. 145, Sm. Sp. T., No. 404.
Figs. 8, 9.   FORMOSUM, W. S., S. B. D., 20, f. 195, Prit., 917, 8, f. 32, M. J., 1856, p. 205, 13, f. 5, Sm. Sp. T., No. 402.
Fig. 10.   AESTUARII, W. S., S. B. D., 31, f. 275, Sm. Sp. T., No. 394. "A pale, thin, small form of P. angulatum." H. L. S.
" 11.   "   FASCICOLA, W. S., S. B. D., 21, f. 211, Prit., 916, 12, f. 60, 61, Sm. Sp. T., No. 401.
" 12.   OBSCURUM, S. B. D., 20, f. 206, Sm. Sp. T., No. 407. Very nearly allied to P. peisonis of Grun.
13.   KUETZINGII, Grun., 1860, p. 561, 4, f. 3, V. II., 21, f. 14,   P. Spencerii, var.
" 14.   "   EXIMIUM, Breb., V. II., 21, f. 2, - Sullivants obtusatum; — formerly Collectonema eximium.

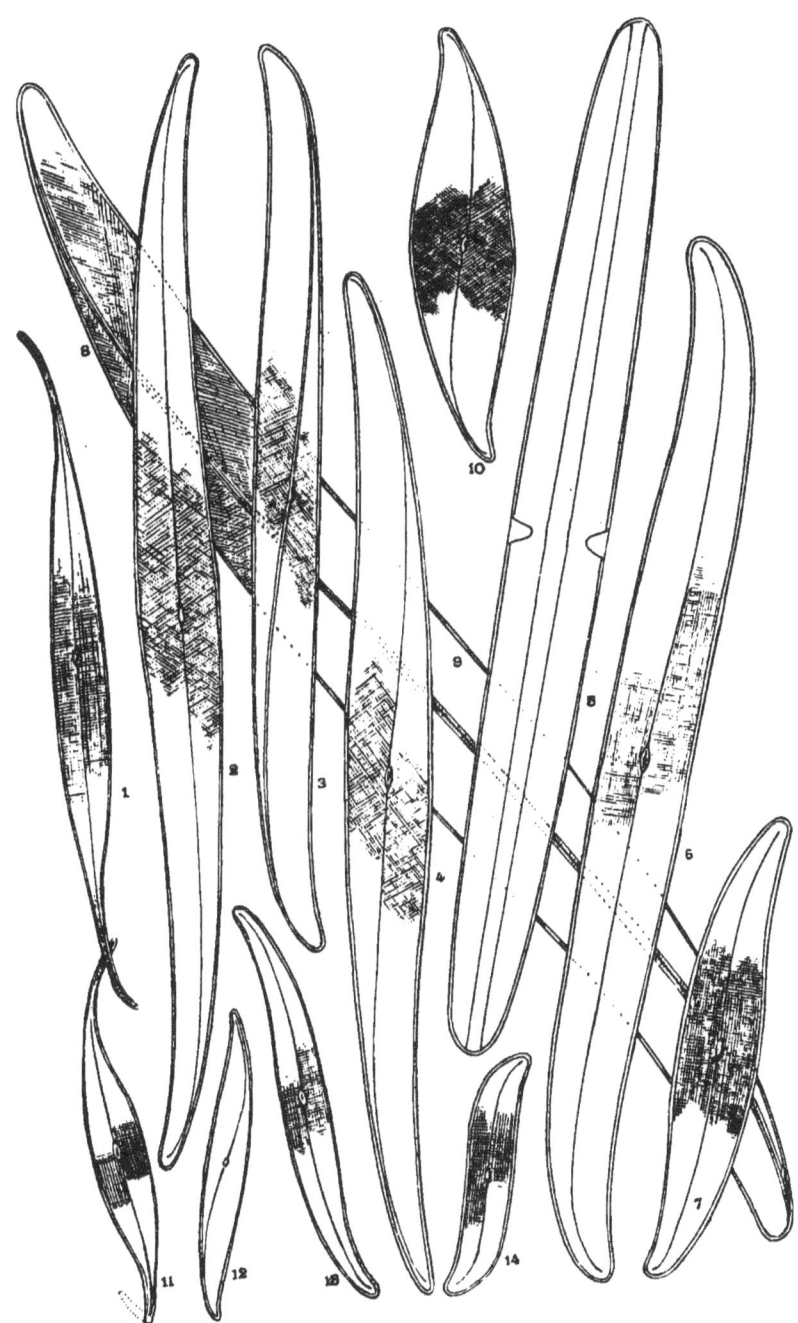

PLATE XXXIII.

Figures magnified 500 diameters.

Figs. 1, 2.	COCCONEIS, PSEUDO-MARGINATA Greg., Prit., 871, 7, f. 39; Jan. and Rab., p. 7, 4, f. 16; V. H., 29, f. 20, 21. Superior valve and inferior valve, Sm. Sp. T., No. 74.
" 3, 4.	" THWAITESII, W: S., S. B. D., 3. f. 33, — Achnanthidium, flexellum, Breb. Fig. 3, inferior valve; Fig. 4, superior valve, V. H., 26, f. 29, 30.
" 5, 11.	" INTERRUPTA, Grun., 1863, p. 144, 13, f. 14; V. H., 30, f. 3, 4. Fig. 5, inferior valve; Fig. 11, superior valve.
" 6, 7.	" LINEATA, Ehrb., lower valves, Mik., 39, 3, f. 11; Abh., 1869, 1, A. f. 8; 1, B. f. 2, etc. Sill. J., 1851, p. 44; V. H., 30, f. 31, 32.
Fig. 8.	" GREVILLEI? small form, S. B. D., 3, f. 35.
Figs. 9, 10.	" COSTATA, Greg., M. J., 1855, p. 39, 4, f. 10; 1857, p. 68, 1, f. 27; V. H., 30, f. 11-17.
Fig. 11.	" INTERRUPTA, superior valve; vide Fig. 5.
Figs. 12, 13, 14.	" SCUTELLUM, Ehrb. Mik., 7, 3 A, f. 57; 15 A, f. 56, 19, f. 33; K. B., 5, f. 6; Rab. S. D., 3, f. 4; S. B. D., 3, f. 34; Sm. Sp. T., No. 78. Fig. 12, superior valve; 13, inferior valve; 14, hoop.
" 15, 16.	DIRUPTA, Greg., D. C., p. 19, 1, f. 25, Jan. Guan., II, B., f. 14; V. H., 29, f. 13-15, inferior valves.
" 17, 18.	" PLACENTULA, Ehrb., 17, superior valve; 18, inferior valve, Mik., 5, 1, f. 24; 7. 1, f. 46, K. B., 28, f. 13; Rab. S. D., 27, 3, f. 3; S. B. D., I, p. 21, 3, f. 32; Sm. Sp. T., No. 77.
" 19, 20.	" CALIFORNICA, Grun., V. H., 30, f. 8, 10.
" 21, 27.	" PEDICULUS, Ehrb. Mik., 34, 12 B, f. 1; K. B., 5, f. 9; S. B. D., 2, f. 31; Rab. S. D., 27, 3, f. 1; Sm. Sp. T., No. 76. Fig. 21, frustule "?"; 27, under lower power, in situ.
" 22, 23.	" SCUTELLUM, Ehrb., same as Figs. 12-14 in different stages.
" 24, 25.	CAMPYLONEIS ARGUS, Grun., 1862, p. 429, 10, f. 9; V. H., 28, f. 15, 16. 24, superior valve; 25, inferior valves, varieties of Coc. Grevillei.
Fig. 26.	COCCONEIS STRIATA, Ehrb. Mik., 58, f. 4; 14, f. 41; 15 A, f. 55; 15 B, f. 16, etc.; Rab. S. D., 3, f. 12.
" 27.	" PEDICULUS, vide Fig. 21.
Figs. 28, 29, 30.	CAMPYLONEI GREVILLEIS, Grun., R. M. S., 1878, p. 245, 14, f. 5; V. H., 28, f. 10-12. Fig. 28, a skeleton; 29, inferior valve; 30, superior valve.
Fig. 31.	COCCONEIS DECUSSATA. Ehrb., E. Amer., 2. 6, f. 12; K. B., 28, f. 17; Rab. S. D., 8, f. 6.
" 32.	" AMERICANA, Ehrb., p. 123; K. B., p. 73; Rab. S. D., p. 27; Prit., Pl. 12, f. 48, — C. Mexicana.
" 33.	" PEDICULUS, Ehrb., same as Figs. 21, 27.

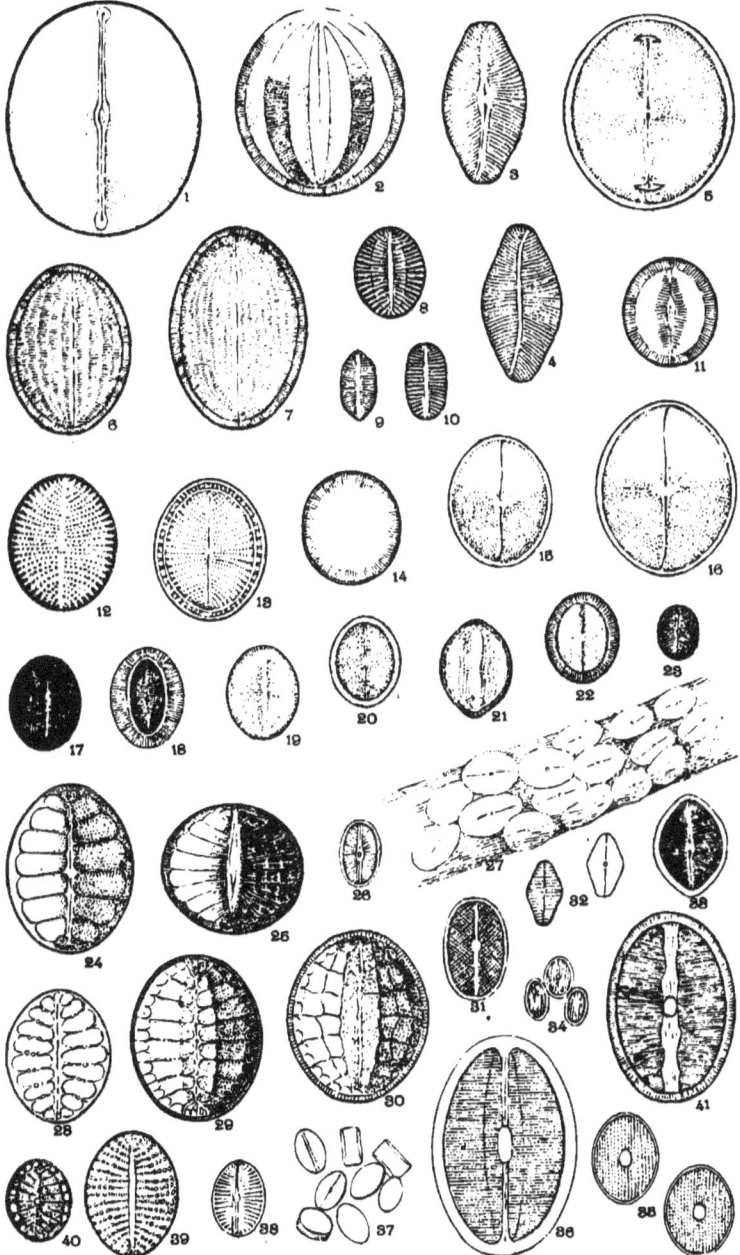

Fig. 34. COCCONEIS OBLONGA, K. B., p. 72, 5, f. 7, 8, small form of C. placentula?
" 35. " OCEANICA, Ehrb., Abh., 1840, p. 65, E. Amer., 1, 1, f. 14; 2-6, f. 11, 12, K. B., 72, 5, f. 5; 8, f. 4, Prit., 868, 12, f. 42, Sm. Sp. T., No. 75   C. dirupta var.
" 36. " PINNICA, Ehrb. Mik., 33, 3, f. 10; 17, 2. f. 19, 38 a, 20, f. 9, etc., Abh., 1870, 2, 1, f. 44, K. B., 28, f. 18, Prit., 938, 12, f. 41,   C. Smithii, var.
" 37. " NITIDULANS, Kg., K. B., 4, f. 16, Prit., 867.
" 38. " MARGINATA, Ehrb., K. B., 5, f. 6, Prit., p. 868, C. sentellum?
Figs. 39, 40. CAMPYLONEIS REGALIS,   C. Grevillei, var., comp. Figs. 28-30, Grev., M. J., 1859, p. 156, 7, f. 1, Prit., p. 870.
Fig. 41. COCCONEIS PINNULARIA, Bail., Sill. J., 1841, f. 2, 31, K. B., 73, probably N. Smithii, 5, f. 34, or Elliptica, which is the fresh-water form.

PLATE XXXIV.

Figures magnified 500 diameters.

Figs. 1, 2. NITZSCHIA SCALARIS, var. undulata, W. S., S. B. D., 39, 14, f. 115, Prit. 781, 4, f. 22, Sm. Sp. T., No. 365.
"   3, 5. COCCONEIS, NITIDA, Greg., D. C., p. 20, 1, f. 26, Prit. p. 871.
Fig. 4.   "   LINEATA, inferior valve; Ehrb. Mik., 39, 3, f. 11. Abh., 1869, 1, A, f. 8; 1, B, f. 2; 1, D, f. 7; Sill. J., 1851. f. 44; V. H., 30, f. 31, 32.
Figs. 6, 7. REGALIS, Grev., M. J., 1859, p. 156; 7, f. 1, Prit., p. 870.
"   8, 9, 10. "   GREVILLEI, W. S., S. B. D., 3, f. 35; Sm. Sp. T., No. 72; Grun., 1862, 4, f. 32.
Fig. 11. "   DISRUPTA, Greg.. D. C.. p. 19, 1, f. 24; V. H., 29, f. 13-15.
"   12. EPITHEMIA SUCCINCTA, Breb., J. Q. C., 1870, p. 42, 1, f. 7; Sm. Sp. T., No. 154; V. H., 32, f. 16-18.
"   13. "   SEREX, Kg., K. B., 5, f. 12; Rab., S. D., 1, f. 7; S. B. D., 1, f. 9; Sm. Sp. T., No. 153; V. H., f. 10.
"   14. COCCONEMA ASPERUM, var. of C. lanceolatum, Ehrb. Mik., 16, 3, f. 30; 6, 1, f. 1; 5, 1, f. 1; Sill. J., 1838, 2, f. 2.
Figs. 15, 16. GEPHYRIA JAPONICA.
"   17, 18, 19. GONIOTHECIUM ROGERSII, Bail., Sill. J., March, 1844, p. 301; Mik., 18, f. 92, 93; M. J., 1856, 7, 43-46; M. D., 42, f. 30.
Fig. 20. RHABDONEMA, side views of single frustules.
"   21. SURIRELLA FASTUOSA, var. Ehrb., (S. recedens, A. S., Schm. At., 19, f. 2-4), E. Amer., 2-4, f. 7; 2-6, f. 14; K. B., 28, f. 19; S. B. D., 9, f. 66; M. J., 1855, p. 40, 4, f. 12.

Plate XXXIV.

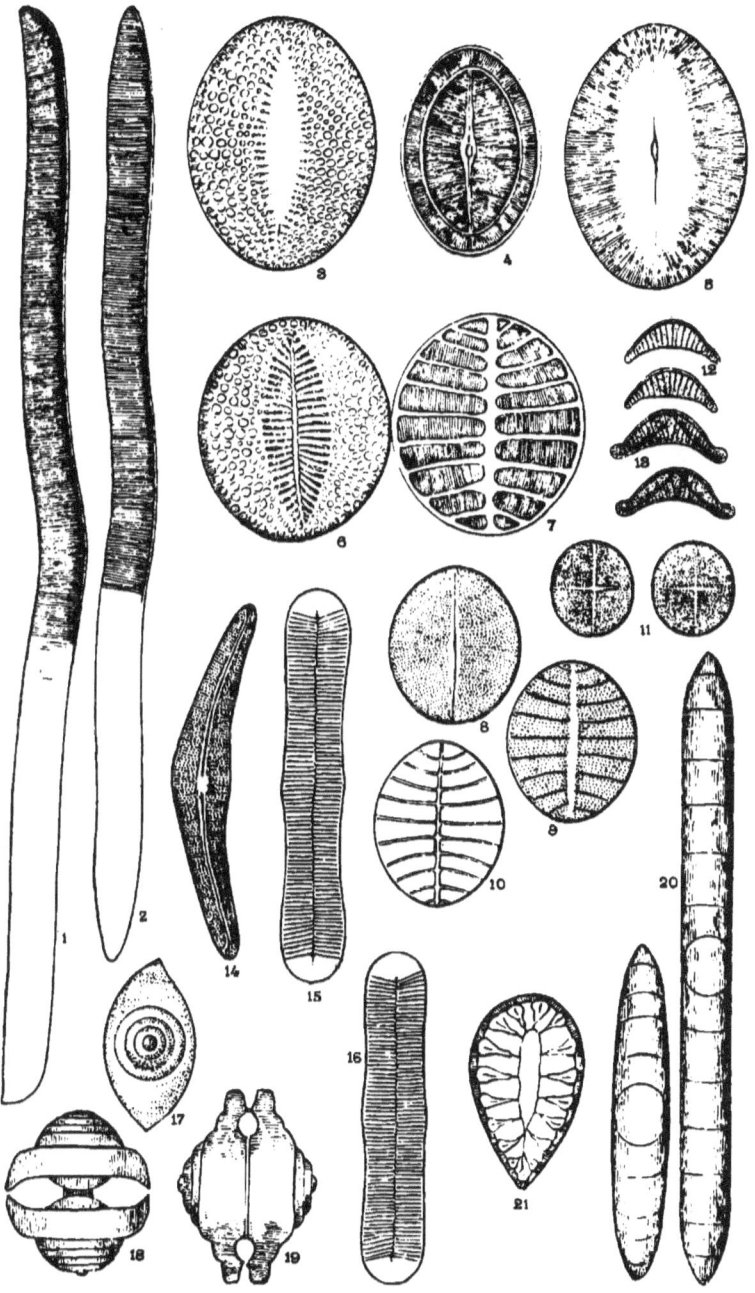

PLATE XXXV.

Figures magnified 500 diameters.

Figs. 1-3.	EPITHEMIA GIBBA, Kg., K. B., 4, f. 22; Rab., S. D., 1, f. 3; S. B. D., 15, 1, f. 13; Prit. p. 759, 12, f. 27; Sm. Sp. T., No. 150.	
" 4, 5.	" GRANULATA, Kg., K. B., 5, f. 20; S. B. D., 1, f. 3; Prit., 761, 9, f. 165.	
" 6, 7.	" HYNDMANII, W. S., S. B. D., 12, 1, f. 1; V. H., 31, f. 34.	
" 8, 9.	" GIBBA, var. ventricosum, Grun.	
" 10-13.	" TURGIDA, vars., Kg., K. B., 8, f. 14; Rab. S. D., 1, f. 11; Prit., 761, 4, f. 1; 9, f. 156-161 and 11, f. 1-8; Sm. Sp. T., No. 155.	
" 14-17.	ARGUS, Kg., K. B., 29, f. 55, 56; Rab. S. D., 1, f. 33; S. B. D., 12, f. 5; Prit., 759, 15, f. 11; Sm. Sp. T., No. 149.	
18, 19.	MUSCULUS, Kg., K. B., 30, f. 6; S. B. D., 14, 1, f. 10; Prit., 760, 13, f. 18; Sm. Sp. T., No. 151.	
" 20, 21.	" CONSTRICTA, S. B. D., 20, f. 248; V. H., 32, f. 16-18.	
" 22, 23.	" OCELLATA, Kg., K. B., 29, f. 57; Rab. S. D., 1, f. 25; S. B. D., p. 13, 1, f. 6.	
" 24, 25.	SOREX, Kg., K. B., 5, f. 12; Rab. S. D., 1, f. 7; S. B. D., p. 13, 1, f. 9; V. H., 32, f. 6-10.	
" 26-28.	" GIBBERULA, Kg., K. B., 29, f. 54; Sm. Sp. T., No. 652.	
" 29-33.	" ZEBRA, K. B., 5, f. 11; Rab. S. D., 1, f. 8; S. B. D., 1, f. 4; V. H., 31, f. 9-11-14. E. zebra is only a var. of E. argus.	
Fig. 34.	" TENTRICULA, K. B., 29, f. 53; Rab. S. D. 1, f. 13.	

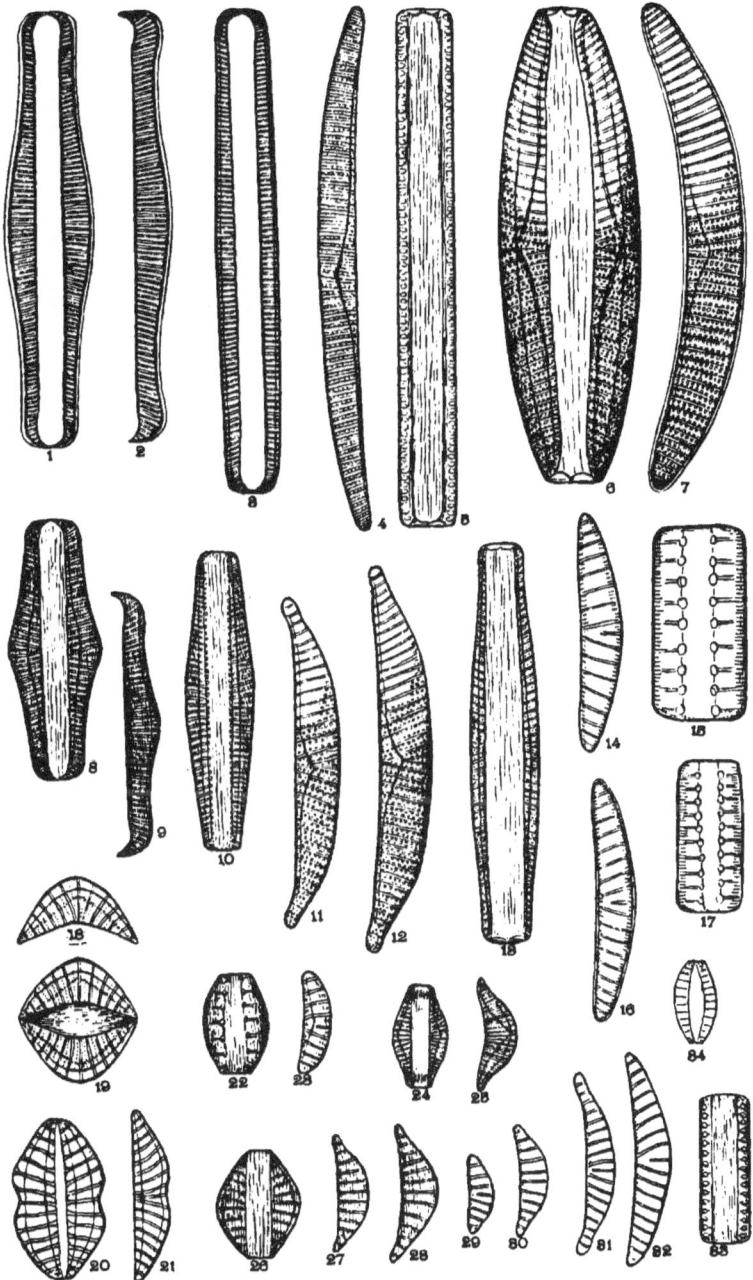

PLATE XXXVI.

Figures magnified 500 diameters.

Fig. 1-3. EPITHEMIA ARGUS, large forms. Compare Pl. 34, f. 14-17.
Figs. 4, 12. EUNOTIA MONODON, Ehrb. Mik., 39, 3, f. 46; Sill. J., May, 1846, 2, f. 38; May, 1851, f. 27.
"   5, 6. "   DIODON, Ehrb. Mik., 17, f. 28ab; 16, 2, f. 18, K. B., 5, f. 24. Variety of robusta.
"   7, 8. "   TRIODON, Ehrb. Mik., 39, 3, f. 5; 16, 2, f. 20; Prit. 763, 4, f. 4. Var. of robusta.
Fig. 9. "   ROBUSTA, var. tetrodon, Ralfs., Prit., p. 763. Striae moniliform.
"   10. "   ROBUSTA, var. papilio, (Ehrb.) Ralfs., Grun., 1868, p. 94.
"   11. "   ROBUSTA, var. diadema, (Ehrb.) Ralfs., Mik., 16, 3, f. 17; 16, 1, f. 33; K. B., 5, f. 28.
"   12. "   Compare Fig. 4, monodon, Ehrb.
"   13. "   ROBUSTA, var. hendecaodon, (Ehrb.) Ralfs., Mik., 16, 2, f. 30; 17, 1, f. 38; Prit., 764.
"   14. "   DECAODON, Ehrb. Mik., 16, 2, f. 29; 17, 1, f. 37; Sill. J., Jan., 1842; K. B., 5, f. 29.
Figs. 15, 16. "   PECTINALIS, Kg., var. curta; V. H., p. 143, 33, f. 15; C. and M., No. 234.
"   17, 18. "   PECTINALIS, Kg., var. stricta, Rab.
"   18, 20. "   PECTINALIS, Kg., var. elongata; Kg., Rab., striae moniliform.
"   21, 22. "   DIODON, Ehrb., large form, Sill. J., May, 1842, 2, f. 29, K. B., 5, f. 24, S. B. D., 2, f. 17; compare Figs. 5, 6.

Plate XXXVI.

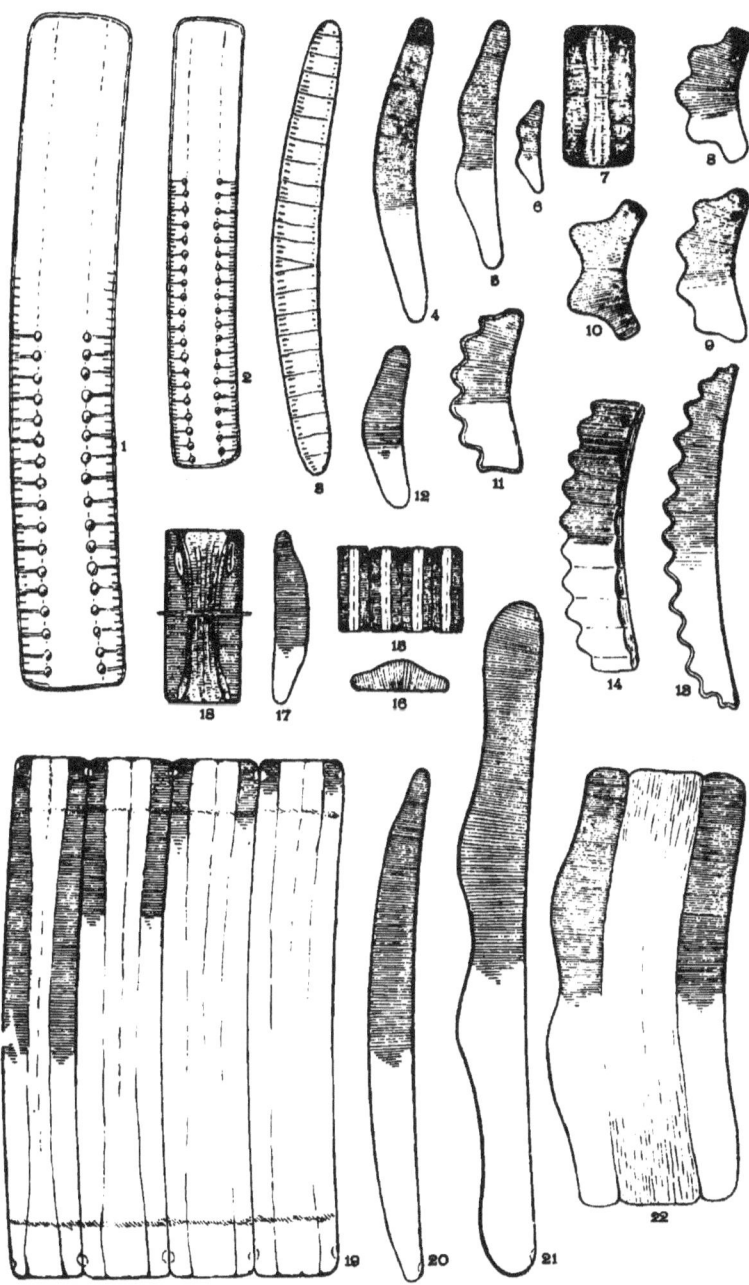

PLATE XXXVII.

Figures magnified 500 diameters.

Fig. 1.   EUNOTIA POLYDON, Ehrb. Mik., 17, 1, f. 45; Prit., 764.
" 2.   "   SERRA, Ehrb. Mik., 16, 1, f. 35, a, b; 2, f. 31, 17; 1, f. 39, 40, Sill. J., May, 1842, p. 100, 2, f. 33, K. B., 5, f. 30.
" 3.   "   MONODON, var. depressa, Ehrb. Mik., 14, f. 59; 15, a, f. 6, 8, b, Abh., 1869, 2, 1, f. 8; 1, G., f. 3, K. B., 29, f. 39.
" 4.   FRAGILARIA DIMEREGRAMMA? dubia, Grun., 1862, p. 873, 7, f. 28; V. H., 36, f. 18.
" 5.   EUNOTIA CYGNUS, Ehrb., Abh., 1869, 1, d, f. 3; 1, a, f. 2; 1871, 1, b, f. 28.
Figs. 6, 7, 8, 10.   ACTINELLA PUNCTATA, Lewis, N. and J. forms, p. 343, 1, l. 5, V. H., 35, f. 18, 21.
Fig. 9.   DIODESMIS.
" 11.   RHAPHONEIS COCCONEIS, Ehrb., Abh., 1869, p. 52, 2, 1, f. 7.
" 12.   SCEPTONEIS, NITZSCHIOIDES, Grun., V. H., 37, f. 4.
" 13.   "   CADUCEUS, Ehrb. Mik., 33, 17, f. 15; Sill. J., April, 1845, f. 11, Prit., 772, 4, f. 11; V. H., 37, f. 5.
" 14.   "   GEMMATA, Grun., J. Q. C., 1871, p. 170, 14, f. 4, 5.
" 15.   RHAPHONEIS, PRETIOSA, Ehrb., T. M. S., 1854, 6, f. 9; V. H., 36, f. 25.
" 16.   "   SCALARIS, Ehrb. Ber., 1844, p. 271. V. H., 36, f. 32.
" 17.   "   GEMMIFERA, Ehrb. Ber., 1844, p. 87; T. M. S., 1854, 6, f. 7; V. H., 36, f. 31. Probably a var. of amphoaros.
" 18.   "   AMPHICEROS, Ehrb. Mik., 33, 20, f. 20; 33, 14, f. 22; Prit., p. 791, 14, f. 21; Sm. Sp. T., 698; V. H., 36, f. 22, 23.
" 19.   "   AMPHICEROS, var. Californica, Grun., V. H., 36, f. 24.
" 20.   "   "   Ehrb., typical form.
Figs. 21, 22.   "   "   var. rhombica, Grun., V. H., 36, f. 20, 21.
Fig. 23.   COCCONEIS MORMONORUM, Ehrb., Abh., 1870, 53, 2, 1, f. 45.
Figs. 24, 25.   MERIDION CIRCULARE, Ag., K. B., 7, f. 16; Rab. S. D., 1, t. 1; S. B. D., 32, f. 277; Prit., 767, 9, f. 177-179.
Fig. 26.   "   INTERMEDIUM, var. constrictum, H. L. S., A. Q. J. M., 1878, p. 12, 3, f. 2.
Figs. 27, 28.   "   CIRCULARE, side views.
" 29, 30.   "   Side views of Intermedium. Fig. 26 showing the constriction.

Plate XXXVII.

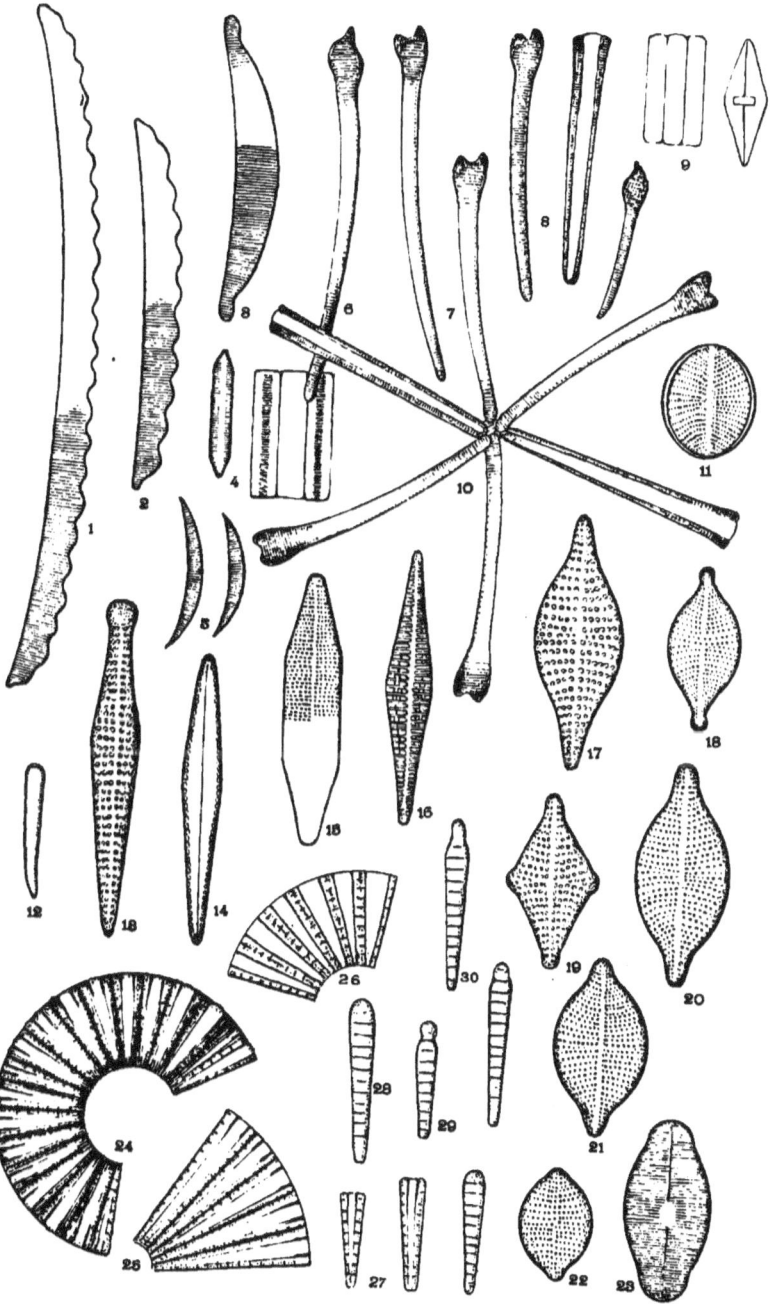

PLATE XXXVIII.

Figures magnified 500 diameters.

Figs. 1-4. EUNOTIA MAJOR, Rab., V. H., 34, f. 14; Grun., 1865, p. 5, 1, f. 8.
Fig. 5. " PRAERUPTA, Ehrb. Mik., 5, 1, f. 5; 3, 2, f. 14; Rab. S. D., p. 16; V. H., 34, f. 19. Striae granulate.
" 6. " PRAERUPTA, var. monodon.
" 7. " DIZYGA, Ehrb., E. Amer., 2, 1, f. 8; K. B., 29, f. 51; Rab., S. D., 1. f. 30, B. J. N. H., p. 333, f. 27.
" 8. " GIBBOSA, V. H., 8, 35, f. 13.
" 9. " BIDENTULA, S. B. D., II, p. 83; V. H., 34, f. 28.
" 10. " BACTRIANA, Ehrb. Mik., I, f. 29; 16, 11. f. 19, Lewis, W., M.D., 13, 2, f. 13; V. H., 34, f. 32.
" 11. " POLYGLYPHIS, Grun., var. hexaglyphis, Ehrb.; V. H., 34, f. 33.
Figs. 12, 13. " PECTINALIS, var. Rab., comp. V. H., 33, f. 15, 16.
Fig. 14. " MAJOR, var. bidens; V. H., 34, f. 15.
" 15. " PARALLELA, Ehrb. Mik., 15, f. 58; 3, 2, f. 11; Sill. J., May, 1842, p. 100, 2, f. 32; V. H., 34, f. 16. Not separable from E. major, Fig. 4.
Figs. 16, 18, 19. " LUNARIS, vars. Ehrb., V. H., 35, f. 3, 6a, 4.
Fig. 17. " LUNARIS, var. excisa; V. H., 35, f. 6c.
Figs. 20, 21. " FORMICA, Ehrb. Mik., 4, 3, f. 19; 3, 4, f. 18; 4, 1, f. 13; Abh., 1869, 2, 1, f. 12; V. H., 34, f. 1.
Fig. 22. " IMPRESSA, Ehrb. Mik., 15, f. 66; 3, 4, f. 20; 5, 2, f. 8; 2, 2, f. 30.
" 23. " DECLIVES, Ehrb., E. Amer., 2, 1, f. 3; K. B., 29, f. 62; Rab., 1, 26.
" 24. " INCISA, Greg., M. J., 1854, p. 94, 4, f. 4; Lewis, W., M.D., p. 13, 2, f. 12; V. H., 34, f. 35a.
Figs. 25, 26. " TRIDENTULA. Ehrb. Mik., 39, 3. f. 51; Sill. J., 1851, f. 38; K. B., 29, f. 60, Rab., S. D., 1, f. 16; V. H., 34, f. 2, 7, E. Ehrenbergii, Ralfs.
" 27, 28. " ZYGODON, Ehrb. Mik., 16, 13, f. 15; 17, 1, f. 27, 28; K. B., 29, f. 49.
" 29. " PENTAGLYPHIS, Ehrb. Mik., 16, 2, f. 22; 17, 1, f. 32; Lewis, W., M.D., 13, 2, f. 4.
Figs. 30, 31, 32. " MOSIS, Ehrb., Abh., 1870, 54, 3, f. 7-10; 1871, 1C, f. 13, 1870, 2, 1, f. 55.
" 33, 34. " SERRULATA, Ehrb. Mik., 16, 1, f. 36; 16, 2, f. 32; 17, 1, f. 40. (Vars. of E. robusta.)
Fig. 35. " UNDENARIA, Ehrb. Mik., 33, 10, f. 12, Prit., 764.

NOTE.—Figs. 25-29, Ralfs unites as varieties of E. Ehrenbergii, varying only in having one, two or more undulations.

Plate XXXVIII.

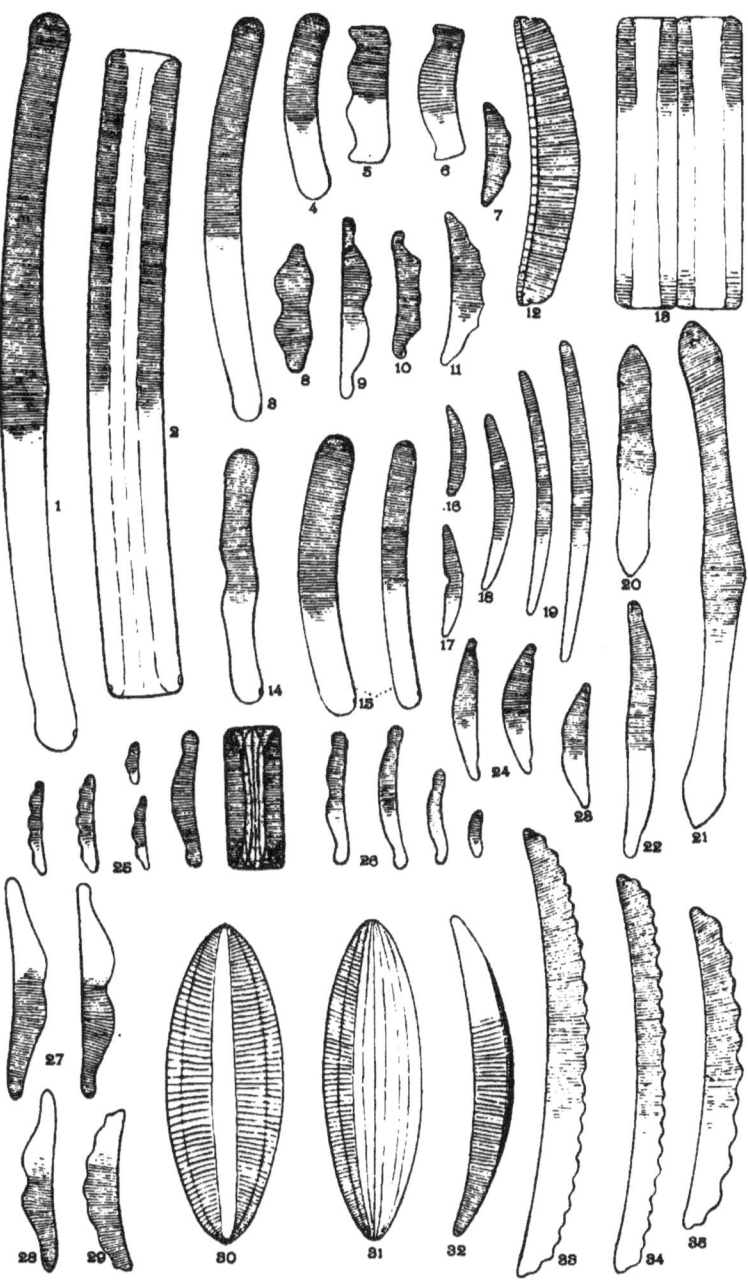

PLATE XXXIX.

Figures magnified 500 diameters.

Fig. 1.	SYNEDRA	ROBUSTA, Ralfs., Prit., 789, 8, f. 3; V. H., 42, f. 67.
Figs. 2, 3.	"	CRYSTALLINA, Kg., K. B., 16, f. 1; S. B. D., 74, 12, f. 101; V. H., 42, f. 10.
Fig. 4.	"	FULGENS, W. S., S. B. D., 74, 12, f. 103; Prit., 789, 13, f. 20; M. D., 13, f. 24; Sm. Sp. T., No. 556.
" 5.	"	CAPITATA, Ehrb. Mik., 39, 3, f. 114; 17, f. 25b, 33; K. B., 14, f. 19; Rab. S. D., 5, f. 6; S. B. D., 72, 12, f. 93; Sm. Sp. T., No. 547. Striae of ends too radiate in figure.
" 6.	"	UNDULATA, Greg., S. B. D., II, p. 97; V. H., 42, f. 2; Sm. Sp. T., 586.
" 7.	"	LONGISSIMA, W. S., S. B. D., 72, 12, f. 95; Sm. Sp. T., Nos. 564, 565.
" 8.	"	ULNA, var. longissima, W. S., S. B. D.
" 9.	"	FULGENS, W. S., (Ardissonia fulgens, Kg.), S. B. D., 74, 12, f. 103; Prit., 789, 13, f. 20; M. D., 13, f. 24; V. H., 43, f. 42.
Figs. 10, 11.	"	GALLIONII, Ehrb., K. B., 30, f. 42; 28, f. 36; S. B. D., 30, f. 265; Prit., 788, 12, f. 34, 36.
" 12.	"	ONYRRHYNCHUS, (vide small form), Kg., K. B., 14, f. 82; Rab. S. D., 5, f. 23; S. B. D., 11, f. 91; Sm. Sp. T., No. 570.
Figs. 13, 14.	"	RUMPENS, Kg., K. B., 16, f. 6; Rab. S. D., 5, f. 8; V. H., 40, f. 14.
Fig. 15.	"	INVESTIANS, W. S., S. B. D., p. 98; Sm. Sp. T., No. 560; V. H., 40, f. 33.
" 16.	"	VAUCHERIE, K. B., 14, f. 4, Rab. S. D, 5, f. 15; S. B. D., 11, f. 99; V. H., 40, f. 19.
" 17.	"	RADIANS, magnified only about one-half as much as the other figures, Kg., K. B., 14, f. 7; Rab. S. D., 4, f. 40; Sm. Sp. T., No. 574; V. H., 30, f. 11.

Plate XXXIX.

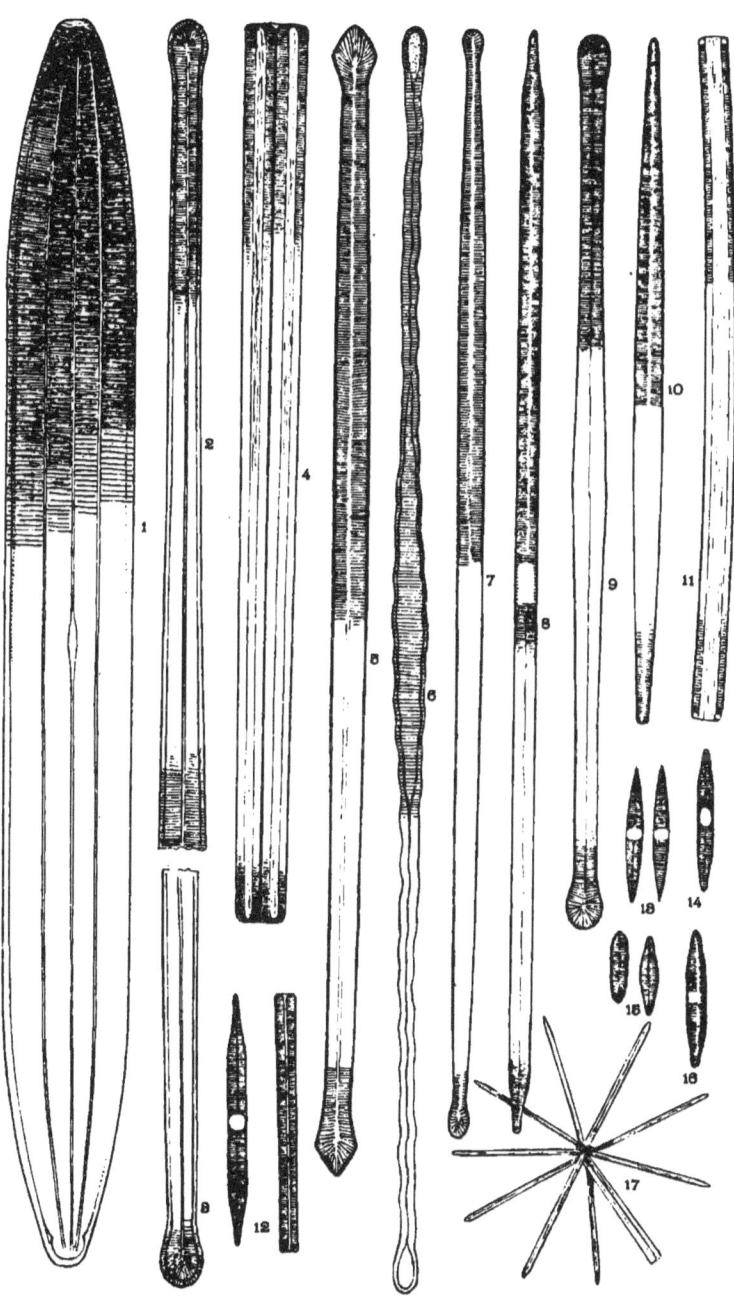

PLATE XI.

Figures magnified 500 diameters.

Fig. 1.   NITZSCHIA SPECTABILIS, Ralfs., Prit., p. 782. More properly
          N. ampionys; var. *spectabilis*, H. L. S.; striae
          moniliform, V. H., 67, f. 8.
"   2.   "   LONGISSIMA, Grun., 1862, p. 581; M. D., 13, f. 11;
          V. H., 70, f. 1-3 — birostrata, W. S.
"   3.   "   LONGISSIMA, forma parva; V. H., 70, f. 3.
"   4.   "   SIGMA, S. B. D., 13, f. 108; Prit., 781, 4, f. 21; Sm.
          Sp. T., No. 367; V. H., 65, f. 7, 8.
"   5.   "   SIGMA, formae elongatae; V. H., 66, f. 7.
Figs. 6, 7. "   SIGMA, vars. rigida, Grun., and rigidula, Grun.;
          V. H., 66, f. 5, 8.
"   8, 9.   "   LANCEOLATA, S. B. D., 14, f. 118; T. M. S., 1860,
          p. 48, 2, f. 20; M. D., 13, f. 10; V. H., 68, f. 1-4;
          Sm. Sp. T., No. 352.
"   10-13. "   CLOSTERIUM, S. B. D., 15, f. 120; V. H., 70, f. 5,
          7, 8.
"   14-17. "   FASCICULATA, Grun., 1878, p. 21; V. H., 66, f.
          11-13,   Homoeocladia sigmoidea; S. B. D., 55,
          f. 349; Sm. Sp. T., Nos. 199, 200.
Fig. 18. "   CLOSTERIUM, small form, *vide* Figs. 10, 13.
Figs. 19, 20. "   TENUIS, S. B. D., 13, f. 111; V. H., 67, f. 16.
Fig. 21. SYNEDRA LANCEOLATA, Kg., K. B., 30, f. 31; Rab., S. D., 4, f.
          14, 18. Compare more elongated form, Plate
          41, f. 9.
Figs. 22, 23. NITZSCHIA FRUSTULUM, H. L. S., p. No. 342; V. H., 68, f.
          28, 29.
"   24, 25. "   COMMUNIS, Rab., Alg., Sachv., 949 and 843; Grun.,
          1862, p. 578, 12, f. 18; V. H., 69, f. 32.
"   26, 27. "   PALEA, N. S., S. B. D., p. 89, Grun., 1862, p. 579,
          12, f. 3; V. H., 69, f. 22, b. c., Sm. Sp. T., No. 361.
"   28, 29, 30. SYNEDRA DANICA, K. B., 14, f. 13; Sm. Sp. T., 549; V. H.,
          38, f. 14.
Fig. 31. "   LONGICEPS, Ehrb., Ber., 1845, Prit., 788; Sm. Sp.
          T.; No. 564 = capitata, with stiliform produced
          apices.
"   32.   LONGISSIMA, S. B. D., 12, f. 95; Sm. Sp. T., No. 564.

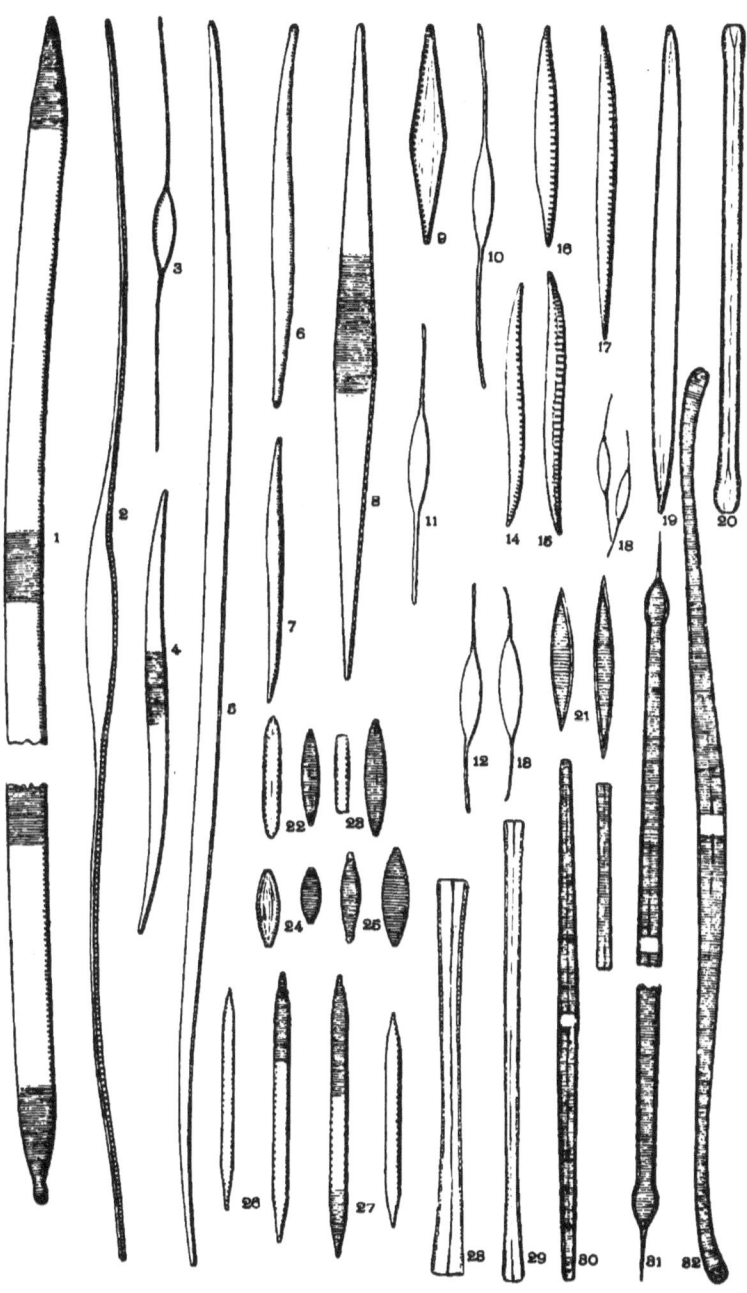

Plate XL.

PLATE XLI.

Figures magnified 500 diameters.

Fig. 1. SYNEDRA SPATHULIFERA, Grun; *S. ulna*, var.; V. H., 38, f. 4.
" 2-4. " ACUS, Kg., K. B., 15, f. 7; Rab. S. D., 5, f. 42; V. H., 39, f. 4. Might all pass for S. ulna, var.
Figs. 5, 6. " DANICA, (*ulna*, var.), K. B., 14, f. 13; Rab., S. D., p. 54; Sm. Sp. T., 549; V. H., 38, f. 14.
Fig. 7. " OBTUSA, S. B. D., 11, f. 92; Sm. Sp. T., No. 569.
" 8. " VITREA, K. B., 14, f. 17; Rab. S. D., 5, f. 24. *S. ulna*, var., V. H., 88, f. 12.
" 9. LANCEOLATA, *S. ulna*, var., V. H., 38, 10. Compare Plate, 36, f. 21.
" 10–12. " AFFINIS, K. B., 15, f. 6, 24, f. 1, 5; S. B. D., 12, f. 97; V. H., 41, f. 13, 14. Fig. 10, *genuina forma parva*; Fig. 11, var. *acuminata*, Grun.; Fig. 12, var. *gracilis*, Grun.
Figs. 13, 14. " BICEPS, f. and s. views, K. B., 14, f. 18, 21; Rab. S. D., 5, f. 9; Sm. Sp. T., No. 545. Fig. 13a, smaller form.
Fig. 15. " AMPHIRRHYNCHUS, Ehrb., K. B., 14, f. 15; Rab. S. D., 5, f. 57. V. H., 38, f. 5.
" 16. " PULCHELLA, forma major, K. B., 29, f. 87; Rab. S. D., 5, f. 17; S. B. D., 30, f. 84; Prit., p. 786, 4, f. 28; Sm. Sp. T., No. 573.
Figs. 17, 18. " PULCHELLA, var., *genuina*, Kg.; V. H., 40, f. 28, 29.
" 19, 20. " CROTONENSIS, (Edw.), Grun.; V. H., 40, f. 10; var. *prolongata*, Grun. Usually hyaline, often flexuose.
" 21-23. " ACUTA, Ehrb., = (*Ulna*, var. *acuta*, H. L. S.); Mik., 39, 3, f. 118, 119, 6, 1, f. 3a, b, etc.; K. B., 30, f. 49; Rab. S. D., 5, f. 23; Sm. Sp. T, 541.
" 24, 25. " SUPERBA, K. B., 15, f. 13; S. B. D., 12, f. 102; Sm. Sp. T., No. 578.
Fig. 26. " PULCHELLA, Compare Figs. 17, 18.
Figs. 27-29. " INVESTIENS, S. B. D., p. 98; V. H., 40, f. 33c.
Fig. 30. " ACUTA, Compare Figs. 21-23, above.
" 31. " ULNA, var. *vitrea*, Compare Fig. 8; V. H., 38, f. 11, 12.
" 32. " SUPERBA, Compare Figs. 24, 25. Two small clusters seen under lower power.
Figs. 33, 34. " AFFINIS, var. with ends more than usually blunt, and more than ordinarily inflated.

Plate XLI.

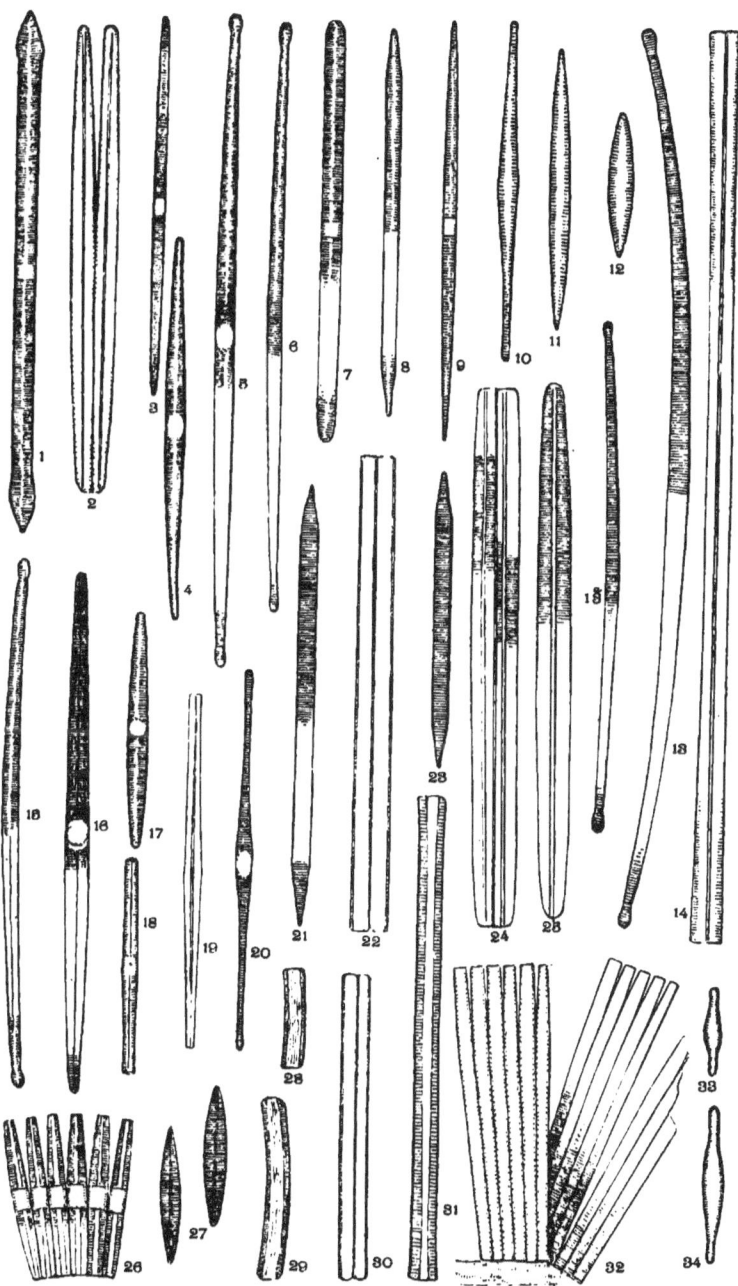

PLATE XLII.

Figures magnified 500 diameters.

Fig. 1.   NITZSCHIA VALIDA, forma longissima, Cleve. and Grun., 1878, p. 12, 3, f. 19; J. M., 1879, p. 33, 2, f. 19; V. H., 65, f. 4, 5.
Figs. 2, 3.   "   parts of same seen under high power, about $\frac{1000}{1}$.
Fig. 4.   "   MAJOR, Grun., V. H., 65, f. 6, (N. Sigma var.).
"   5.   "   parts of same under higher power.
Figs. 6, 8.   "   BREBISSONII, S. B. D., 31, f. 266; V. H., 64, f. 4, 5. Figs. should be beaded along the edge.
Fig. 7.   "   OBTUSA, smaller form, compare Fig. 18.
"   9.   "   SIGMOIDEA, S. B. D., 13, f. 104; Prit., 781, 7, f. 148; V. H., 63, f. 5-7.
Figs. 10, 11.   "   SIGMA, S. B. D., 13, f. 108; Prit., 781, 4, f. 21; V. H., 65, f. 7, 8; Sm. Sp. T., No. 367.
"   12-15.   "   VERMICULARIS, Hantz., Prit., 781; V. H., 64, f. 1, 2. All are small forms of sigmoidea.
"   16, (21).   "   VITREA, Norm., T. M. S., 1861, p. 7, 2, f. 4; V. H., 67, f. 10; Sm. Sp. T., No. 376.
Fig. 17.   "   LINEARIS, S. B. D., 13, f. 110 and 31, f. 110; V. H., 67, f. 13-15; Sm. Sp. T., Nos. 353, 354. Striae fine and light. Fig. too heavy.
"   18.   "   OBTUSA, S. B. D., 13, f. 109; V. H., 67, f. 1; Sm. Sp. T., No. 358. Vide Fig. 7.
Figs. 19, 20.   "   DISSIPATA, Rab. Alg. Sachv., No. 948; V. H., 63, f. 1-3.
Fig. 21.   "   VITREA, Norm., forma major. Vide Fig. 16.
"   22.   "   LINEARIS, Vide Fig. 17.
"   23.   "   VITREA, Normal form. Vide Figs. 16, 21.

Plate XLII.

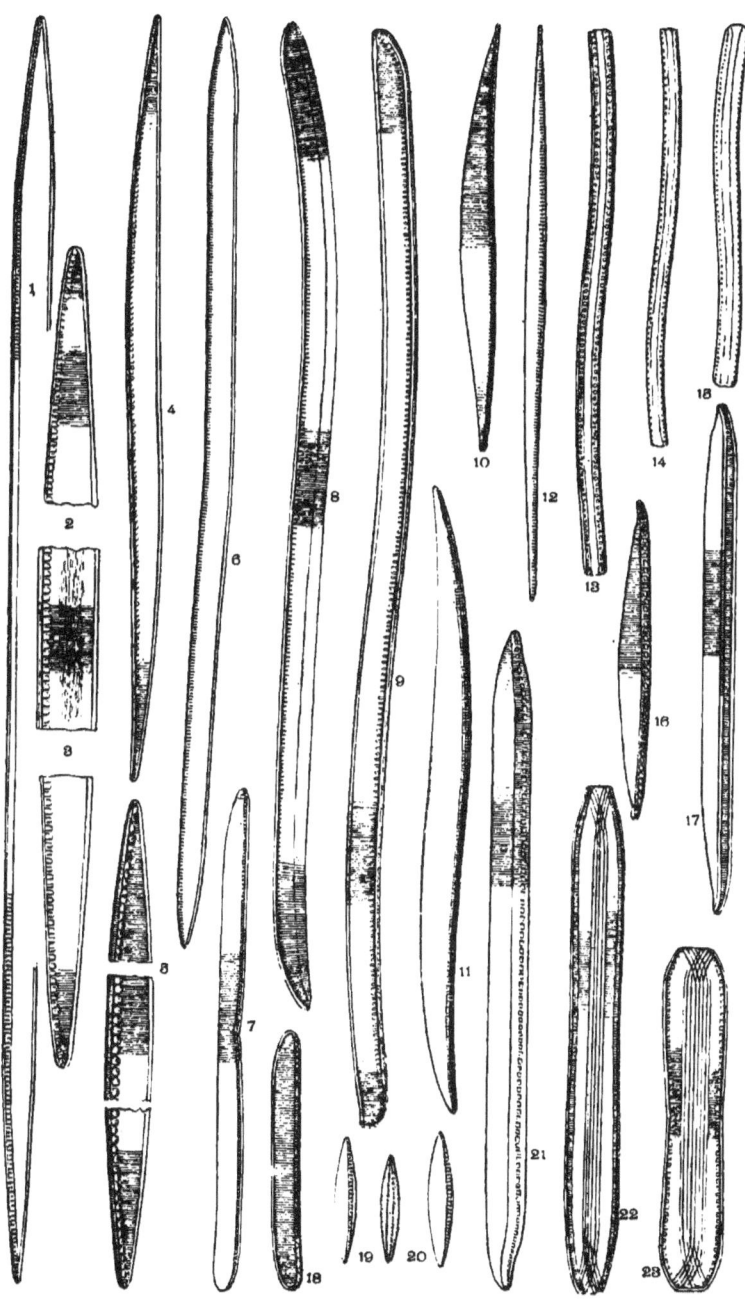

PLATE XLIII.

Figures magnified 500 diameters.

Fig. 1. NITZSCHIA FEDIGERII, Grun., J. R. M. S., 1880, p. 396, 13, f. 15.
" 2, 3. " CAMPEACHEANA, Grun.; J. R. M. S., 1880, p. 395, 13, f. 16.
" 4. " SCALIGERA, Grun.; J. R. M. S., 1880, p. 395, 12, f. 3. Allied to *N. Grundleri*. Fla.
Figs. 5, 6. " BILOBATA, S. B. D., 15, f. 113; V. H., 60, f. 13; Sm. Sp. T., No. 335.
" 7, 8. " SPATHULATA, Breb.; S. B. D. 31, f. 268; Sm. Sp. T., No. 370; V. H., 62, f. 7, 8.
Fig. 9. " SCALARIS, W. S.; S. B. D., 14, f. 115; Prit., 781, 4, f. 22; V. H., 60, f. 14, 15; Sm. Sp. T., No. 365.
" 10. " LIMNICOLA, Grun.; M. M. J., 1880, p. 395, 13, f. 10; = *Tryblionella punctata*, S. B. D., 10, f. 76 a.
" 11. " GRANULATA, Grun., M. M. J., 1880, p. 395, 12, f. 7. Not *Nitzschia*; Bailey's PYXIDICULA COMPRESSA, Mic. Obs., p. 40, 2, f. 13, 14; Sm. Sp. T., Nos. 430, 431. (Savannah).
" 12-14. " EPITHEMOIDES, Grun.; Cleve., 1880, p. 82; V. H., 60, f. 6-8.
" 15. " BILOBATA, small form, *vide* Figs. 5, 6.
Figs. 16, 17. " LITTOREA, Grun.; V. H., 59, f. 24; = *N. Thermalis*, var.
Fig. 18. " CURSORIA, Grun.; Cleve., 1880, p. 89; V. H., 62, f. 19.
" 19. " MAJUSCULA, Grun.; Cleve., 1880, p. 87; V. H., 62, f. 5. Closely allied to *fluminensis*. Campeachy Bay.
" 20. " FLUMINENSIS, Grun.; 1862, p. 581, 12, f. 35; V. H., 62, f. 3, 4. Striae moniliform.
" 21. " SINUATA, Grun.; Cleve., 1880, p. 82; V. H., 60, f. 11; = *Denticula sinuata*, W. S.
Figs. 22, 23. " TABELLARIA, Grun., Cleve., 1880, p. 82; = *Denticula tabellaria*, = *N. Sinuata*, var.
Fig. 24-26. " ANGULARIS, S. B. D., 13, f. 117; V. H., 62, f. 11-14. Fig. 25, var. *occidentalis*, striae oblique. Camp. Bay.
" 27-29. " PARADOXA, (Gmel), Grun.; V. H., 61, f. 6, 7, = *Bacillaria paradoxa*.
" 30. " KITTONII, H. L. S.; A. Q. J. M., 1878, p. 14, 3, f. 5.
" 31-33. " DENTICULA, Grun.; (Kg.), W. S.; V. H., 60, f. 10; S. B. D., 34, f. 292. Not really a *Nitzschia* but *Denticula tenuis*, S. B. D., 34, f. 293.

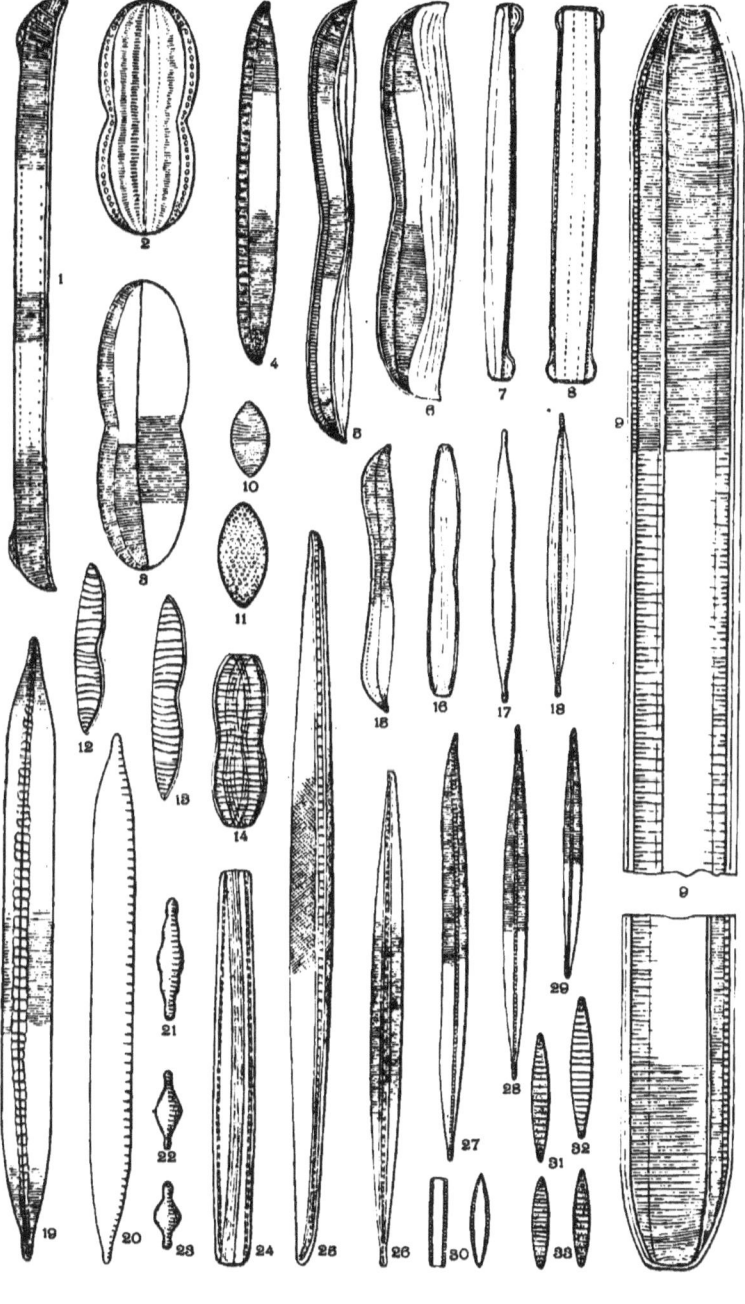

PLATE XLIV.

Figures magnified 500 diameters.

Figs. 1, 2.   NITZSCHIA PLANA, S. B. D., 15, f. 114; Sm. Sp. T., No. 363, V. H., 58, f. 10, 11.
" 3, 4, (9).   "   PANDURIFORMIS, Greg., D. C., p. 57, 6, f. 102; V. H., 58, f. 1-3.
" 5, 6, (11). "   DUBIA, S. B. D., Pl. 13, 14, f. 112; V. H., 59, f. 9-12; Sm. Sp. T., No. 341.
Fig. 7.   "   CIRCUMSUTA, Grun., Cleve., 1880, p. 77; V. H., 91. S. Tryblionella circumsuta, Ralfs.
" 8.   "   LEVIDENSIS, W. S., V. H., 57, f. 15-17; 59, f. 7, Tryblionella levidensis, granulata.
" 9.   "   Compare Figs. 3, 4.
" 10.   "   COARCTATA, Grun., V. H., 59, f. 4. Punctæ should be coarser. Tryblionella, var. coarctata.
" 11.   "   DUBIA, Compare Figs. 5, 6.
Figs. 12, 13.   "   HUNGARICA, Grun., 1862, p. 568, 12, f. 31; V. H., 58, f. 19-22; Sm. Sp. T., No. 347.
" 14, 15.   "   VIVAX, S. B. D., 31, f. 267; V. H., 62, f. 1, 2; Nit. amphioxys, var.
" 16, 17.   "   MARINA, Grun., V. H., 57, f. 26, 27.
" 18, 19.   "   ANGUSTATA, Greg., Cleve., 1880, p. 70; V. H., 57, f. 22, 24; Tryblionella angustata; S. B. D., 30, f. 262.
" 20, 21, 22. "   PUNCTATA, S. B. D., 10, f. 76; 30, f. 261; B. J. N. H., p. 344, 2, f. 76 = Tryblionella punctata.
Fig. 23.   "   TRYBLIONELLA, Hantz, V. H., 57, f. 9, 10; probably the same as Tryblionella gracilis; W. S.
" 24.   "   SCUTELLUM, S. B. D., 10, f. 74; M. D., 13, f. 30; (markings rather too distinct) Tryblionella scutellum; W. S., the same as T. circumsuta; V. H., and called by Bailey Surirella circumsuta.

Plate XLIV.

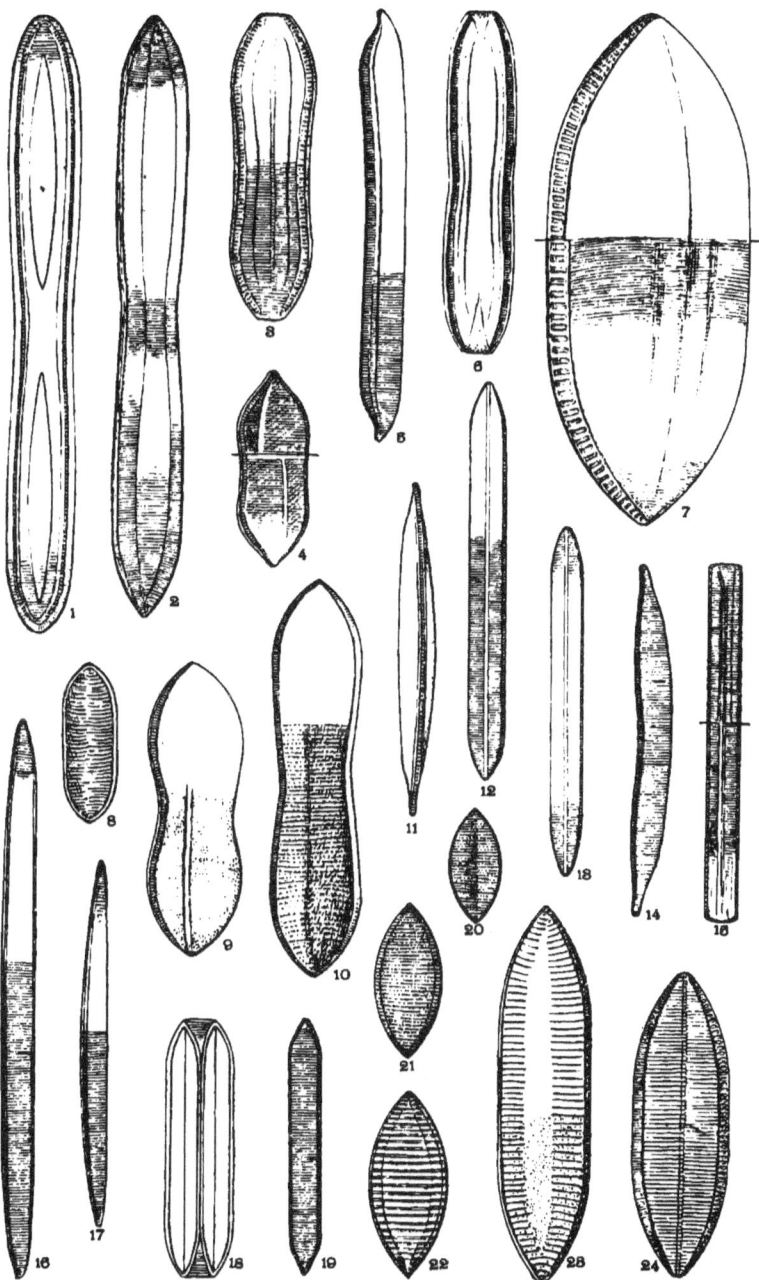

PLATE XLV.

Figures magnified 500 diameters.

Figs. 1, 2.   PLAGIOGRAMMA PULCHELLUM, Grev.; M. J., 1859, p. 209, 10, f. 4-6; Prit., 774, 4, f. 32. Costae are beaded. Cal. Guano.
Fig. 3.   "   PULCHELLUM, a very large specimen.
Figs. 4, 5.   DIMMEREGRAMMA MARINUM, Ralfs.; Prit., p. 790; V. H., 36, f. 9.
" 6, 7.   PLAGIOGRAMMA PYGMAEUM, Grev.; M. J., 1859, p. 210, 10, f. 11. Camp. Bay.
" 8, 9.   "   VALIDUM, Grev.; M. J., 1859, p. 209, 10, f. 8, Cal. and Camp. Bay.
" 10, 11.   "   GREGORIANUM, Grev.; M. J., 1859, p. 208, 10, f. 1, 2; Jan. and Rab., p. 10, 2, f. 8. Costae are beaded. Mobile.
" 12-14.   "   OBESUM, Grev.; M. J., 1859, p. 211, 10, f. 12, 13. Charleston.
" 15.   "   SPINOSUM, Cleve.; N. L. K. D., p. 18, 4, f. 55. Camp. Bay.
Figs. 16, 17. "   CALIFORNICUM, Grev.; M. J., 1859, p. 211, 1 D, f. 15-17. (Striae moniliform.)
" 18, 19.   "   TESSELLATUM, Grev.; M. J., 1859, p. 208, 10, f. 7.
Fig. 20.   "   ORNATUM, Grev.; M. J., 1859, p. 209, 10, f. 9; V. H., 36, f. 3. Costae moniliform as in all of this genus, H. L. S.
Figs. 21, 22. GLYPHODESMIS DISTANS, V. H.; 36, f. 15, 16; Dimeregramma, Greg.
" 23, 24.   "   WILLIAMSONII, W. S.; V. H., 36, f. 14.

Plate XLV.

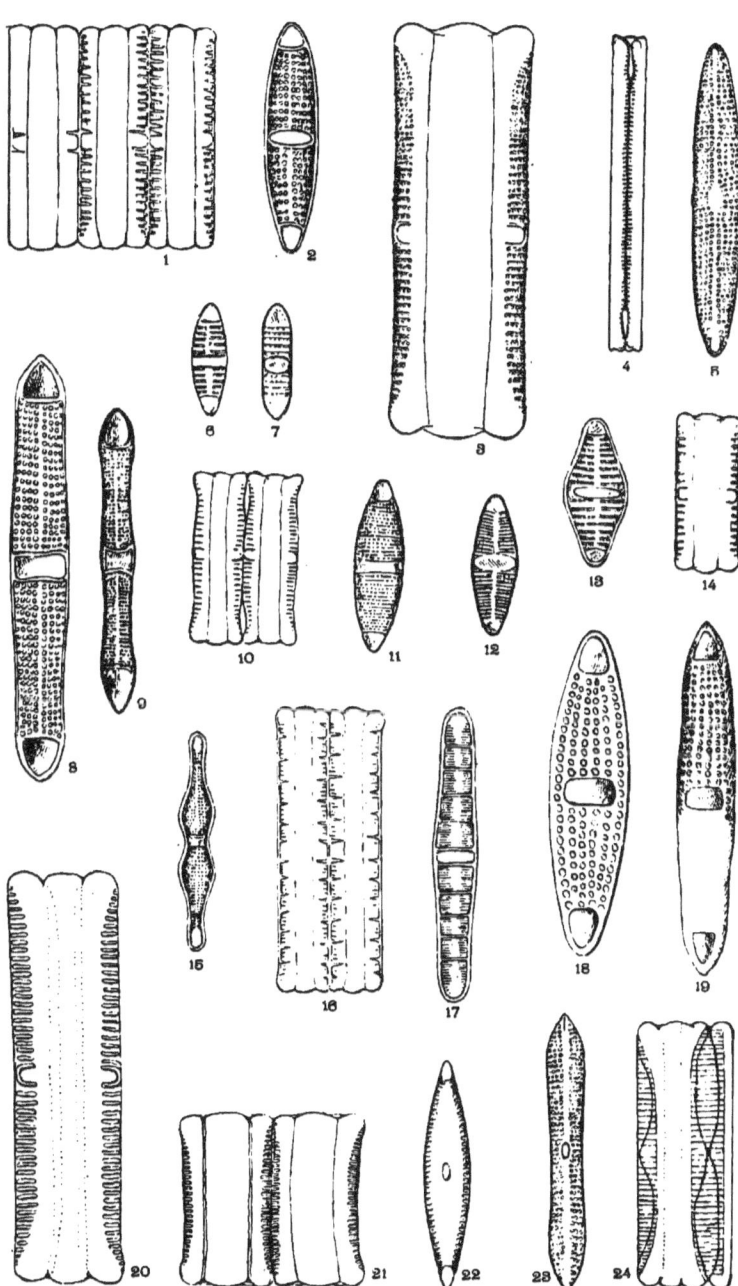

PLATE XLVI.

Figures magnified 500 diameters.

Fig. 1.   Asterionella formosa, var. subtilissma, N. America?
" 2.   "   " var. subtilis, Grun. Lake Erie.
" 3.   "   notata, (var. of Bleakleyi) Grun.; V. H., 52, f. 3.
" 4.   "   formosa, var. gracillima, Hantz; V. H., 51, f. 22.
Figs. 5, 6.   "   formosa, var. Ralfsii; W. S., a large and a miniature form; S. B. D., p. 81; T. M. S., 1860, p. 150, 7, f. 9.
Fig. 7.   "   Bleakeleyi, W. S., a var. of *A. formosa*; S. B. D., p, 82; T. M. S., 1860, p. 150, 7, f. 10; Lewis, N. and R. Sp., p. 10, 2, f. 9.
" 8.   "   formosa, var. gracillima; V. H., 51, f. 22; S. B. D., 11, p. 81; T. M. S., 1860, p. 149, 7, f. 8; Prit., 779, 4, f. 17; M. D., 43, f. 14; Sm. Sp. T., No. 46.
" 9.   Denticula thermalis, K. B., p. 43, 17, f. 6; Rab. S. D., 1, f. 3; V. H., 49, f. 17, 18; Sm. Sp. T., No. 128.
" 10.   "   lauta, Bail, N. Sp., 9, f. 1, 2; V. H., 49, f. 1, 2.
" 11.   Diatom anceps, front view; Figs. 17, 18, end view; Ehrb. V. H., 51, f. 5-8.
Figs. 12-14.   "   vulgaree, front view; Bory.; K. B., 17, f. 15; Rab. S. D., 2, f. 6; S. B. D., 39 and 40, f. 309; Sm. Sp. T., No. 127.
Fig. 15.   "   elongatum, end view; Ag., K. B., 17, f. 18; Rab. S. D., 35, 2, f. 1; S. B. D., 40 and 41, f. 311.
Figs 16,(21,22).   "   Ehrenberghi, end view; K. B., 17, f. 17; Sm. Sp. T., No. 133.
" 17, 18.   "   anceps, end view; V. H., 51, f. 5-8.
" 19, 20.   "   pectinale, end and front views; K. B., 17, f. 11; Rab. S. D., 2, f. 2; V. H., 50, f. 23-26.
" 21, 22.   "   Ehrenberghi, front view, compare Fig. 16.
" 23-25.   "   tenue, front view; Ag. Syst.; K. B., 17, f. 9, 10; Sill. J., 1836, 2, f. 12.
" 26, 27.   "   elongatum, front view, compare Fig. 15.

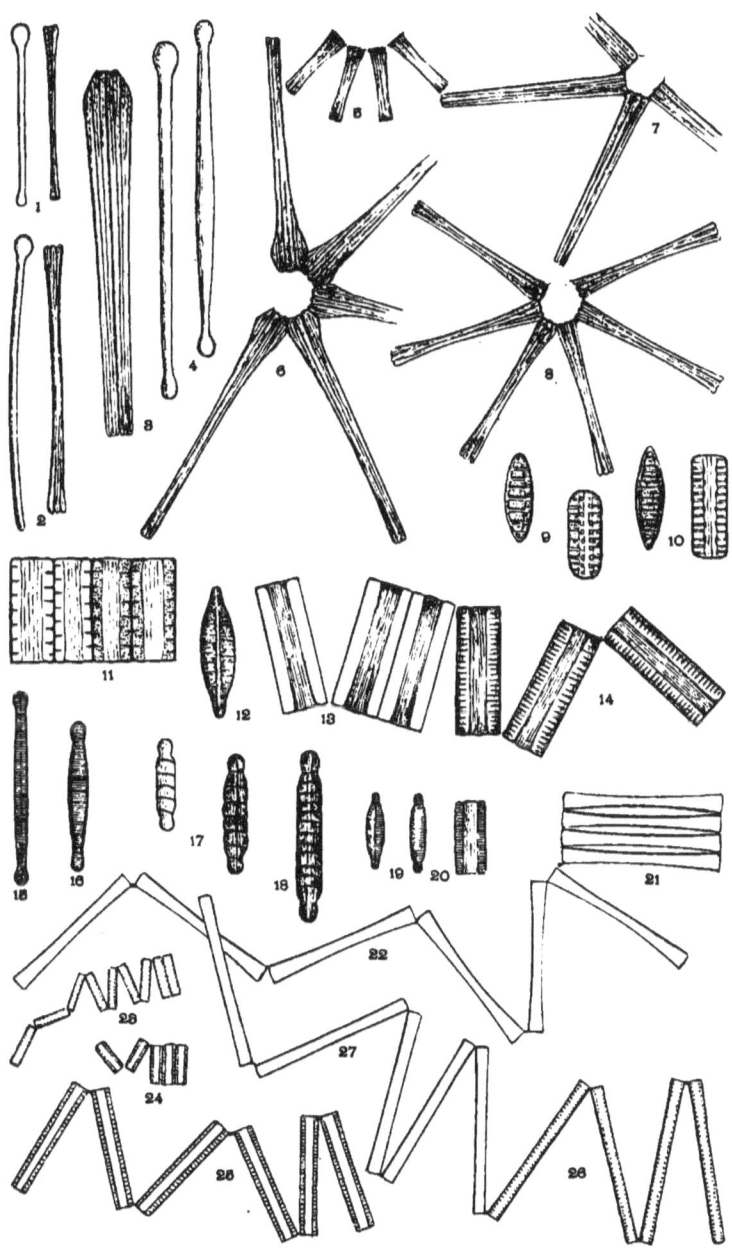

Plate XLVI.

PLATE XLVII.

Figures magnified 500 diameters.

Fig.	1-3.	FRAGILARIA CAPUCINA, Desmaz.; K. B., 16, f. 3; Rab., S. D., 1, f. 2; S. B. D., 34, f. 266; Prit., 776, 9, f. 173; Sm. Sp. T., No. 165.
"	4-7.	" VIRESCENS, Ralfs.; A. M. D., 1843, p. 110, 2, f. 6; K. B., 16, f. 4; Rab., S. D., 1, f. 1; S. B. D., 35, f. 297.
"	8.	" BINOIDES, Ehrb.; Abh., 1841, p. 415; Mik., 16, 2, f. 36; Mik., 6, 1, f. 4; this and Figs. 9, 10, rather doubtful species.
"	9.	" BICEPS, Ehrb. Mik., 7, 2, f. 9, 10; 7, 3, f. 23, 24; 14, f. 49-51; K. B., p. 46.
"	10.	" PARISITICA, Grun.; V. H., 45, f. 30; var. 45, f. 29; 46, f. 14.
"	11.	" TURGENS, Ehrb. Mik., 7, 1, f. 8; 1, 3, f. 8; Abh., 1869, 1a, f. 4.
"	12.	" HARRISONII, Grun.; V. H., 45, f. 28.
Figs.	13, 14.	" CONSTRUENS, Grun.; V. H., 45, f. 27.
Fig.	15.	" BREVISTRIATA, Grun.; var. Mormonorum; V. H., 45, f. 31-34.
Figs.	16, 17.	" PARADOXA, Ehrb. Mik., 33, 15, f. 13; 33, 14, f. 10. Somewhat doubtful species.
"	18, 19.	" MUTABILIS, Grun.; V. H., 45, f. 12; Bran. Alp., p. 119, 4, f. 8. Frustules cohere strongly and form bands.
Fig.	20.	" CONSTRUENS, Grun. Same as Figs. 13, 14.
"	21.	" AMPHICEPHALA, Ehrb. Mik., 35a, 14, f. 1-3; Mik., 37, 2, f. 5, 6.
"	22.	" CALIFORNICA, Grun.; V. H., 44, f. 13; a var. of striatula. Marine.
Figs.	23-25.	" PACIFICA, Grun.; 1862, p. 373, 8, f. 19; V. H., 33, f. 20-23. N. Amer.? No Fragilaria, cuneate and allied to Trachysphaenia, H. L. S.
Fig.	26.	" ENTOMEN, Ehrb. Mik., 5, 3, f. 50; K. B., p. 46.
Figs.	27-29.	LICMOPHORA FLABELLATA, S. B. D., 26 and 32, f. 234; J. R. M. S., 1879, p. 683.
"	30, 31.	" GRACILIS, Grun.; Hedw., VI, p. 34; V. H., 46, f. 13.
"	32, 32a.	" CALIFORNICA, Grun.; V. H.. 47, f. 14. Striae oblique.
"	33, 34.	" JURGENSII, Ag.; Cleve., 1880, p. 110, 7, f. 125; V. H., 46, f. 10, 11. Frustules not bent.
Fig.	35-37.	" TINCTA, Grun.; Hedw., VI, p. 35; V. H., 48, f. 13-15. Very hyaline.

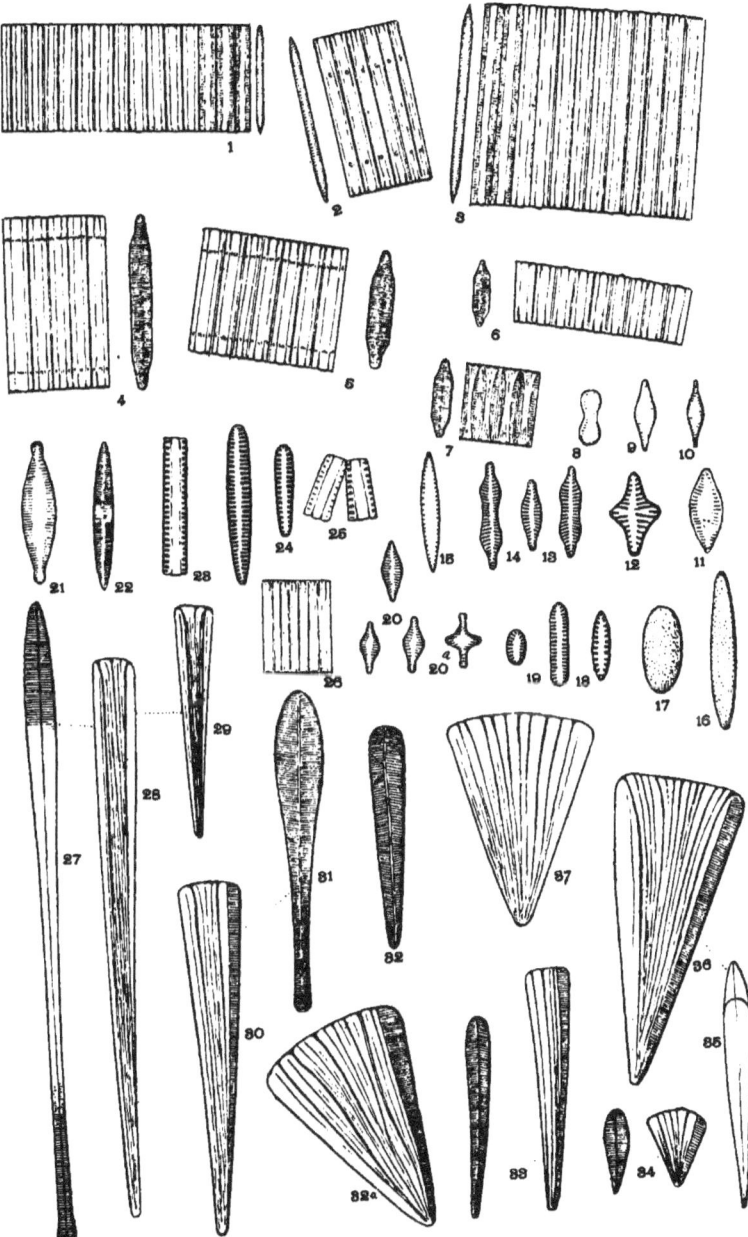

Plate XLVII.

PLATE XLVIII.

Figures magnified 500 diameters.

Figs. 1-4. ODONTIDIUM MESODON, K. B., 17, f. 1; S. B. D., 34, f. 288, Sm. Sp. T., No. 382.
" 5, 6. " HYEMALE, K. B., 17, f. 4; S. B. D., 34, f. 289; Sm. Sp. T., No. 381.
" 7-12. " MUTABILE, S. B. D., 34, f. 290; Sm. Sp. T., Nos. 383, 692.
" 13, 14. TESSELA (*Striatella*) INTERRUPTA, Ehrb.; K. B., 18, f. 4. M. D., 14, f. 35; Sm. Sp. T., No. 591.
" 15-20. ODONTIDIUM TABELLARIA, S. B. D., 34, f. 291; Lewis, W., M. D., p. 13, 2, f. 1, 2. These figures represent abnormal or sporangial forms collected by Dr. Lewis in White Mountain pools.
Fig. 21. AMPHIPRORA ORNATA, Bail. M. O., p. 38, 2, f. 15, 23; V. H., 22 bis f. 5. Compare plate, Figs. Bail.'s figure, but too small.
" 22-24. " CALUMETICA, List of diatoms of Lake Michigan, B. W. Thomas, Chicago.
" 25-28. DICTYOCHYA forms frequently found in diatomaceous material; no longer accepted as diatoms.
" 29-31. AMPHICAMPA many similar siliceous forms found, but cannot be classed with diatoms.
Figs. 32, 33. CLIMACOSPHAENIA ELONGATA Bail. N. Sp., p. 8, f. 10, 11; Prit., p. 772.

Plate XLVIII.

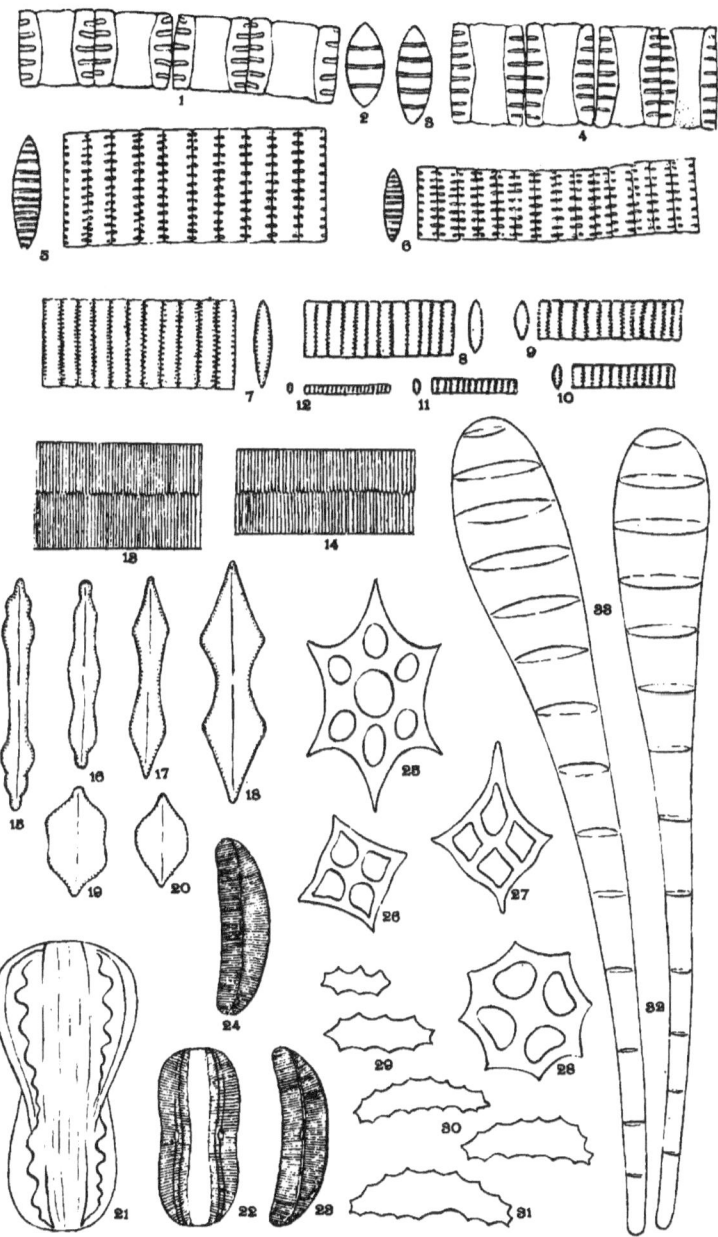

PLATE XLIX.

Figures magnified 500 diameters.

Fig. 1. GRAMMATOPHORA SERPENTINA, Ehrb. Mik., 35 A, 22, f. 14, K. B., 29, f. 82; S. B. D., 42, f. 315; V. H., 53, f. 1-3.
Figs. 3-5. " UNDULATA, Ehrb. Mik., 18, f. 87; 19, f. 37; K. B., 29, f. 68.
" 6-7. " GIBBA, Ehrb., Amer., 2, 6, f. 8; K. B., 29, f. 77; Prit., p. 808, 11, f. 48. Fig. 6 rather too much inflated at center. Fig. 7 may be a var. of marina.
" 8-10. " MARINA, K. B., 17, f. 24; S. B. D., 42, f. 314; Prit., 808, 4, f. 47; Sm. Sp. T., No. 188.
" 11-13. " ANGULOSA, Ehrb. Amer., 1, 3, f. 12; Mik., 21, f. 18; 18, f. 88; K. B., 29, f. 79; 30, f. 79; = hamilifera, Kg.
" 14-16. " MARINA, M. D., 12, f. 35; compare Figs. 8-10.
" 17, 18. " STRICTA, Ehrb.; Amer., 1, 1, f. 22; 3, 7, f. 31; K. B., 29, f. 76.
Fig. 19. " CARIBÆA, Cleve., 1878, p. 14, 4, f. 27; V. H., 53², f. 19.
Figs. 20-23. " SERPENTINA, same as Fig. 1. Smaller form; dots on this species more regularly quincuna than illustrated.
" 24-25. " MEXICANA, Ehrb.; K. B., 18, f. 1; 29, f. 67, 68; V. H., 53², f. 11, = marina, vide Figs. 14-16.
Fig. 26. " MAXIMA, Grun., 1862, 416, 8, f. 5; V. H., 53²; f. 12.

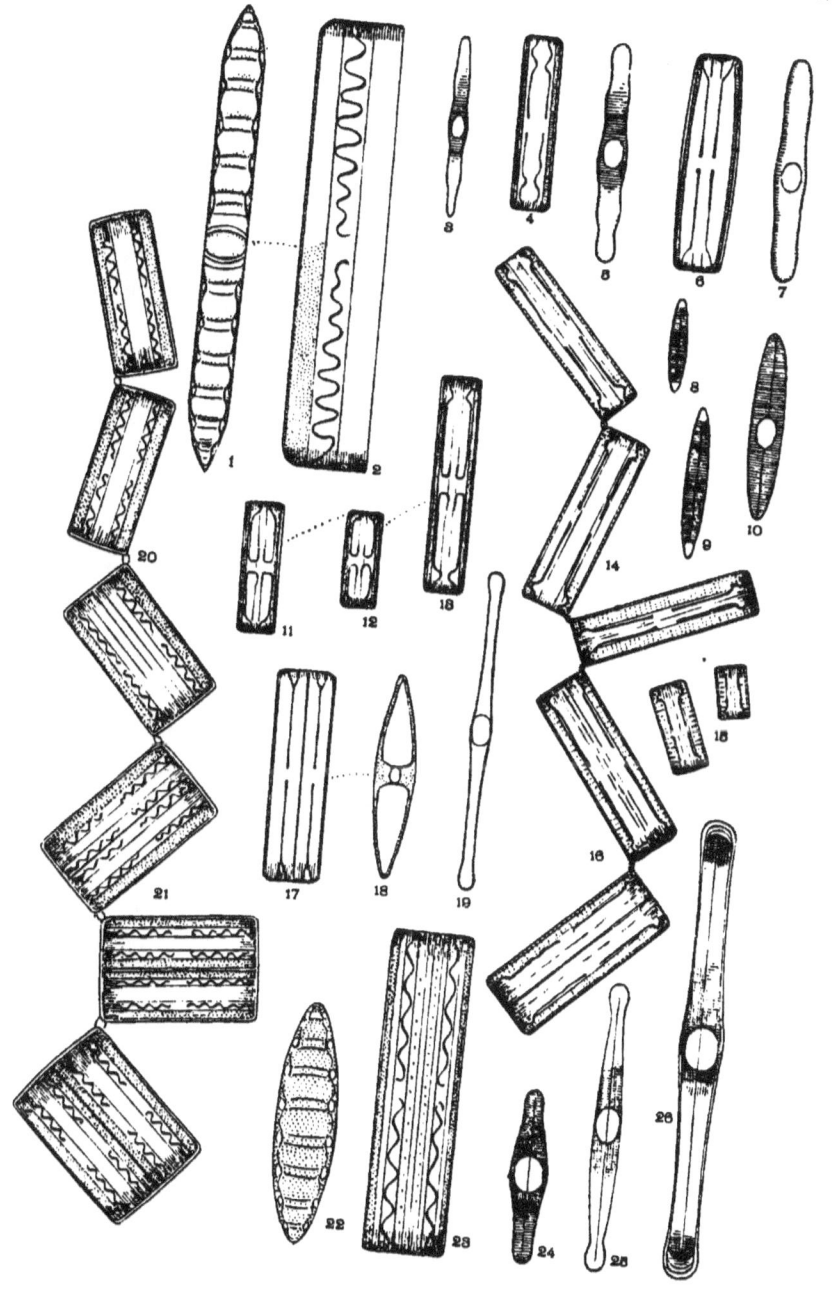

Plate XLIX.

PLATE L.

Figures magnified 500 diameters.

Figs. 1-3.   TABELLARIA FENESTRATA, K. B., 17, f. 22; 18, f. 2; 30, f. 73; S. B. D., 43, f. 317; Prit., 807, 13, f. 29; Sm. Sp. T., No. 588.
"   4-6.   "   FENESTRATA, Kg.; front views.
"   7-11.   "   FLOCULOSA, K. B., 17, f. 21; front and side views; Rab., S. D., 10, f. 2; S. B. D., 43, f. 316; Prit., p. 807, 13, f. 29; Sm. Sp. T., No. 589.
"   12, 15, 16.   TETRACYCLUS EMARGINATUS, W. S.; S. B. D., II, p. 38; Prit., 806.
"   13, 14, 17, 18.   "   LACUSTRIS, Ralfs; A. N. H., 1843, p. 105, 2, f. 2; K. B., 29, f. 70; S. B. D., 39, f. 308; Prit., p. 806, 11, f. 24, 25. All side views.
"   19, 20.   "   LACUSTRIS, another form; three front views and one side view.

Plate L.

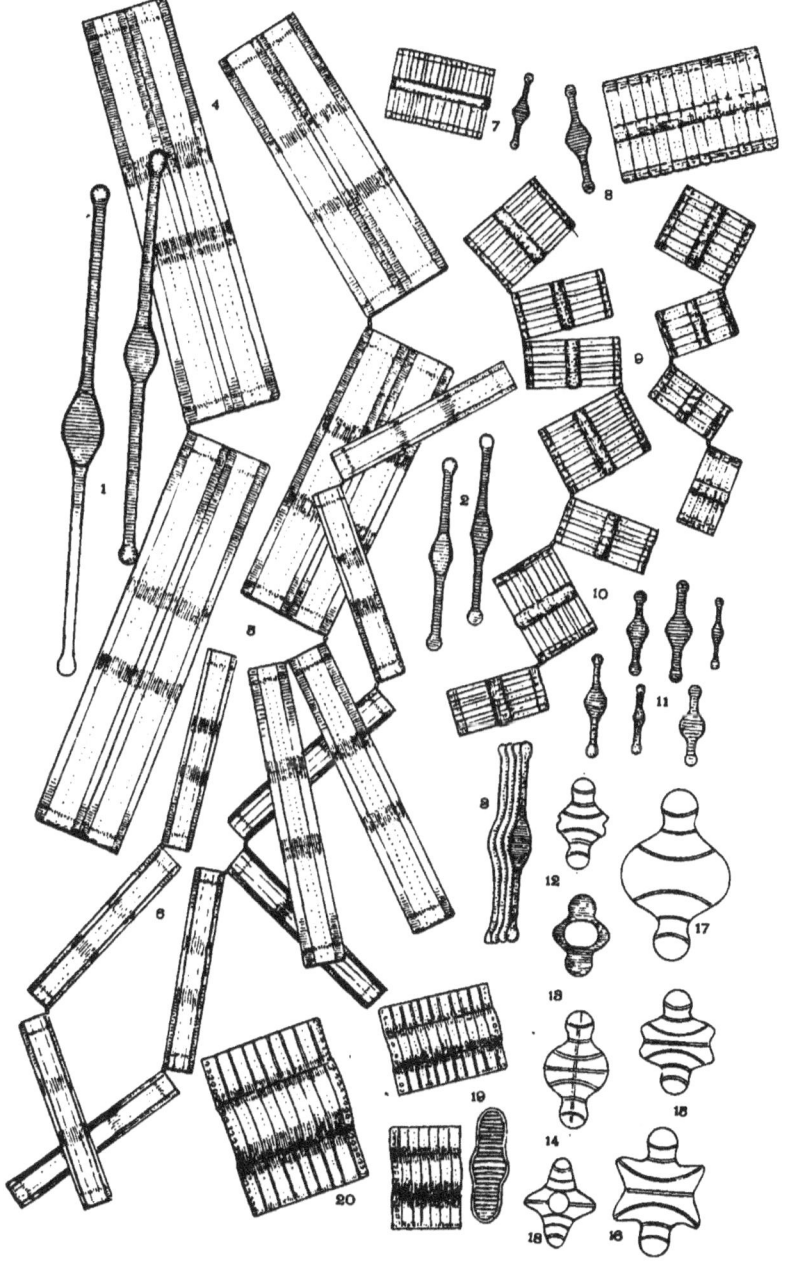

PLATE LI.

Figures magnified 500 diameters.

Figs. 1, 12, 13.   STRIATELLA UNIPUNCTATA, Ag.; K. B., 18, f. 5; S. B. D., 39, f. 307; Prit., 4, f. 40; Sm. Sp. T., No. 510.
"   2, 3, 4.   RHABDONEMA ADRIATICUM, K. B., 18, f. 7; Prit., 805, 13, f. 27; S. B. D., 38, f. 305, Sm. Sp. T., No. 432.
"   5, 6, 7.   "   ARCUATUM, K. B., 18, f. 6; S. B. D., 38, f. 305; Prit., 804, 10, f. 203, 204; V. II., 54, f. 14-16. Fig. 7 dotted all over; Fig. 6 perfectly smooth.
"   8-11.   "   MINUTUM, K. B., 21, f. 2; S. B. D., 38, f. 306; Prit., 4, f. 41; Sm. Sp. T., No. 435.
Figs. 12, 13.   STRIATELLA, end views of Fig. 1, margins quite smooth.
Fig. 14.   PODOCYSTIS AMERICANA, Bail.; N. Sp., p. 11, f. 38; M. D., 42, f. 21; S. B. D., II, p. 101.
Figs. 15-17.   HANTZSCHIA AMPHIONYS, = *Nitzschia amphionys;* hardly a need for the newer genius Hantzschia: V. II., 56, f. 1, 2.   Compare Fig. 20.
"   18.   PODOCYSTIS AMERICANA, Bail., same as Fig. 14.
"   19.   "   ADRIATICA, K. B., p. 62; Prit., 772, 4, f. 10; Sm. Sp. T., No. 418; V. II., 55, f. 8.
"   20.   NITZSCHIA AMPHIONYS, S. B. D., 13, f. 105; Sm. Sp. T., No. 334; Prit., p. 780. Same as Figs. 15-17.

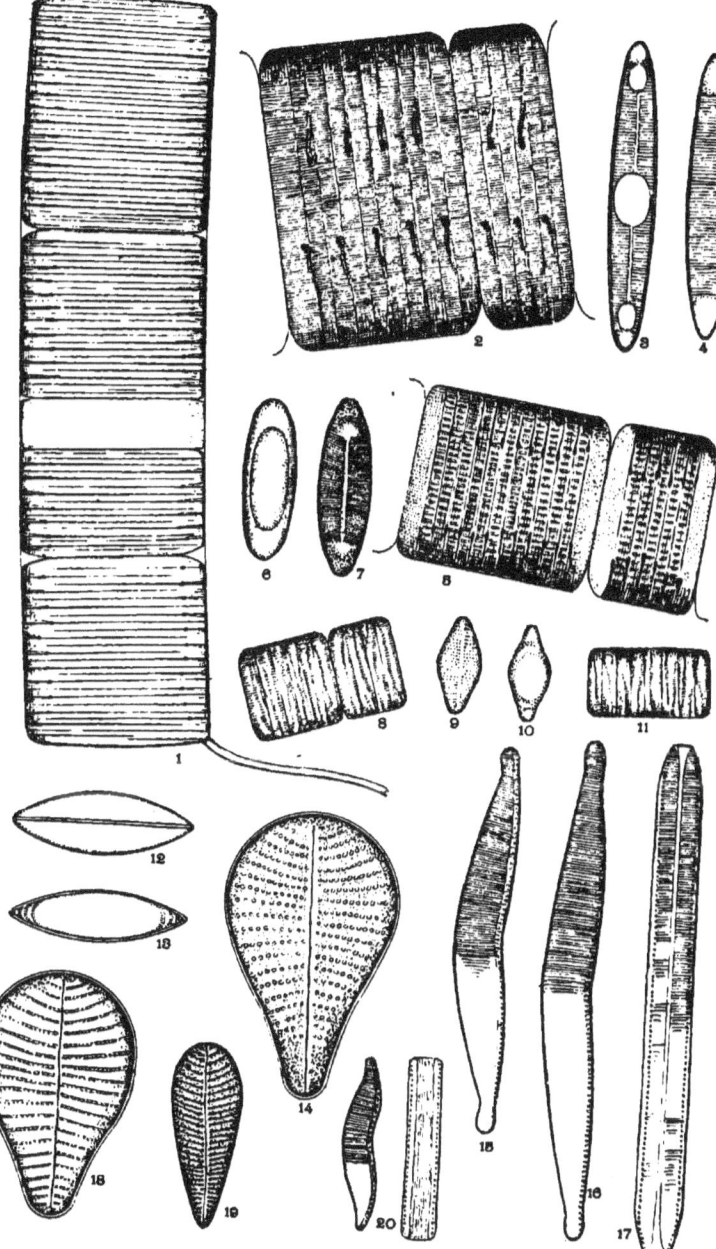

PLATE LII.

Figures magnified 500 diameters.

Figs. 1, 2. SURIRELLA BISERIATA, W. S.; S. B. D., 8, f. 57. Breb. has a *S. biseriata* which is *Sur. linearis*, W. S.
Fig. 3. " RECEDENS, A. S.; Schm. At., 19, f. 2-4; 24, f. 28. Var. of *S. fastuosa*.
Figs. 4, 5. " NORWEGICA, Eulenst.; Schm. At., 21, f. 17. Only a *Sur. elegans*, Ehrb.
" 6, 7. " CRUCIATA, A. S.; Schm. At., 56, f. 15, 16.
Fig. 8. " RECEDENS, A. S. Probably a variety of *fastuosa*. Compare Fig. 3.
9. " FLUMINENSIS, Grun., 1862, p. 463; Schm. At., 5, f. 6; 4, f. 9. Probably a variety of *S. fastuosa*.
" 10. " FASTUOSA, Ehrb.; K. B., 28, f. 19; S. B. D., 9, f. 66; M. J., 1855, p. 40, 4, f. 12; Schm. At., 5, f. 7, 8, 11. This figure is one of A. S. not W. S. A variable form differing greatly in size. Marine. Compare Plate 53, figures 1-4.
11. " GEMMA, Ehrb. Abh., 1840, p. 76, 14, f. 5; S. B. D., 9, f. 65; Prit., 795, 12, f. 2-4; Schm. At., 24, f. 26, 27, etc.
Figs. 12, 13. " OBLONGA, Ehrb. Mik., 17, 2, f. 1; 2, 3, f. 15, etc.; K. B., 29, f. 38; Schm. At., 22, f. 6-8; near Kg.'s *splendida*; S. B. A., 8, f. 63.
14, 15. " MOLLERIANA, Grun.; Schm. At., 23, f. 36, 37; 56, f. 21-23. Pensacola.
" 16, 17. " ANGUSTA, K. B., 30, f. 52; Rab. S. D., 3, f. 17; Schm. At., 23, f. 39-41; Sm. Sp. T., No. 512.

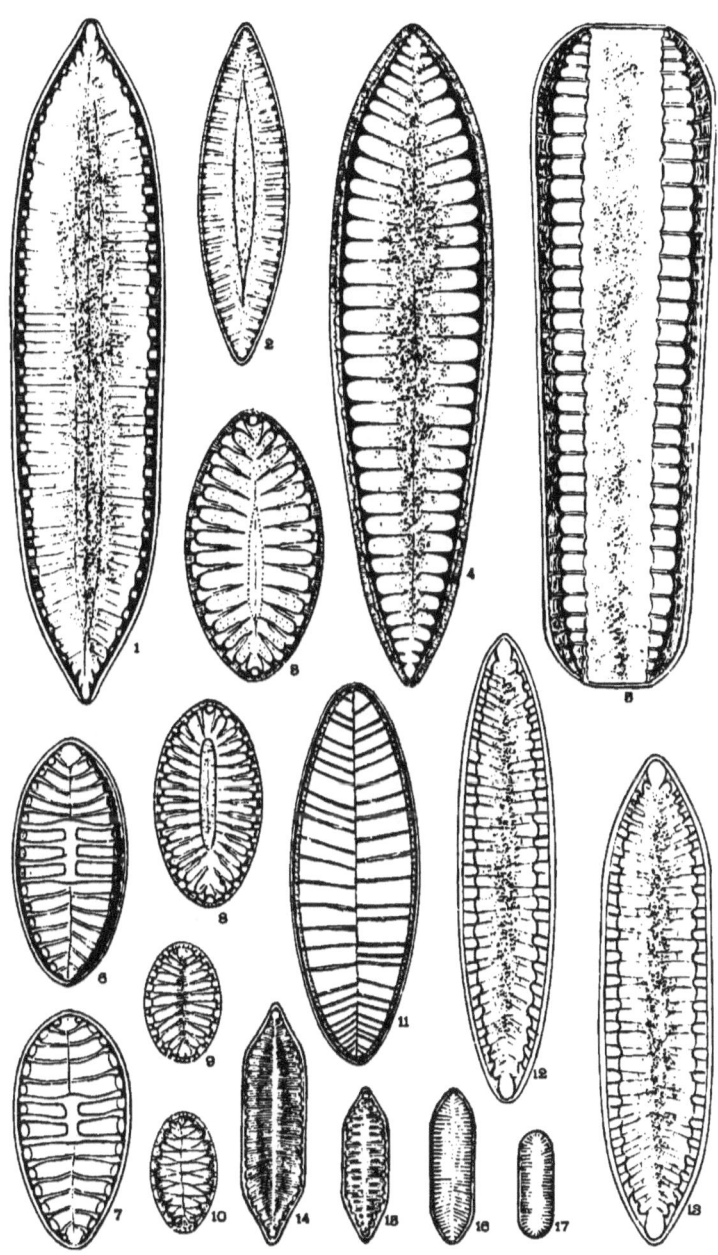

PLATE LIII.

Figures magnified 500 diameters.

Fig. 1–3.   SURIRELLA MEXICANA, A. S.; Schm. At., 4, f. 10-12. These and the following, Fig. 4, evidently only varieties of *S. fastuosa*, Ehrb. Compare pl. 52. Gulf of Mexico.
"   4.   "   BELDJECKII, Norman; Schm. At., 4, f. 22. Camp. Bay.
Figs. 5, 6.   "   CRUMENA, Breb., Schm. At., 24, f. 7-9; V. H., 73, f. 1; Sm. Sp. T., No. 518. Breb. suggested they are varieties of *S. Brightwellii*, H. L. S., considers them nearer to *S. ovalis*.
"   7, 8.   "   OVALIS, K. B., 30, f. 64; Rab. S. D., 3, f. 34; S. B. D., 9, f. 68; Schm. At., 24, f. 1-6. Salt Lake.
"   9, 10.   "   APICULATA, S. B. D., II, p. 88. Schm. At., 23, f. 31, 35.
Fig. 11.   "   MICROCORA, Ehrb., K. B., 29, f. 15; Rab. S. D., 3, f. 26; Sm. Sp. T., No. 526.
Figs. 12, 13.   "   SPIRALIS, K. B., 3, f. 64; Rab. S. D., 3, f. 5; Schm. At., 56, f. 25, 26; better, *Campylodiscus spiralis*.
Fig. 14.   "   INDUCTA, A. S.; Schm. At., 20, f. 10; 24, f. 15, 25. Probably the same as Fig. 8, *S. ovalis*: var.
Figs. 15, 16.   "   OVATA, Ehrb., K. B., 7, f. 1-4; S. B. D., 9, f. 70; Schm. At., 23, f. 49-55; V. H., 73, f. 5-7. Probably a var. of *S. ovalis*; H. L. S.
Fig. 17.   "   DELICATISSIMA, Lewis, N. and I. Forms, p. 343, 1, f. 4.
Figs. 18, 19.   "   MINUTA, Breb.; Rab. S. D., 3, f. 28; S. B. D., 9, f. 73; Schm. At. 23, f. 42-48; Sm. Sp. T., Nos. 527, 528.
"   20, 21.   "   EUGLYPTA, Ehrb., Amer., 3-5, f. 2, 4; K. B. 28, f. 27; Rab. S. D., 3, f. 23. Doubtful forms, probably small varieties of *Sur. splendida* or may be *Sur. elegans*, H. L. S.
"   22, 23.   "   BAYLEYI, Lewis, N. and I. Forms, p. 338 1, f. 1.
"   24, 25.   "   ANCEPS, Lewis, N. and I. Forms, p. 342, 1, f. 3.
Fig. 26.   "   INTERMEDIA, Lewis, N. and I. Forms, p. 339, 1, f. 2.
"   27.   "   ARCTISSIMA, A. S., Schm. At., 56, f. 13, 14.

Plate LIII

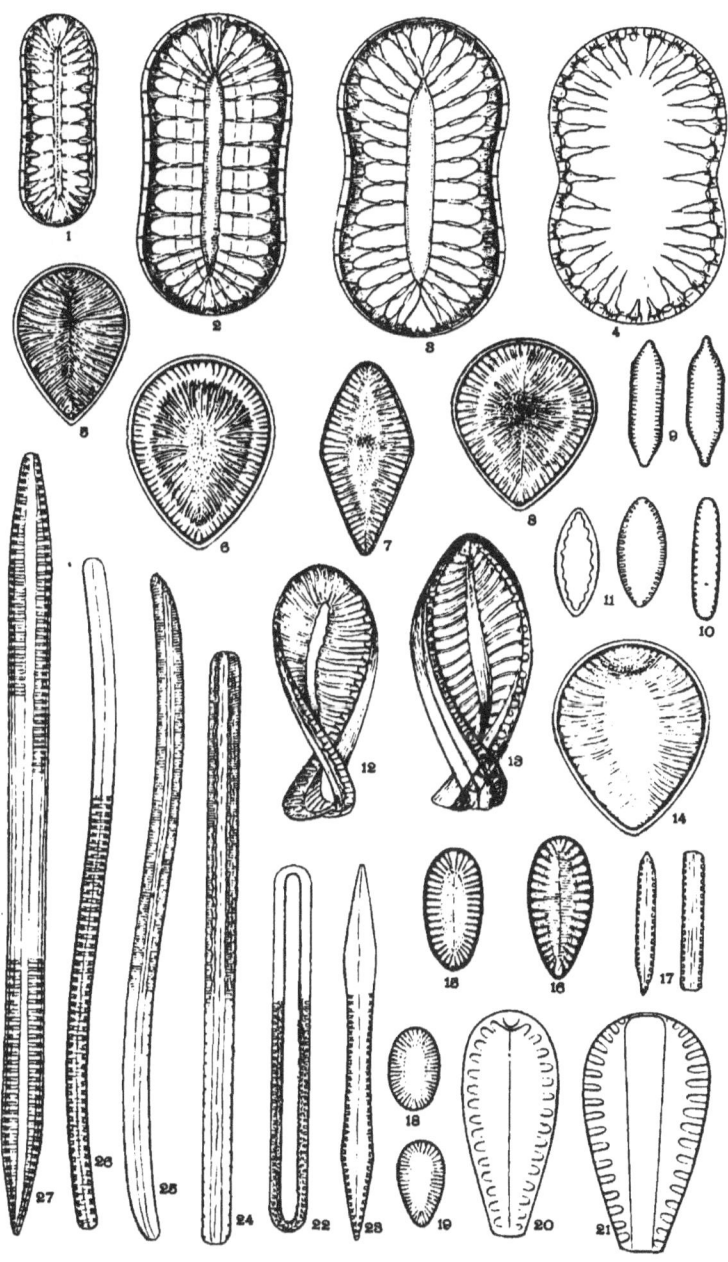

PLATE LIV.

Figures magnified 500 diameters.

Fig. 1.   SURIRELLA ELEGANS, Ehrb.; K. B., 28, f. 23. Rab. S. D., 3, f. 2; Schm. At., 21, f. 18. Not the typical form. Usually more like Figs. 7, 8 in outline.

" 2.   "   SPLENDIDA, var. bifrons; A. S., — (biseriata, W. S., not Breb.'s form): Mik., 7, 3 A, f. 17, 20; 7, 1, f. 2; 14, f. 36; K. B., 7, f. 10; Rab. S. D., 3, f. 21.

Figs. 3, (7).   "   SPLENDIDA, Ehrb. Mik., 5, 1, f. 22; 14, f. 35; S. B. D., 8, f. 62; V. H., 72, f. 4.

Fig. 4.   "   SENTIS, A. S.; Schm. At., 19, f. 9, 11. Has appearance of a very large form of Sur. fastuosa. Camp. Bay.

" 5.   "   VALIDA, A. S.; Schm. At., 23, f. 3. Looks like an exaggerated form of S. splendida.

" 6.   "   GUATIMALENSIS, Ehrb., — (S. cardinalis, Kitton; — S. limosa, Bail.); Mik., 33, 14, f. 24; Sm. Sp. T., No. 523; S. cardinalis, Schm. At., 21, f. 11–14; S. limosa, Mik., 1869, 179, 9, f. 5.

" 7.   "   SPLENDIDA, vide Fig. 3.

" 8.   "   ROBUSTA, Ehrb. Mik., 16, 3, f. 31; 17, 2, f. 1; 17, 1, f. 14; Schm. At., 22, f. 3. 4. Marine. Probably a variety of splendida; Sur. nobilis of Grun. is a sporangial form of splendida.

Plate LIV.

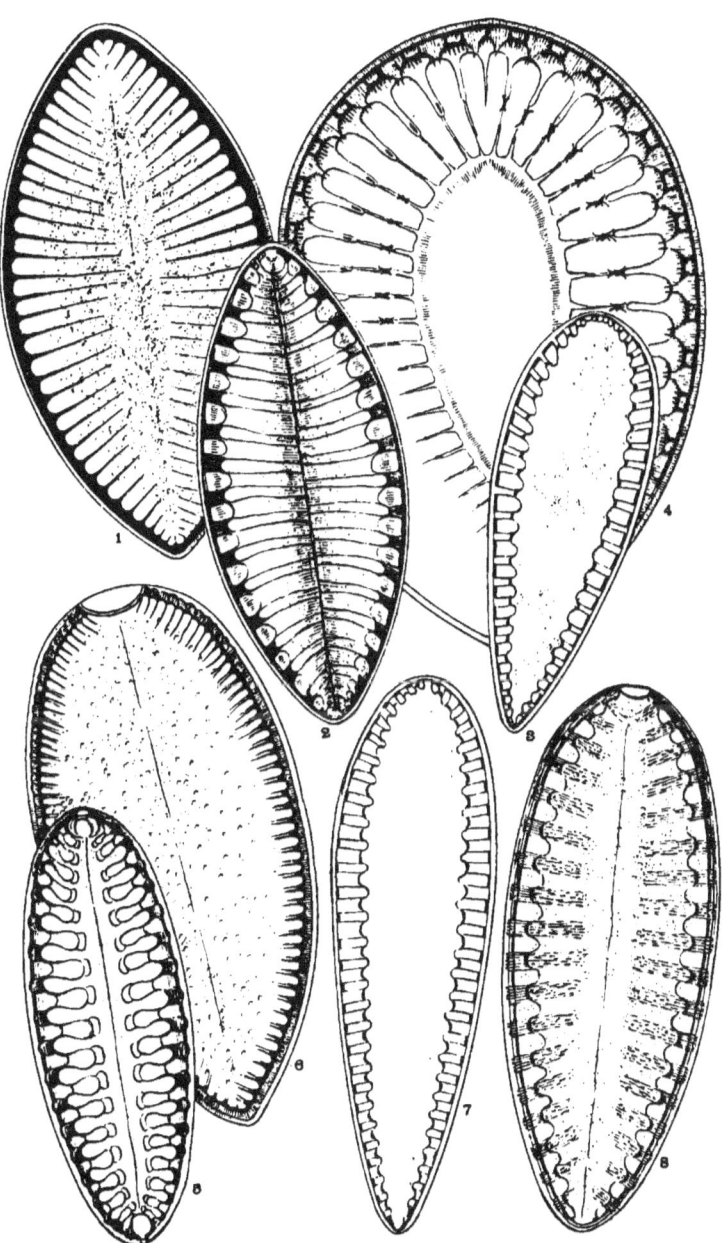

PLATE LV.

Figures magnified 500 diameters.

Fig. 1. SURIRELLA FEBIGERII, Lewis N. and R., Sm. Sp. T., 1, f. 2; Schm. At., 20, f. 9; 21, f. 1; Sm. Sp. T., No. 520.
" 2. " INDUCTA, or Sur. striatula, var. inducta; A. S.; Schm. At., 20, f. 10; 24, f. 15, 25.
" 3. " ELEGANS, Ehrb.; K. B., 28, f. 23; Rab. S. D., 3, f. 2; Schm. At., 21, f. 18.
" 4. " SAXONICA, Auers.; Schm. At., 22, f. 1. Probably only a variety of S. splendida.
" 5. " FASTUOSA, Ehrb.; Schm. At., 5, f. 7, 11; compare Pl. 49, f. 10; 52, f. 1-3.
" 6. " STRIATULA, Turp.; K. B., 7, f. 6; Rab. S. D., 3, f. 22; S. B. D., 9, f. 64; Schm. At., 24, f. 17-22; Sm. Sp. T., No. 536, = Sur. testudo, Ehrb. Salt Lake.
" 7. " REGINA, Janisch; Schm. At., 21, f. 5, 6. Cal.
" 8. " SPLENDIDA, var. Oregonica, H. L. S., = Sur. Oregonica, Ehrb. Mik., 33, 12, f. 27; Schm. At., 22, f. 9.
" 9. " SPLENDIDA, var. turgida, H. L. S., = Sur. turgida; S. B. D., 9, f. 60; Schm. At., 22, f. 10. Lake Erie.
Figs. 10, 11. " ELEGANS, var. Davidsonii, H. L. S., = Sur. Davidsonii, A. S.; Schm. At., 21, f. 7-10.
Fig. 12. " LAEVIGATA, Ehrb., Ber., 1845, p. 81; Schm. At., 24, f. 23, 24. Probably a variety of Sur. regina.

Plate LV.

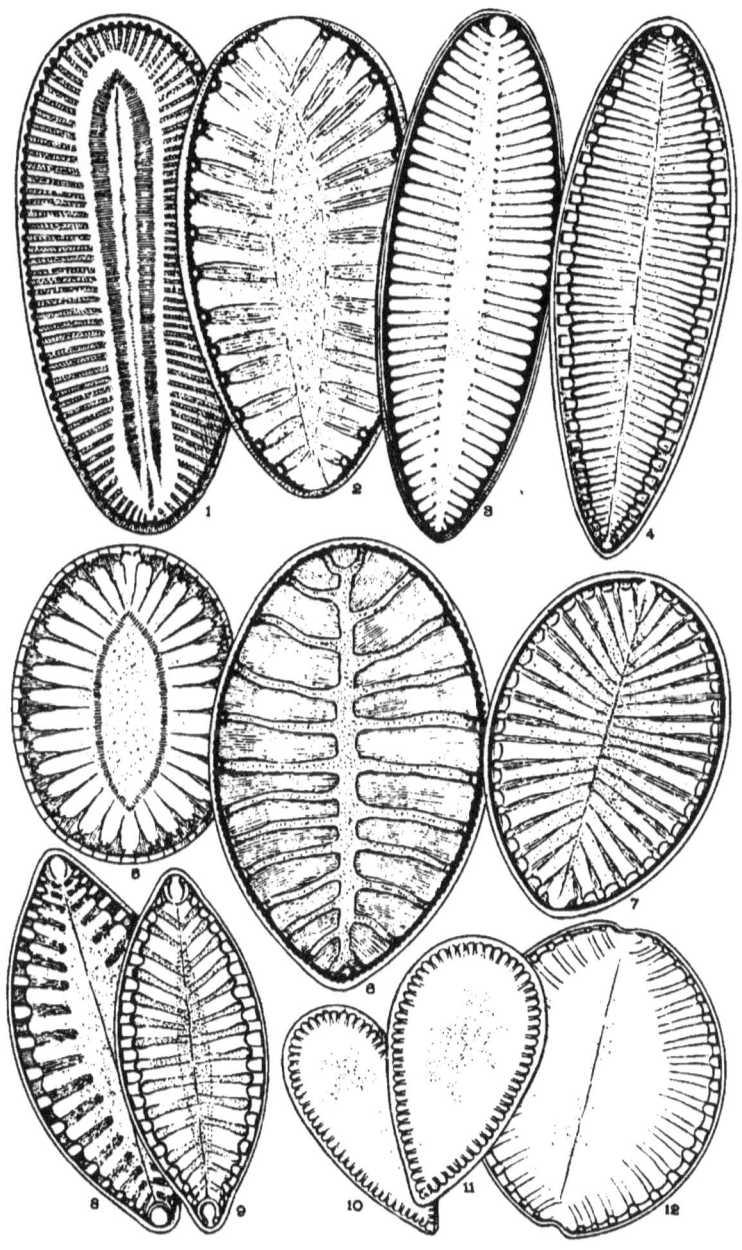

PLATE LVI.

Figures magnified 500 diameters.

Fig. 1. SURIRELLA REFLEXA, Ehrb. Mik., 33, 11, f. 13; Abh., 1870, 3, 2, f. 5. Probably a variety of Sur. splendida. Oregon.
" 2. " LEPTOPTERA, Ehrb. Abh., 1870, 3, 2. f. 7. Oregon.
Figs. 3, 4. " GEROLTII, Ehrb. Abh., 1869, p. 53; Abh., 1870, 2, 1, f. 7; Sm. Sp. T., No. 517. Utah.
Fig. 5. " CRENULATA, Ehrb. Mik., 33, 14, f. 23; Abh., 1870, 2, 1, f. 9; Sm. Sp. T., No., 517. Utah.
" 6. CAMPYLODISCUS EHRENBERGII, Ralfs.; Prit., 802, 12, f. 12, 13, 22, 23, = Surirella campylodiscus; Ehrb. Amer., 3, 5, f. 6; K. B., 28, f. 26; Rab. S. D., 3, f. 4.
" 7. " AMERICANUS, Ehrb. Abh., 1870, p. 52, 3, 2, f. 1.
" 88. SURIRELLA MINUTA, Breb.; Rab., S. D., 3, f. 28; S. B. D., 9, f. 73; Schm. At., 23, 42–48; V. H., 73, f. 9, 10, 14.
" 9. CAMPYLODISCUS CASTILLII, Ehrb. Abh., 1869, 1, f. F., 9; Abh., 1870. Mexico.
" 10. " LATUS, Shadb.; T. M. S., 1854, p. 16, 1, f. 13.
" 11. SURIRELLA PANDURIFORMIS, S. B. D., 30, f. 258; V. H., 73, f. 11; Sm. Sp. T., No. 533.
" 12. CAMPYLODISCUS HUMBOLDTII, Ehrb. Abh., 1869, p. 46, 1E, f. 3. Oregon.

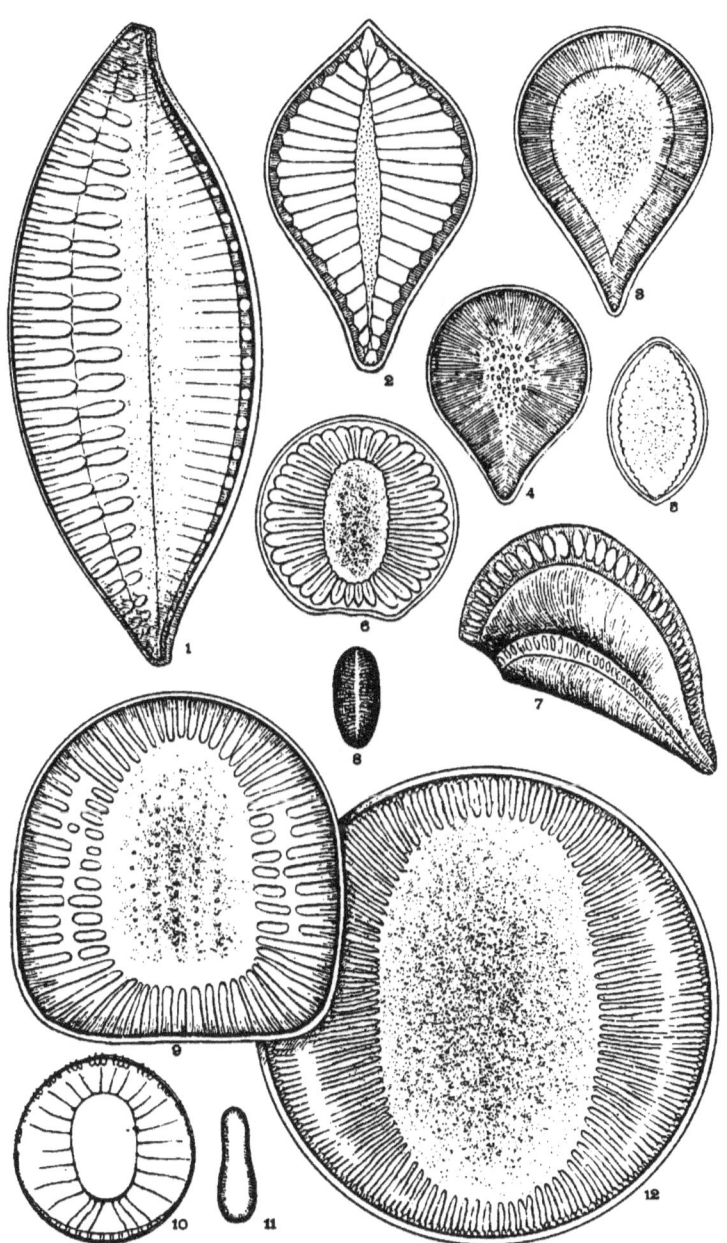

PLATE LVII.

Figures magnified 500 diameters.

Figs. 1-5. MELOSIRA NUMMULOIDES, Ag.; K. B., 3, f. 3; S. B. D., 49, f. 329; Prit., p. 816, 5, f. 64; 11, f. 14; Abh., 1870, 2, f. 67. M. moniliformis, Kg. This name was given by Keitzing, but others applied it indiscriminately to Borreri, nummuloides, salina, discigera, etc. It is of no value.

Fig. 6. " CROTONENSIS, Bail., crenulata, Kg. K. B., 2, f. 8; V. H., 88, f. 3-5.

Figs. 7-9. " GRANULATA, L. W. Bail.; B. J. N. H., p. 332, 1, f. 7, V. H., 87, f. 10-12, — Mel. punctata, — Gallionella granulata; G. marchica; procera; tenerrima.

Fig. 10. " AMERICANA, Rab. Of very doubtful value. No one able to tell what this is.

Figs. 11-14. " VARIANS, Ag.; A. N. H., 1843, p. 357, 9, f. 5; K. B., 2, f. 10; S. B. D., 51, f. 332; V. H., 85, f. 10-15.

Fig. 15. " VARIANS, Fig. from Rab., S. D.

Figs. 16-20. " CRENULATA, Bail.; 18, var. laevis; 19, 20, var. tennis. Compare Fig. 6.

" 21, 22. " GRANULATA, Bail. Vide Figs. 7-9.

" 23-25, 28. " SCALARIS, Grun.; V. H., 86, f. 30.

" 26, 27, 29. " SPIRALIS, Kg.; V. H., 87, f. 19-22; Sm. Sp. T., No. 231, — vars. of granulata.

" 30-32, 36. " DISTANS, Kg.; K. B., 2, f. 12; S. B. D., 61, f. 385; Rab. S. D., 2, f. 9; V. H., 86, f. 21-23.

Fig. 33. " GRANULATA, var. = M. carconensis, Grun.; V. H., 87, f. 27.

Figs. 34, 35. " SCULPTA, K.; V. H., 91, f. 13, 14, - Orthosira marina. Side and front views.

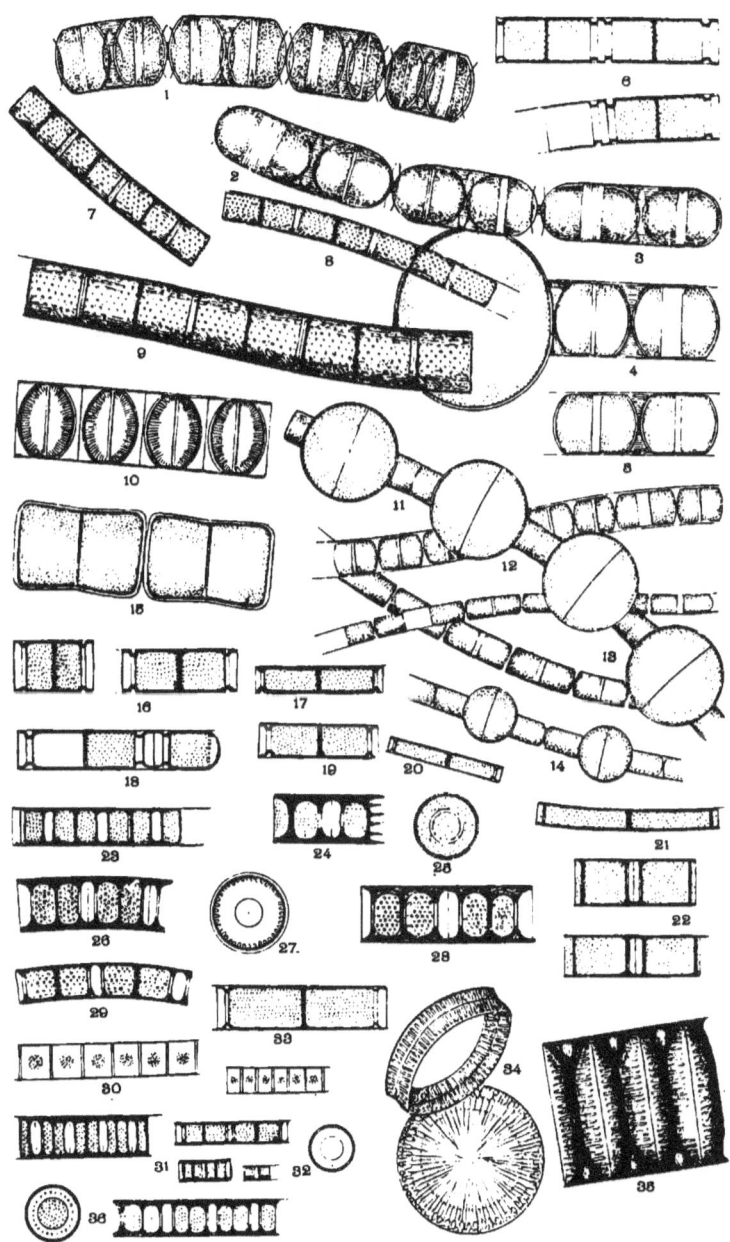

Plate LVII.

PLATE LVIII.

Figures magnified 500 diameters.

Figs. 1, 2. MELOSIRA SOL., Kg.; V. H., 91, f. 7-9; Witt., 1888, p. 17, 4, f. 18. Syn. Gallionella and Orthosira sol.
"   3, 4.   "   SCULPTA, Ehrb. Abh., 1870, 2, 1, f. 71; vide Pl. 56, f. 34, 35.
"   5-7.   "   ARENARIA, Moore.; A. N. H., 1843, p. 349, 9, f. 4; K. B., 21, f. 27; Rab. S. D., 2, f. 15; Prit., 819, 8, f. 17.
"   8-11. "   BORRERI, Grev.; S. B. D., 50, f. 330; V. H., 85, f. 5-7. Ordinary and sporangial frustules.
"   12-15. "   SULCATA, (= Marina) Ehrb.; K. B., 2, f. 7; Abh., 1870, 2, 1, f. 72; V. H., 91, f. 16; Jan. Guan., p. 26, 1A, f. 3, 4,   Orthosira marina.
Fig. 16.   "   CLAVIGERA, Grun.; V. H., 91, f. 1, 2.
Figs. 17, 18. "   ORICHALCEA, Kg.; S. B. D., II, p. 61, 53, f. 337, = Orthosira orichalcia; W. S., = Gallionella, orichalea, Ehrb.

Plate LVIII.

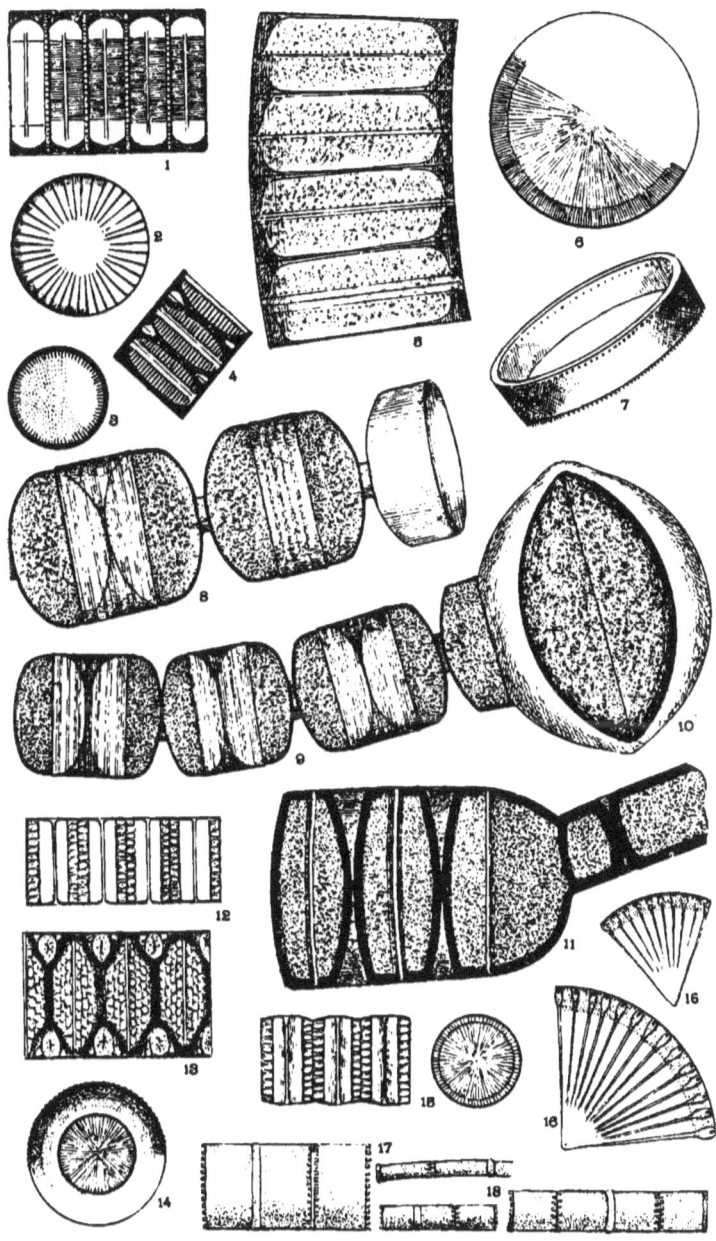

PLATE LIX.

Figures magnified 500 diameters.

Fig. 1.   MASTOGONIA CRUX, Ehrb. Mik., 33, 18, f. 8; M. D., 43, f. 23a; V. H., 83°, f. 1.
" 2.   HYALODISCUS LAEVIS, Ehrb. Mik., 33, 15, f. 17; Cast. 1887, p. 140, 24, f. 4. This Fig. is taken from Castracane. The granules should be more dense and the lines like Fig. 3. Pl. 24, f. 4.
" 3.   "   SUBTILIS, var. Japonica, Cast., 18, f. 4; Bail., N. Sp., 10, f. 12; Prit., 815, 5, f. 60.
" 4.   PODOSIRA ARGUS, Grun., 1878, p. 35; J. R. M. S., 1879, p. 691, 21, f. 6.
" 5.   HYALODISCUS WHITNEYI, Ehrb. Abh., 1870, p. 57, 2, 1, f. 21.
" 6.   "   STELLIGER. Bail. N. Sp., p. 10; V. H., 84, f. 12, Podosira maculata; Pyxidicula, etc.
Figs. 7, 8.   "   MAXIMUS, Eulenst.; J. R. M. S., 1878, p. 230, 14, f. 7.
Fig. 9.   MASTOGONIA HEPTAGONA, Ehrb. Ber., 1844, p. 269; Sill. J., March, 1845, p. 326, 4, f. 12.
" 10.   PODOSIRA FEBIGERII, Grun.; V. H., 84, f. 22-24.
" 11.   "   STELLILIFERA, Grun.; J. R. M. S., 1879, p. 690, 21, f. 3; V. H., (var.) 84, f. 25.
Figs. 12, 13.   "   FEBIGERII, Grun.; Cal., vide Fig. 10.
" 14, 15.   "   HORMOIDES, $\frac{100}{1}$, Kg., K. B., 28, f. 5; 29, f. 84.
Fig. 16.   "   " $\frac{2}{1}°$, S. B. D., 49, f. 237; Prit., 815, 2, f. 45; J. R. M. S., 1879, p. 689, 21, f. 7; Sm. Sp. T., No. 419.
Figs. 17, 18.   "   MONTAGNEI, K. B., 29, f. •85; S. B. D., 40, f. 326; Prit., 815, 5, f. 61; Sm. Sp. T., No. 422; V. H., 84, f. 11, 12. Parasitic in Polysiphonia, Fig. 19.

Plate LIX.

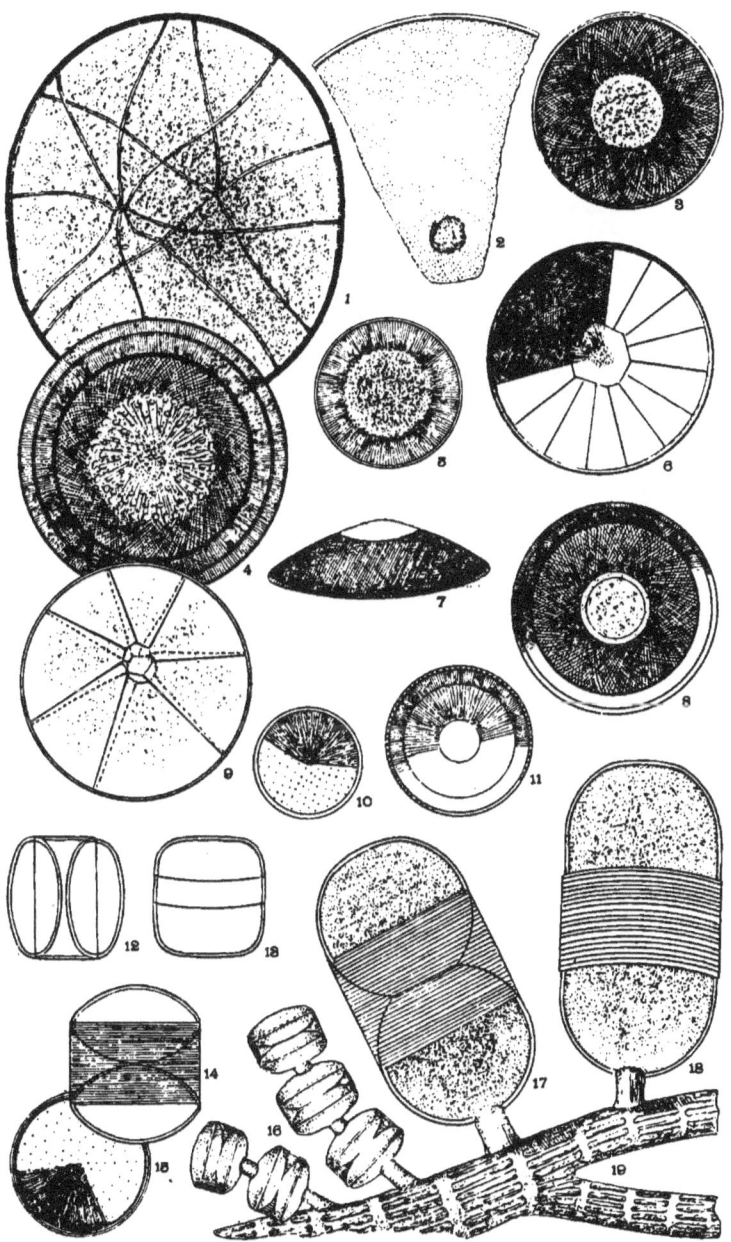

PLATE LX.

Figures magnified 500 diameters.

Figs. 1-4.   CYMATOPLEURA SOLEA, S. B. D., 10, f. 78; Prit., 793, 9, f. 155; 16, f. 9; M. D., 12, f. 24; Sm. Sp. T., No. 114.
"   5-7.   "   ELLIPTICA, S. B. D., 10, f. 80; Prit., 793, 9, f. 149; M. D., 12, f. 23; V. H., 55, f, 1-4.
"   8, 10, 11.   "   HIBERNICA, W. S., S. B. D., 10, f. 31; Sm. Sp. T., No. 113; V. H., 55, f. 34. Front and side views.
"   9, 12.   "   APICULATA, W. S., S. B. D., 10, f. 79; Grun., 1862, p. 466. A var. of solea.
Fig. 13.   "   SOLEA, another side view. Compare Figs. 1-4.
Figs. 14, 15.   "   MARINA, Lewis, N. and R. Sp., p. 5, 1, f. 4. Lewis, U. S. Sea Board.
Fig. 16.   "   ANGULATA, Grev., T. M. S., 1862, p. 89, 9, f. 1. Cal. Guano.

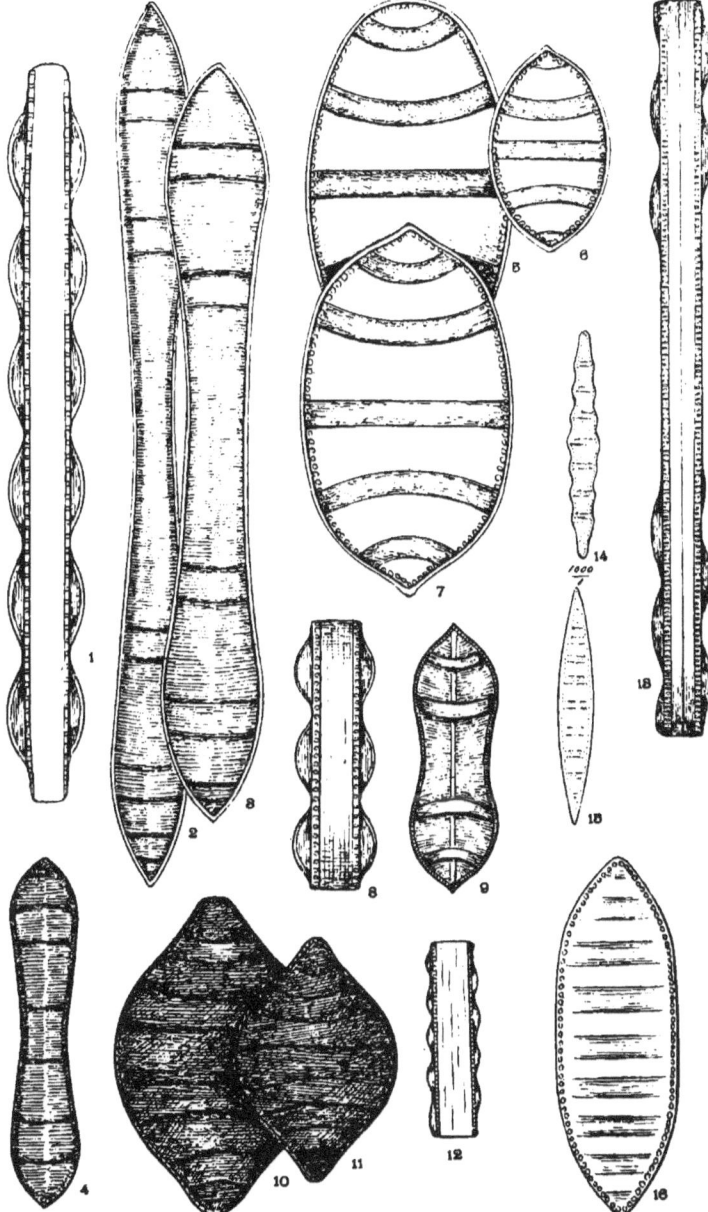

PLATE LXI.

Figures magnified 500 diameters.

Figs. 1, 2.　　GEPHYRIA GIGANTEA, Grev., T. M. S., 1866, p. 122, 11, f. 7, 8;
　　　　　　　　　　Cast., 1887, p. 42, 15, f. 10.
Fig. 3.　　　　　"　　CONSTRICTA, Grev., T. M. S., 1866, p. 77, 8, f. 2.
Figs. 4, 5.　　"　　MEDIA, M. J., 1860, p. 20; Prit., p. 809, 4, f. 49; Sm.
　　　　　　　　　　Sp. T., No. 662.
Fig. 6.　　　　TERPSINOE MUSICA, Ehrb. Mik., 34, 6A., f. 8, 34, 5A., f. 10; K.
　　　　　　　　　　B., 30, f. 72; Rab. S. D., Pl. 10; Prit., 859, 11, f. 47.
"　7.　　　　　"　　MAGNA, L. W. Bailey, B. J. N. H., 340. 2, f. 46-48.
Figs. 8, 9.　　"　　TETRAGAMMA, L. W. Bailey, B. J. N. H., 340, 2, f.
　　　　　　　　　　50, 51. Front and end views.
"　10, 11.　　"　　End views of varieties. Beaufort, N. C.
Fig. 12.　　　"　　MINIMA, L. W. Bailey, B. J. N. H., 340, 2, f. 54.
"　13.　　　　"　　MUSICA, End view of large form, from Florida.
Figs. 14, 15.　　"　　　　End views of larger forms.

Plate LXI.

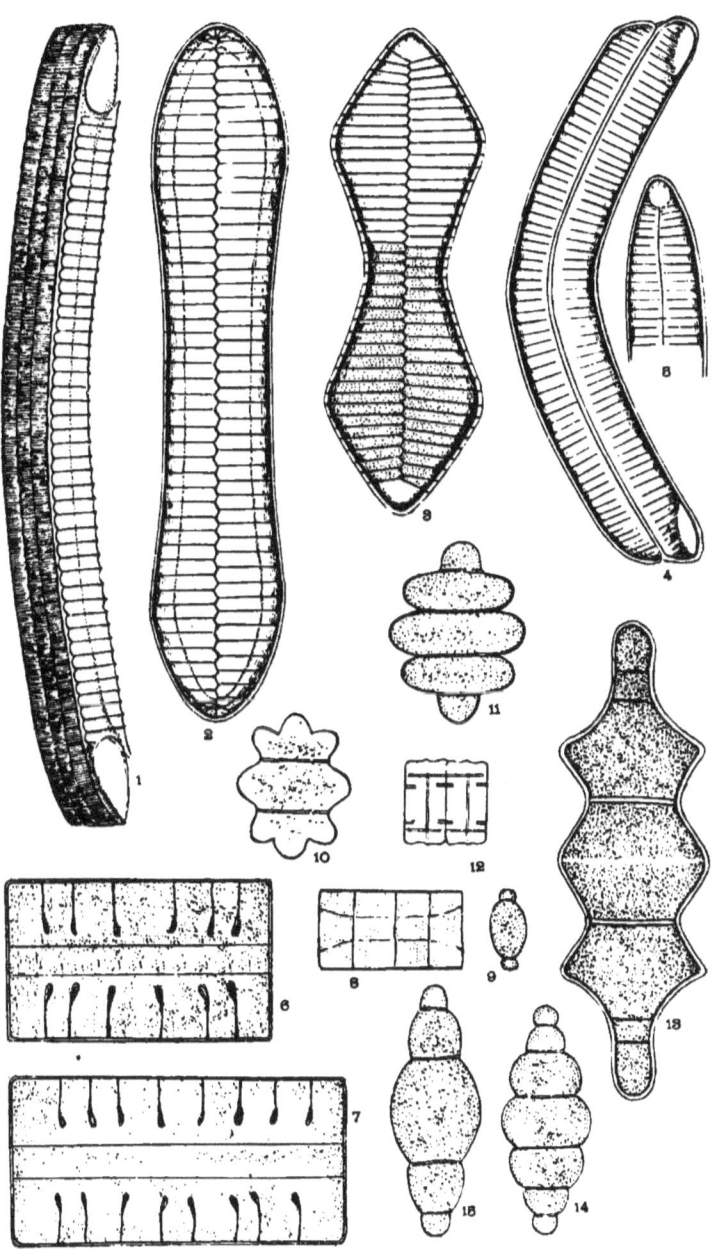

PLATE LXII.

Figures magnified 500 diameters.

Figs. 1, (6). STEPHANOPYXIS CORONA, Grun.; V. H., 83², f. 10, 11; Schm. At., 123, f. 19; 130, f. 36.
Fig. 2. " SPINOSISSIMA, Grun.; Schm. At., 123, f. 18.
" 3. " TURRIS, var. crassipina, Grun.; Schm. At., 130, f. 37, 11. Monterey.
" 4. STEPHANODISCUS LINEATUS, Ehrb. Mik., 33, 13, f. 22, = Coscinodiscus lineatus.
" 5. STEPHANOPYXIS ACULEATA, Ehrb.; Schm. At., 130, f. 12.
" 6. " CORONA, vide Fig. 1.
" 7. PYXIDICULA (Dictyopyxis), CRUCIATA, Ehrb. Mik., 18, f. 2ad. Ehrbs'. figure imperfect, Pyx. Hellenica, var.
" 8. " (Stephanopyxis), LIMBATA, Ehrb. Mik., 18, f. 7; Prit., p. 825.
" 9. " (Stephanopyxis), CRISTATA, Ehrb.; E. Ber., 1844, p. 86, Mik., 18, f. 6.
" 10. " (Dictyopyxis), LENS, E. Ber., 1845, p. 86; Mik., 18, f. 5; Prit., p. 825.
" 11. " CRUCIATA, vide Fig. 7.
Figs. 12-15. STEPHANOPYXIS APPENDICULATA, Ehrb. Mik., 18, f. 4; Schm. At., 130, f. 18, 19, 21, 23, 24.
Fig. 16. " APICULATA, Ehrb. Mik., 19, f. 13, = Pyxidicula, apiculata.
" 17. PYXIDICULA COMPRESSA, Bail. M. O., p. 40, 2, f. 13, 14; Sm. Sp. T., Nos. 430, 431.
" 18. MASTOGONIA SEXANGULATA, Ehrb. Mik., 33, 17, f. 12.
" 19. PYXIDICULA URCEOLARIS, Ehrb. Ber., 1844, p. 86; Mik., 13, f. 3a.
" 20. CYCLOTELLA PHYSOPLEA, Kg.; Prit., p. 811; Mik., 33, 12, f. 28.
" 21. LIOSTEPHANIA COMPTA, Ehrb. Ber., 1847, p. 55; Mik., 36, f. 41; Abh., 1875, 1, f. 5, 6.
" 22. CYCLOTELLA PHYSOPLEA, Ehrb. Mik., 33, 17, 8, = Discoplea physoplea.
Figs. 23, 24. ODONTELLA OBTUSA, Kg., K. B., 18, f. 8, = Biddulphia, aurita and obtusa; Schm. At., 122, 30. Nearly allied to Bid. laevis.
" 25, 26. BIBLARIUM (Stylobiblium) CLYPEUS, Ehrb. Mik., 33, 2, f. 18; Sm. Sp. T., No. 58. Oregon.
Fig. 27. XANTHIOPYXIS CINGULATA, Ehrb. Mik., 33, 17, f. 18. Va.
" 28. LITHODESMIUM CONTRACTUM, L. W. Bailey, B. J. N. H., p. 333, 1, f. 8. Greenport, N. Y. Of doubtful value. H. L. S. says this is not a diatom.
" 29. CHAETOCEROS DISTANS, Cl.; V. H., 82, f. 4. Mobile.
" 30. PYXIDICULA GIGAS, Mik., 33, 13, f. 18. Va.

Plate LXII.

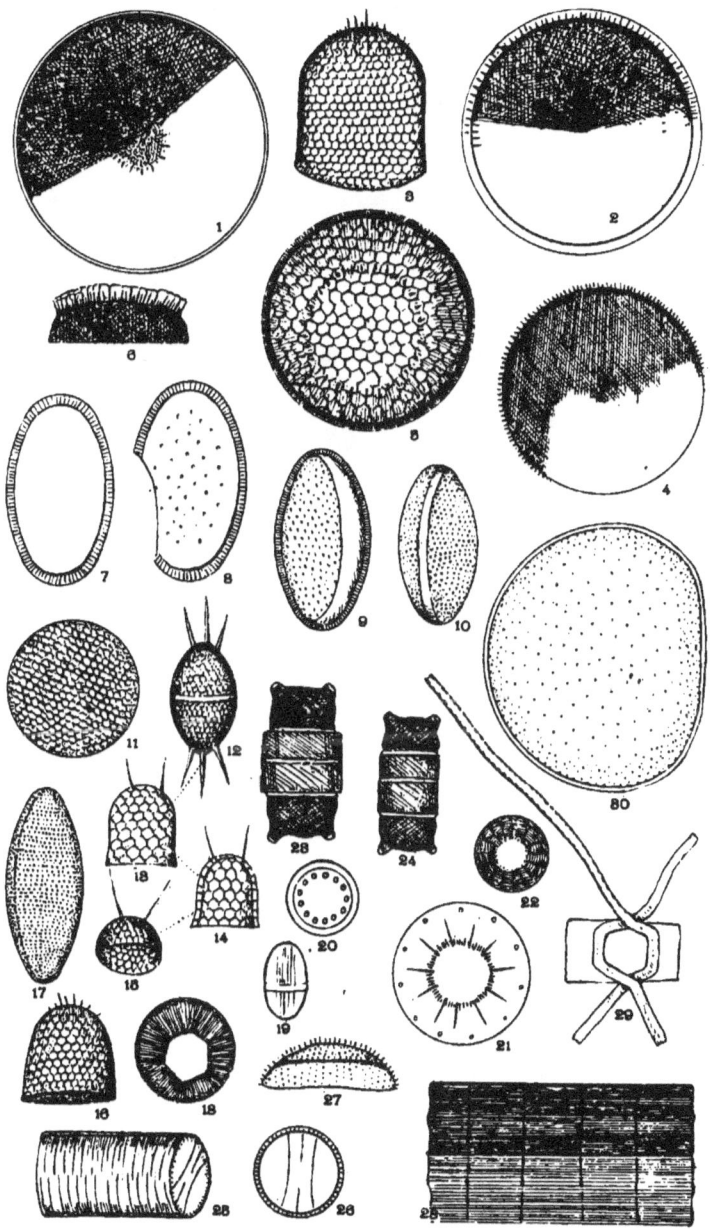

PLATE LXIII.

Figures magnified 500 diameters.

Fig. 1. NAVICULA ENTOMON, Ehrb. K. B., 28, f. 74; Schm. At., 13, f. 43-46, 48, 49; Cast. Chal., 20, f. 10.
" 2. " POLYONCA, Breb.; Lewis, N. and R. Sp., 2, f. 7; V. H., A, f. 14.
Figs. 3-5. TRICERATIUM VARIABILE, Brightw. M. J., 1856, p. 275, 17, f. 18; T. M. S., 1860, p. 149, 7, f. 7.
Fig. 6. TERPSINOE AMERICANA, Ralfs, Prit., p. 859; Grun., 1868, p. 23.
" 7. SYSTEPHANIA (Stephanopyxis) RAEANA, Cast., p. 151-9, f. 11. Va.
Figs. 8, (29). RHAPHONEIS ARCHERI, O'M. M. J., 1867, p. 247, 7, f. 12. Probably only a valve of Cocconeis; C. Costata?
" 9, 10. CHAETOCEROS BACILLARIA, Ehrb., Sill. J., 1845, p. 328, 4. f. 18; M. J., 1856, 7, f. 1, 2.
" 11-13. SYNDENDRIUM DIADEMA, Ehrb. Mik., 35 A, 18, f. 13; M. J., 1856, 7, f. 49-52; M. D., 43, f. 59.
" 14-16. LIRADISCUS MINUTUS, Grev., T. M. S., 1865, p. 47, 5, f. 6.
Fig. 17. RHABDONEMA ATLANTICUM, Kain and Schultze, Bull. Tor. Bot. Club, March, 1889, p. 75, Pl. 89, f. 7, 7a. Artesian well, Atlantic City, N. J.
" 18. COCCONEIS BOREALIS, Ehrb. Mik., 37, 3, f. 2. Of dubious value.
" 19. " RHOMBEA, Ehrb. Mik., 35 A, 7, f. 2. Cannot tell at this day what these two Figs. (18, 19), really are.
Figs. 20, 21. MELOSIRA HOROLOGIUM, Ralfs, Prit., p. 849, 5, f. 62.
" 22-24. STEPHANODISCUS MINUTUS, Grun.; Sm. Sp. T., No. 5, Cyclotella Oregonica, Ehrb. Mik., 37, 2, f. 3.
Fig. 25. RHAPHONEIS AFFINIS, Grun.; Rappahannoc Cliff. Ungarns., Pl. 27, f. 266.
" 26. " PETROPOLITANA, Grun.; Petersburg, Va. Ungarns., 27, f. 268.
" 27. DIMEREGRAMMA FOSSILE, Grun.; Ungarns., 27, f. 265. Md.
" 28. RHAPHONEIS LINEARIS, Grun.; Ungarns., 27, f. 262. Nottingham, Md.
" 29. " ARCHERI, vide Fig. 8.
Figs. 30, 31. CYCLOTELLA COMPTA, Kg., V. H., 92, f. 16-22.
" 32, 33. DENTICULA ANTILLARIA, Cleve., 1878, p. 14, 4, f. 26.
Fig. 34. CYMBELLA STODDERI, Cleve.; N. L. K. D., p. 5, 1, f. 5. H. L. S. says figure is too symmetrical, looks too much like Navicula. Copy from Cleve. is the only one at hand.
Figs. 35, 36. COCCONEIS AMBIGUA, Grun., 1868, p. 14, 1, f. 9; V. H., 30, f. 8-10. -- C. Californica, Gr.
" 37. " ELONGATA, Ehrb. Mik., 7, 38, f. 8; 14, f. 31; 8, 3, f. 13, etc. -- C. placentula, var.
" 38. SYNDENDRIUM DIADEMA, Grev., vide Figs. 11-13.
" 39. RHAPHONEIS FUSCUS? Ehrb.

Plate LXIII.

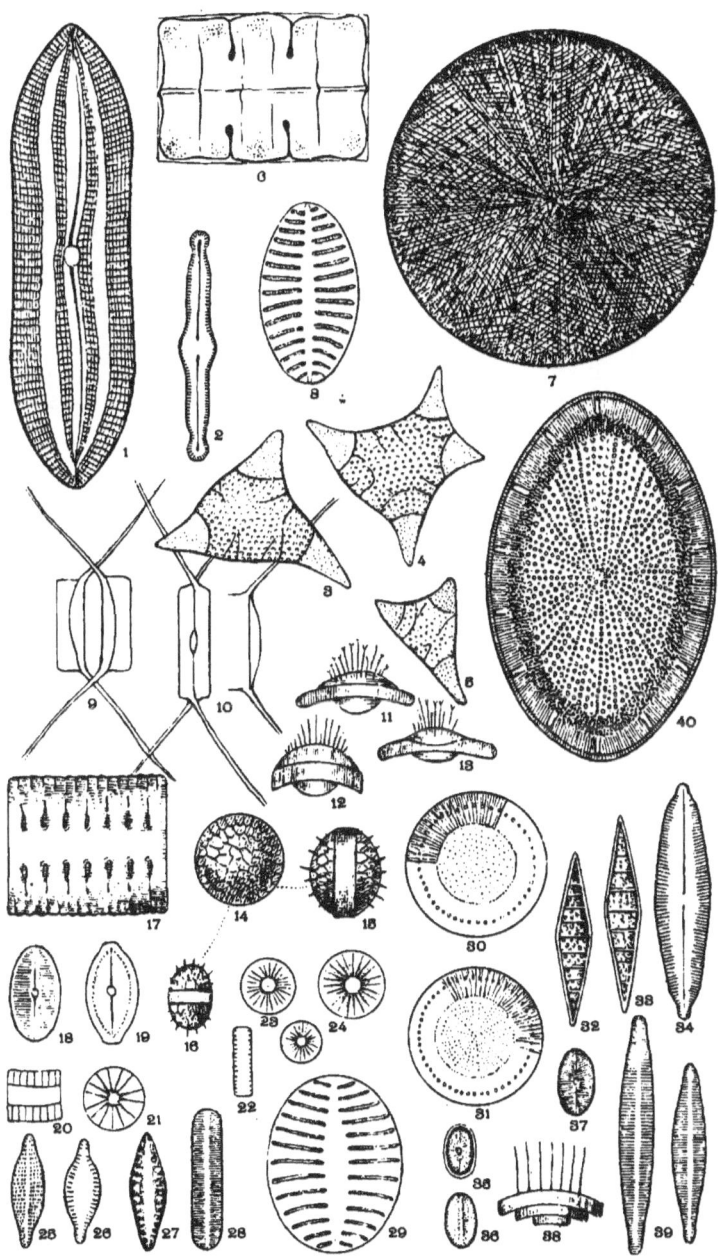

PLATE LXIV.

Figures magnified 500 diameters.

Fig. 1.	BIDDULPHIA (Zygoceros) CIRCINUS, Bail.; N. Sp., p. 11, f. 19, 20; V. H., 105, f. 13.	
" 2.	DICLADIA CLATHRATA, Ehrb. Mik., 18, f. 100; M. D., 43, f. 65.	
Figs. 3, 4.	" MITRA, Bailey, Sill. J., 1856, p. 4, 1, f. 6; V. H., 106, f. 12, 13. Size somewhat exaggerated.	
" 5–7.	" CAPREOLUS, Ehrb. Mik., 35 A; 18, f. 8, 5; 18, f. 101, 102; M. J., 1856, 7, f. 53–60.	
" 8, 9.	BIDDULPHIA (Zygoceros) QUADRICORNIS, Grun.; V. H., 105, f. 5–7.	
Fig. 10	TRICERATIUM SHADBOLTII, L. W. Bailey, B. J. N. H., p. 342, 1, f. 60, 61.	
" 11.	BIDDULPHIA (Zygoceros) OCCIDENTALIS, L. W. Bailey, B. J. N. H., p. 343, 2. f. 66–68. Probably a small variety of Z. Mobilensis.	
Figs. 12–14.	ANAULUS (Biddulphia) BIROSTRATUS, V. H., 22b, f. 15; 103, f. 1–3. Cal.	
" 15, 16.	HEMIAULUS AFFINIS, Grun.; V. H., 106, f. 10, 11.	
" 17, 18.	GONIOTHECIUM ODONTELLA, Ehrb. Mik., 33, 15, f. 16; 18, f. 94; M. J., 1856, Pl. 7, f. 47, 48; Prit., p. 864, 6, f. 29; V. H., 105, f. 11, 12.	
" 19, 20.	HEMIAULUS POLYCYSTINORUM, Ehrb. Abh., 1875, 1, f. 12–15.	
Fig. 21.	RHIZOSOLIMA (?) PILEOLUS, Ehrb. Mik., 18, f. 103; Prit., p. 866.	
Figs. 22, 23.	HERCOTHECA MAMMILARIS, Ehrb. Mik., 33, 18, f. 7; Bot., 867, 7, f. 35; M. D., 43, f. 31.	
" 24, 25, 27.	HEMIAULUS POLYCISTINORUM, Ehrb. Mik., 36, f. 43; Abh., 1875, 1, f. 12–15.	
" 26, 28, 29.	" BIFRONS, (Ehrb.) Grun.; V. H., 103, f. 6–9.	
Fig. 30.	GONIOTHECIUM MONODON, Ehrb. Mik., 18, f. 97; 33, 13, f. 12; M. D., 42, f. 37.	
" 31.	" DIDYMUM, Ehrb. Mik., 18, f. 104; M. D., 42, f. 30.	
" 32.	" ROGERSII, Bail., Sill. J., March, 1844, p. 301 - Mik., 18, f. 92, 93; M. J., 1856, Pl. 7, f. 43–46.	
" 33.	" OBTUSUM, Mik., 18, f. 95; Prit., p. 864. Virginia.	
" 34.	HEMIAULUS CALIFORNICUS, Ehrb., 33, 13, f. 15; Prit., p. 851.	
" 35.	AMPHITETRAS ELEGANS, Grev., T. M. S., 1866, p. 9, 2, f. 24. Monterey.	
" 36.	PERIPTERA CAPRA, Ehrb. Mik., 18, f. 99; M. D., 43, f. 67; Dicladia capra, Ehrb. Ber., 1844, p. 79.	
" 37.	SYRINGIDIUM SIMPLEX, L. W. Bailey, B. J. N. H., p. 343, 2, f. 65.	
" 38.	" AMERICANA, L. W. Bailey, B. J. N. H., p. 342, 2, f. 62–64; Prit., p. 866, 7, f. 34; V. H., 106, f. 2.	
" 39.	AMPHITETRAS ANTEDILUVIANA, Ehrb. Mik., 21, f. 25; 19, f. 19; K. B., 19, f. 3; 29, f. 36; S. B. D., 44, f. 318. Figures only half the usual diameter. Virginia.	
" 40.	" MINUTA, Grev., T. M. S., 1861, p. 77, 9, f. 11. Maryland.	

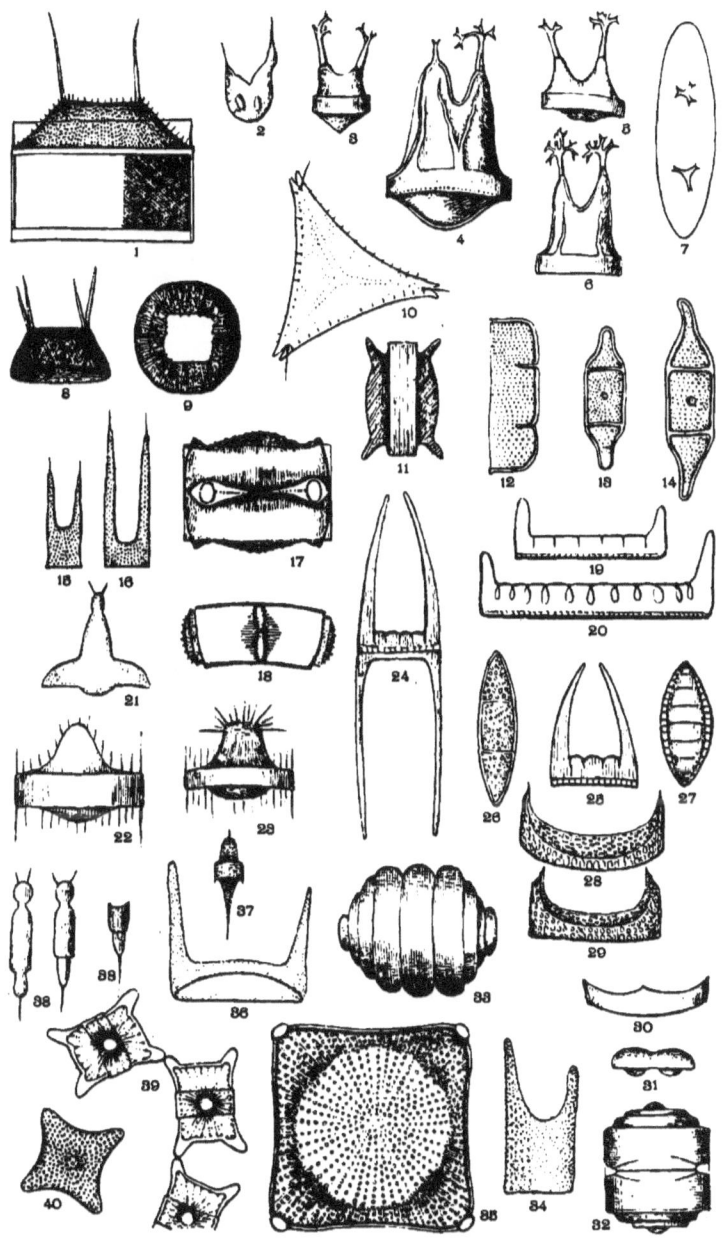

PLATE LXV.

Figures magnified 500 diameters.

Fig. 1.   RHIZOSOLENIA GRACILIS, H. L. S., Proc. A. S. M., 1882, p. 177, 4, f. 166.
" 2.   "   ERIENSIS, H. L. S., Lens, 1, p. 44; A. Q. J. M., 1878, p. 15, 3, f. 7; Sm. Sp. T., No. 447; V. H., 79, f. 9.
Figs. 3–5.   "   AMERICANA, (Ehrb.); Brightwell, M. J., 1858, Pl. 5, f. 3.
Fig. 6.   "   CALYPTRA, Ehrb.; M. J., Pl. 5, f. 2, 1858, Copied by Brightwell.
" 7.   "   ORNITHOGLOSSA, (Ehrb.); Brightwell, M. J., 1858, 5, f. 1. A var. calyptra, Fig. 6.
Figs. 8, (17, 18). PYXILLA AMERICANA, Grun; V. H., 83², f. 1–3.
" 9, 10.   CHAETOCEROS INCURVUM, Bail. N. Sp., 9, f. 30; M. J., 1856, p. 107, 7, f. 9–11.
Fig. 11.   "   BOREALE, Bail. N. Sp., 8, f. 22, 23; M. J., 1856, 7, f. 12–15; T. M. S., 1860, 48, 2, f. 18; 152, 7, f. 13.
" 12.   "   DIDYMUS, Ehrb. Mik., 35A, 17, f. 5; 18, f. 4; M. J., 1856, Pl. 7, f. 3–7.
" 13.   "   CALIFORNICUM, Grun.; V. H., 82, bis f. 8.
Figs. 14–16.   "   DIDYMUS, Ehrb. Same as Fig. 12.
" 17, 18. PYXILLA AMERICANA, vide Fig. 8.
Fig. 19.   "   SUBULATA, Grun.; V. H., 83², f. 6.
Figs. 20, 23.   "   BOREALE, Bail. Same as Fig. 11.
Fig. 21.   "   KITTONIANA, Grun.; V. H., 83, f. 10, 11; 83, bis f. 9–11.
" 22. CHAETOCEROS MONICÆ, Grun.; V. H., 82, bis f. 4.
" 24. PYXILLA DUBIA, Grun.; V. H., 83, f. 7, 8; 83 bis f. 12. A variety from Monterey, Cal.

Plate LXV.

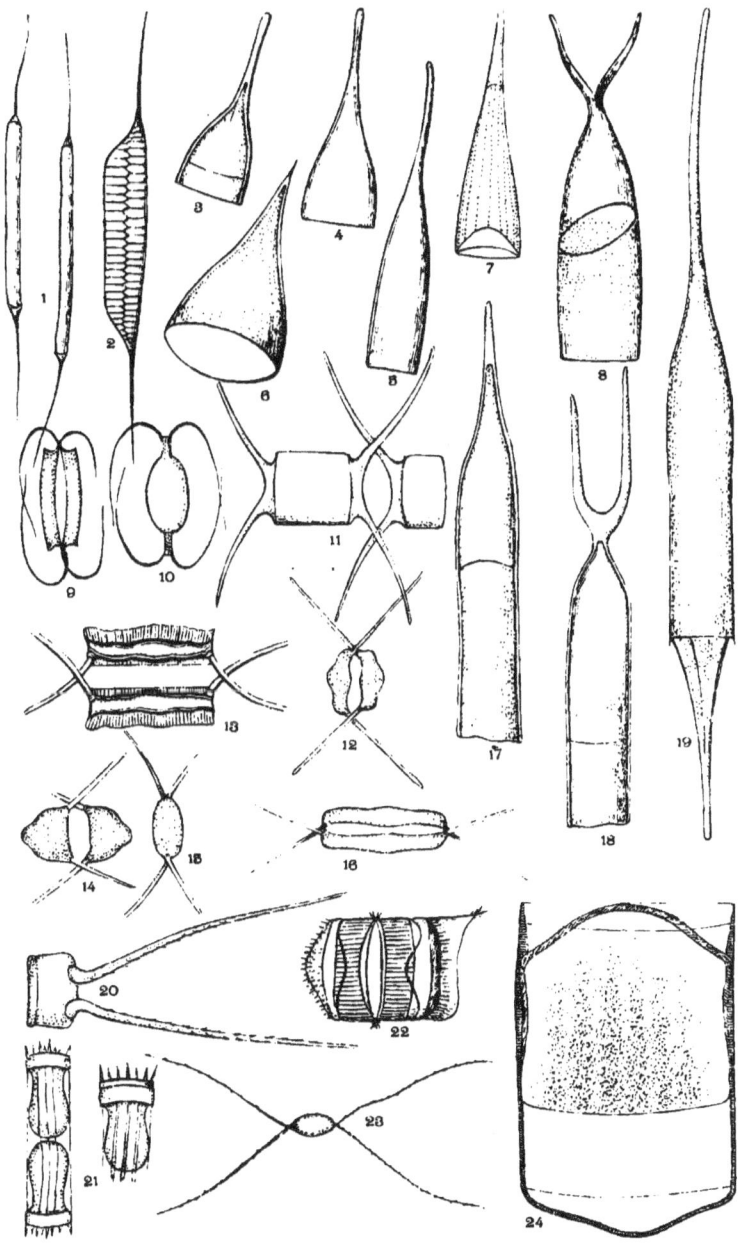

PLATE LXVI.

Figures magnified 500 diameters.

Figs. 1, 2. STEPHANOPYXIS (Creswellia, Grev.) TURGIDA, Ralfs. Prit., p. 826; M. J., 1859, 8, f. 14.
"   3, 4.   "   (Creswellia, Grev.) TURRIS, Ralfs, Prit., 826, 5, f. 74; Sm. Sp. T., No. 507; V. H., 83, tris f. 12.
"   5, 6, 7.   (Creswellia, Grev.) FEROX, Ralfs. Prit., 826, 5, f. 75; M. J., 1859, 8, f. 15, 16.
"   8, 9. CYCLOTELLA KUETZINGIANA, Thw.; S. B. D., 5, f. 47; B. J. N. H., p. 331, 1, f. 45; Sm. Sp. T., No. 103; V. H., 94, f. 1, 4–6.
"   10–12.   "   ROTULA, K. B., 2, f. 4; S. B. D., 5, f. 50; Sm. Sp. T., No. 111.
"   13–15.   "   MENEGHINIANA, Kg. K. B., 30, f. 68; Sm. Sp. T., No. 104; V. H., 94, f. 11–13, — C. Astraea, Kg., — C. Oregonica, Ralfs.
"   16, 17.   "   STRIATA, Grun.; V. H., 92, f. 6–8.
"   18, 19.   "   ANTIQUA, S. B. D., 5, f. 49; V. H., 92, f. 1; Sm. Sp. T., No. 639.
"   20, 21.   "   COMPTA, Kg. V. H., 92, f. 16–22.
"   22–24.   "   OPERCULATA, Kg. K. B., 1, f. 1; Rab., S. D., 2, f. 1; S. B. D., 5, f. 48; M. D., 12, f. 21; V. H., 93, f. 22–24.
Fig. 25.   PORPEIA QUADRICEPS, Bail. Prit., 850, 6, f. 6; T. M. S., 1865, p. 52, 6, f. 18, 19; V. H., 95 bis f. 14. Camp. Bay.
Figs. 26, 27. STEPHANODISCUS CARCONENSIS, Grun., V. H., 95, f. 1, 2. N. Amer.?
"   28, 29.   "   NIAGARA, Ehrb.; V. H., 95, f. 13, 14; Sm. Sp. T., No. 505; Mik., 35a, 7, f. 21, 22.
Fig. 30.   PORPEIA QUADRICEPS, Grev. Trans. M. S., 1865, 58, f. 18.
"   31.   PLAGIOGRAMMA, Grev., T. M. S., 1866, Pl. 1, f. 3, (3).
"   32.   EUCAMPIA VIRGINICA, V. H., 95, bis f. 6. Richmond, Va.
"   33.   PORPEIA QUADRATA, Grev., V. H., 95, bis f. 15; T. M. S., 1863, p. 65, 6, f. 20. Santa Monica.
Figs. 34, 35. EUCAMPIA ZODIACUS, Ehrb., K. B., 21, f. 21; S. B. D., 60, f. 299; Prit., 937, 2, f. 43; V. H., 95, f. 17, 18.

Plate LXVI.

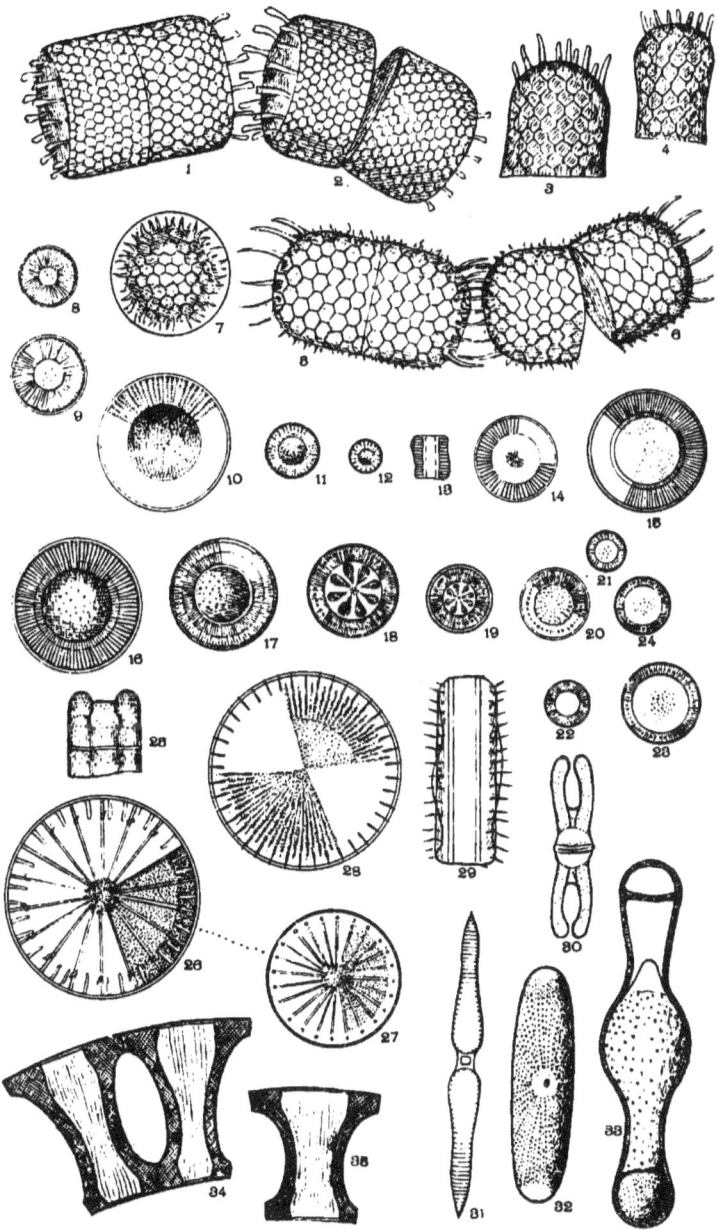

PLATE LXVII.

Figures magnified 500 diameters.

Fig. 1. BACTERIASTRUM FURCATUM, Shad.; T. M. S., 1854, p. 14, 1, f. 1, Prit., 863, 6, f. 26.
Figs. 2, 3, 4. " VARIANS, Lauder, T. M. S., 1864, p. 8, 3, f. 1–6; Sm. Sp. T., No. 57, B. curvatum and furcatum.
Fig. 5. PERIPTERA CHLAMYDOPHORA, Ehrb. Mik., 33, 18, f. 96; Prit., 865, 8, f. 25. A nonentity, evidently only a fragment of Dicladia or the like.
" 6. STEPHANOGONIA ACTINOPTYCHUS, Ehrb.; V. II., 83, ter. f. 2–4. (Mastogonia).
" 7. STEPHANOPYXIS LIMBATA, Ehrb., 18, f. 7; V. II., 83, ter. f. 13, 14. (Pyxidicula limbata?).
" 8. STEPHANOGONIA POLYGONA, Ehrb. Mik., 33, 18, f. 10; Sill. J., April, 1845, p. 326, 4, f. 13; M. D., 43, f. 30.
Figs. 9, 10. STEPHANOPYXIS CAMPEACHIANA, Grun.; Schm. At., 65, f. 19, 20.
" 11, 12. STEPHANOGONIA (Cladogramma) CALIFORNICUM, Grun., 83, bis f. 18, 19.
" 13, 14. TROCHOSIRA SPINOSA, Kitton; V. II. 83, bis f. 14, 15, 17. Nottingham, Md.
" 15, 16. STEPHANOGONIA POLYGONA, Ehrb. Mik., 33, 18, f. 10; Sill. J., April, 1845, p. 326, 4, f. 13; M. J., 1860, p. 68, 5, f. 8.
" 17–19. PERIPTERA TETRACLADIA, Ehrb. Mik., 33, 18; f, 9. A genus of doubtful genuineness. Still no harm in retaining it as does V. II. V. II., 83, ter. f. 7–9; Prit., p. 865, 6, f. 30.
Fig. 20. STEPHANOPYXIS CORONA, Grun.? V. II., 83, ter. f. 10, 11, — Systephanea corona.
" 21. " (Creswellia) RUDIS, Grev.; T. M. S., 1866, Pl. 8, f. 7. Monterey, Cal.

Plate LXVII.

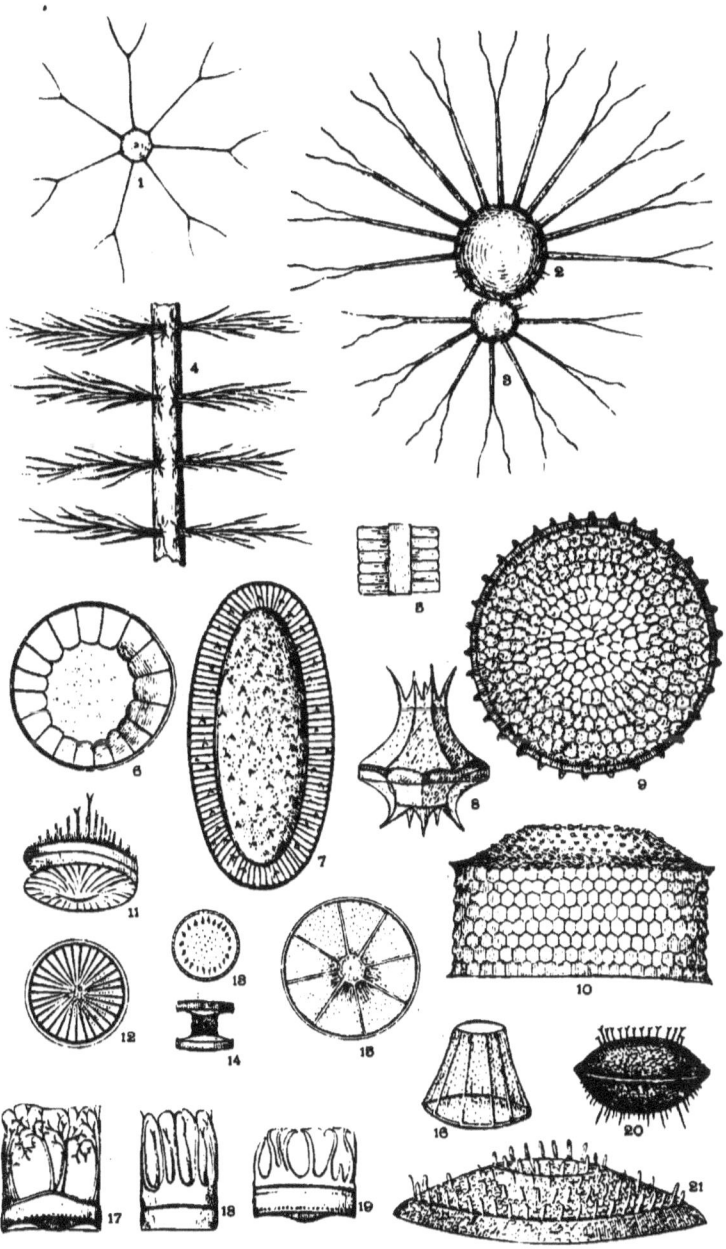

PLATE LXVIII.

Figures magnified 500 diameters.

Fig. 1. RUTILARIA EPSILON, Grev., M. J., 1863, p. 228, 9, f. 1. Monterey.
" 2. " HEXAGONA, Grun.; V. H., 105, f. 8. S. monica.
" 3. RHAPHONEIS BELGICA, Grun.; V. H., 36, f. 25, (R. pretiosa, var.)
" 4. " OREGONICA, Ehrb. Mik., 18, f. 83; Abh., 1870, p. 59; Mik., 37, 2, f. 15.
" 5. " FLUMINENSIS, Grun.; 1862, p. 382, 7, f. 5; V. H., 36, f. 34. Judging by Grun.'s original figure, this may be a Cocconeis.
" 6. STAURONEIS MONOGRAMMA, Kg.; K. B., 29, 18.
" 7. " MONOGRAMMA, Ehrb. One of Ehrb.'s doubtful species, he saw only a fragment; probably S. gracilis.
" 8. NAVICULA GIBBA, (not Stauroptera, Rab., 9, f. 3); K. B., 28, f. 70; Schm. At., 45, f. 45 51; V. H., A, f. 12.
Figs. 9-11. DENTICULA VALIDA, forma major Pedicino; V. H., 49, f. 4-6. Geysers, Cal.
" 12. NAVICULA TRINODIS, S. B. D., II, p. 94; Lewis N. and R., Sp., p. 8, 2, f. 6; V. H., 14, f. 31a.
" 13. " BRAUNII, var., Stauroneiforme, Grun.; V. H., 6, f. 21.
Figs. 14, 15. DENTICULA THERMALIS, K. B., 17, f. 6; Rab. S. D., 1, f. 3; V. H., 49, f. 17, 18.
" 16, 17. " LAUTA, Bail., N. Sp., 9, f. 1, 2; V. H., 49, f. 1, 2; Sm. Sp. T., No. 125.
" 18, 19. COLLECTONEMA SUBCOHAERENS. Thwaites, (Schizonema lacustris, Ag.); V. H., 15, f. 40; S. B. D., 56, f. 353.
Fig. 20. BACTERIASTRUM CURVATUM, Shad., T. M. S., 1854, p. 14, 1, f. 2; M. D., 43, f. 18, — (B. varians).
" 21. TABELLARIA NODOSA, Ehrb. Mik., 14, f. 54; 4, 3, f. 24; 3, 4, f. 31.
" 22. " ROBUSTA, Ehrb., 33, 11, f. 15; Prit., 807.
" 23. EUNOTIA BICEPS, Ehrb. Mik., 17, 1, f. 25; 3, 2, f. 12; K. B., 29, f. 65.
" 24. STAURONEIS PUSILLA, Ehrb. Abh., 1870, p. 59, 2, 1, f. 40. Very nearly if not quite the same as Fig. 6.
" 25. " GRACILE, Ehrb. Mik., 16, 1, f. 4; 17, 2, f. 15; 17, 1, f. 5; K. B., 29, f. 3.
" 26. EUODIA GIBBA, Bail., MSS.; Prit., 852, 3, f. 22.
" 27. EPITHEMIA MUSCULUS, Kg., large form; V. H., 32, f. 14, 15; K. B., 30, f. 6; S. B. D., 1, f. 10.
Figs. 28, 29. RHIZOSOLENIA STYLIFORMIS, Bright, M. J., 1856, p. 95, 5, f. 5; Prit., 865, 7, f. 32; V. H., 78, f. 1-5.
Fig. 30. COSCINODISCUS LIOCENTRUM, Ehrb. Abh., 1870, p. 53, Pl. 2, 2, f. 9; areolation, more or less distinct hexagonal. Humboldt Valley. Oregon.
" 31. HYALODICTYA DIANAE, Ehrb., 1870, p. 57, 3, 17.
Figs. 32-34. BACILLARIA PARADOXA, Gmelin. Sill. J., May, 1842, p. 101, 2, f. 5; S. B. D., 32 and 60, f. 279; Prit., 784, 4, f. 19; 9, f. 166, 167.

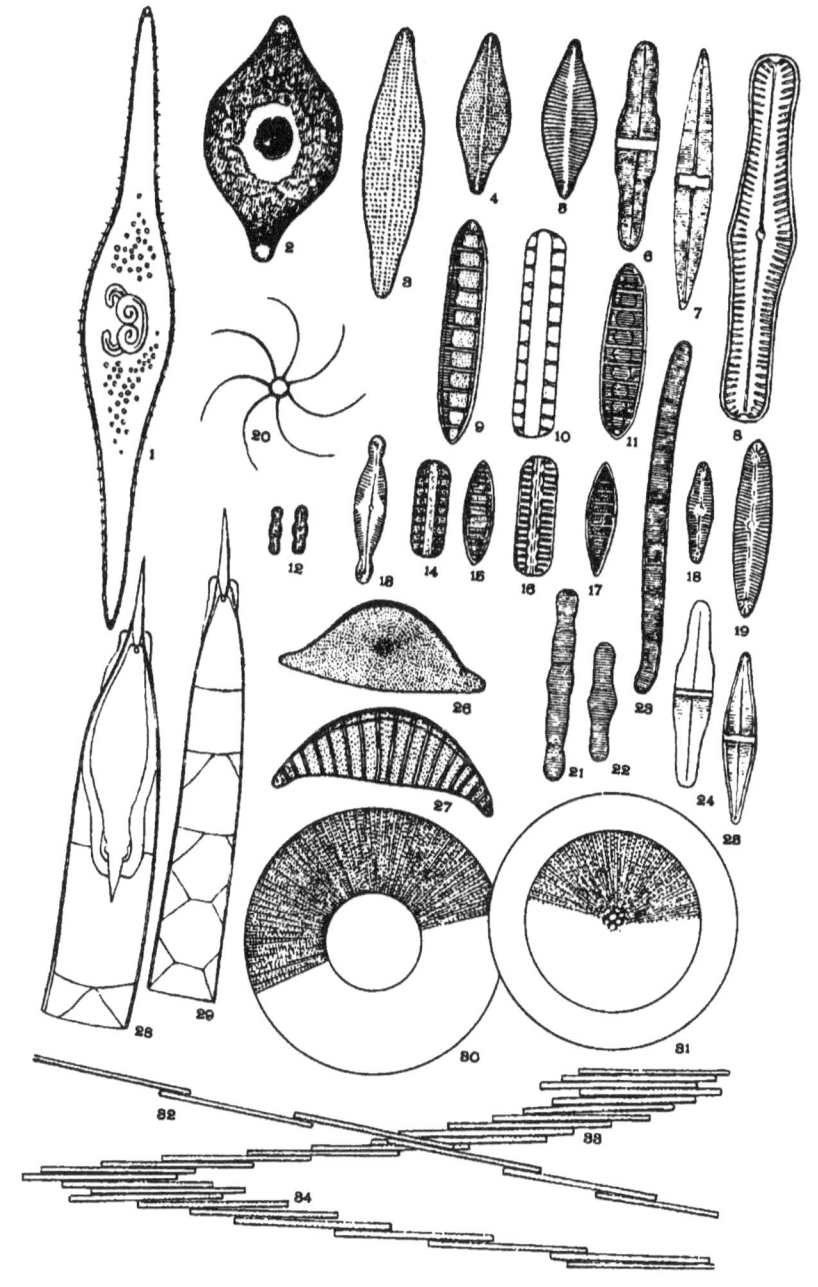

Plate LXVIII.

PLATE LXIX.

Figures magnified 500 diameters.

Figs. 1, 2.   PODOSPHENIA BAILEYI, Edwards; Lewis, N. and R., p. 9, 2, f. 8; Sm. Sp. T., No. 423. New Jersey shore (= Licmophora, Baleyi).

Fig. 3.   PODOSIRA VARIEGATA, A. S.; Schm. At., 140, f. 3. Santa Monica.

Figs. 4, 5.   "   MACULATA, S. B. D., 49, f. 328; Sm. Sp. T., No. 420; O'M. I. D., 26, f. 5a.

" 6, 7.   TRICERATIUM PENTCARINUS, Wallich; M. J., 1858, 12, f. 10–12. (Amphitetras ornata.)

Fig. 8.   ACTINOCYCLUS TRIRADIATUS, Roper, M. J., 1858, p. 23–3, f. 5. 8a, areolation under higher power.

" 9.   ACTINISCUS SIRIUS, Ehrb. Mik., 33, 15, f. 9; T. M. S., 1860, p. 147, 7, f. 14. Carried along for years, proves to be no diatom.

" 10.   SURIRELLA RATRAYI, A. Schm. At., 23, f. 18–21. Vancouvers Island.

" 11.   STICTODESMIS CRATICULA, Grev.; (Surirella craticula), H. L. S. Sp. T., No. 508. This is a sporangial sheath. There is a variety of this form, perhaps different species; also a marine form found at New London.

" 12.   TOXONIDIA GREGORIANA, Donk; T. M. S., 1858, p. 19, 3, f. 1; B. J. N. H., 1879; M. D., 42, f. 42.

Figs. 13, 14.   CAMPYLODISCUS COSTATUS, W. S., S. B. D., 6, f. 52; M. D., 12, f. 6. (C. Hibernicus.)

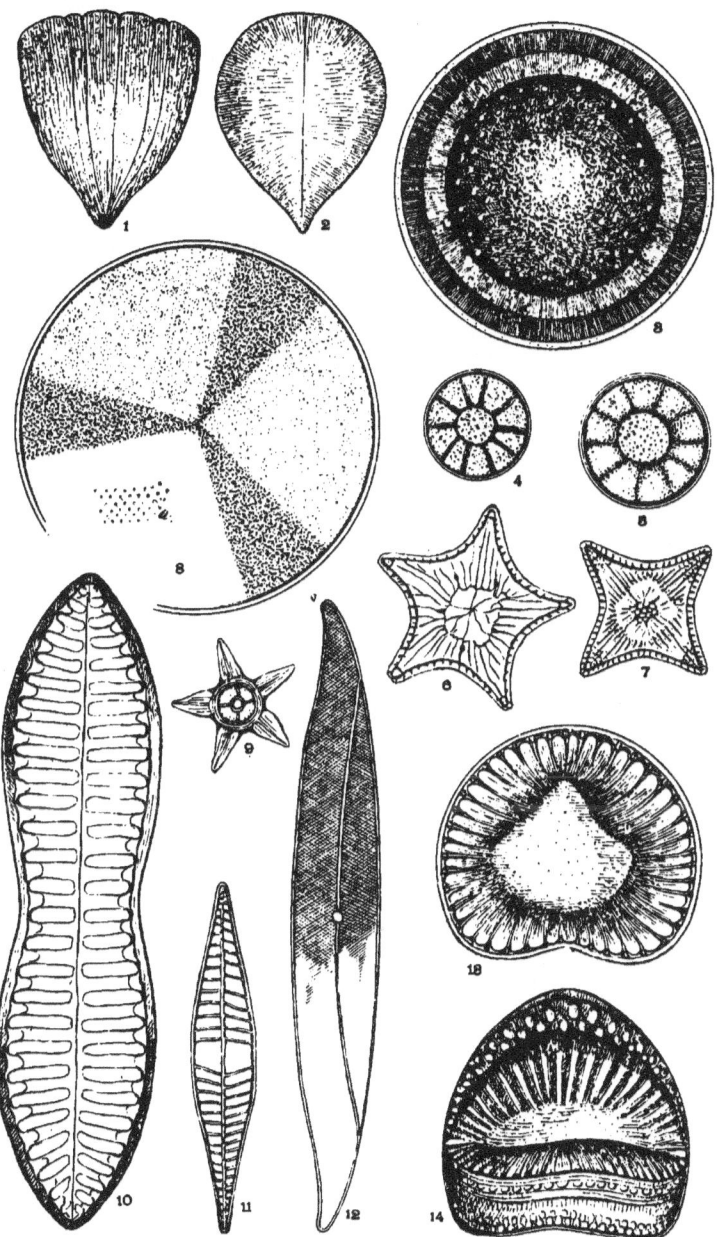

PLATE LXX.

Figures magnified 500 diameters.

Figs. 1, 2.   CAMPYLODISCUS CRIBROSUS, W. S., S. B. D., 7, f. 55; A. N. H., 1851, 1, f. 3.
Fig. 3.   "   HODGSONII, S. B. D., 6, f. 63; Schm. At., 53, f. 5. Same as Camp. Imperialis, Grev.; Schm. At., 52, f. 7; 53, f. 6–7. Figs. 15, 16, are probably only vars. of Hodgsonii.
" 4.   "   SAMOENSIS, Grun.; Schm. At., 15, f. 18–20.
Figs. 5, 6, 7.   "   ARGUS, Bail. M. O. p. 39, 2, f. 24, 25. Hudson River.
" 8, 9.   "   PARVULUS, W. S., S. B. D., 6, f. 56; Prit., p. 801, 15, f. 22, 23; V. H., 77, f. 2.
" 10, 14.   "   LIMBATUS, Breb.; Schm. At., 17, f. 1–3; Grun., 1862, p. 440, 9, f. 4.
Fig. 11.   "   RADIOSUS, Ehrb.; E. Amer., p. 122, 3, 7, f. 14; K. B., 28, f. 12. Of dubious value.
" 12.   "   HUMBOLDTII, Ehrb. Abh., 1862, E, f. 3.
" 13.   "   MEXICANUS, Ehrb. Abh., 1872, 25, f. 19.
" 15.   "   MARGINATUS, Johnst. M. J., 1860, p. 12, 17, 1, f. 11; T. M. S., 1860, p. 30, Pl. 1, f. 2. Cal.
" 16.   "   SAMOENSIS, Schm. At., 19, 20. Compare Fig. 4. The latter two appear closely allied to C. Hodgsonii, Fig. 3, probably merely varieties.

Plate LXX.

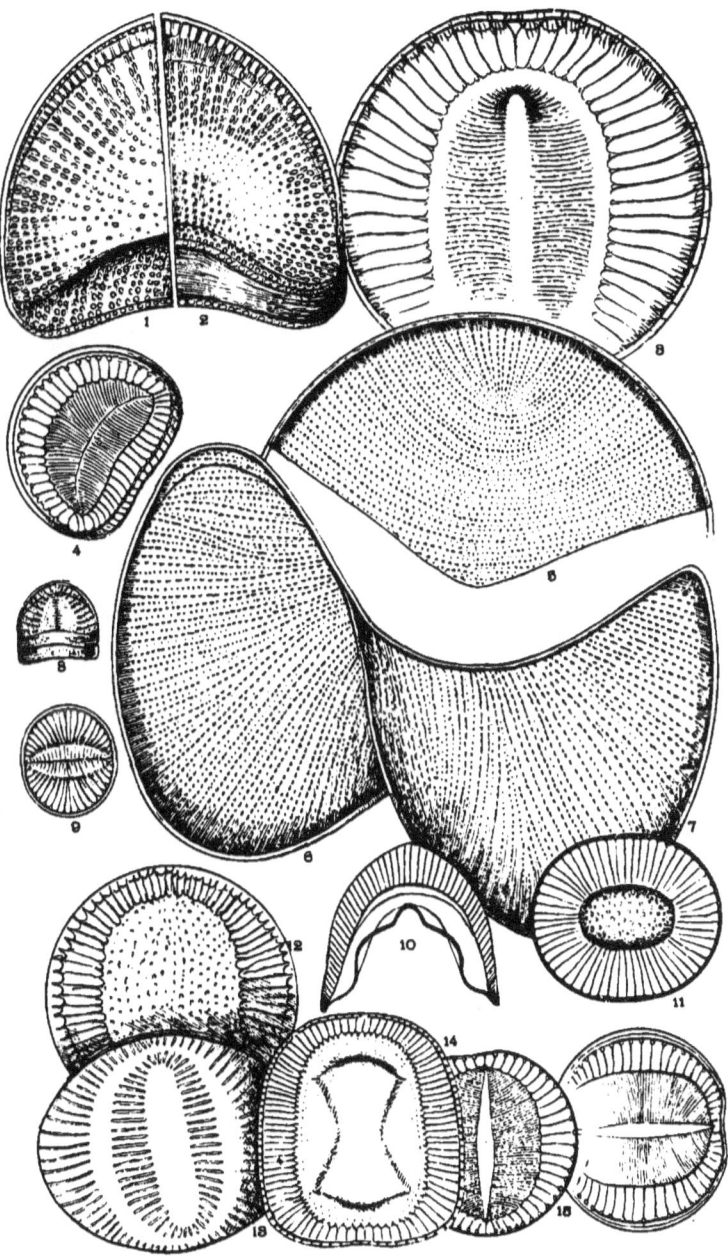

PLATE LXXI.

Figures magnified 500 diameters.

Fig. 1.   AULACODISCUS PULCHER, Norm. Prit., 845, 8, f. 25; granules radiate and rather coarse.
Figs. 2, 3.   CAMPYLODISCUS HIBERNICUS, Ehrb. Mik., 15, A f. 9; Schm. At. 55, f. 9-16.   C. costatus, W. S. ?
Fig. 4.   HELIOPELTA NITIDA, Grev.; T. M. S., 1860, p. 5, 2, f. 18.
Figs. 5-7.   CAMPYLODISCUS NORICUS, Ehrb. Ber., 1840, p. 205; Schm. At., 55 f. 8; V. H., 77, f. 4-6.
Fig. 8.   STICTODISCUS GREVILLIANUS, W. and C., I., p. 5, 1, f. 4.
" 9.   CYMATOPLEURA CAMPYLODISCUS, Bail. B. J., p. 350, Fig. J, q. F; Sm. Sp. T., No. 64.

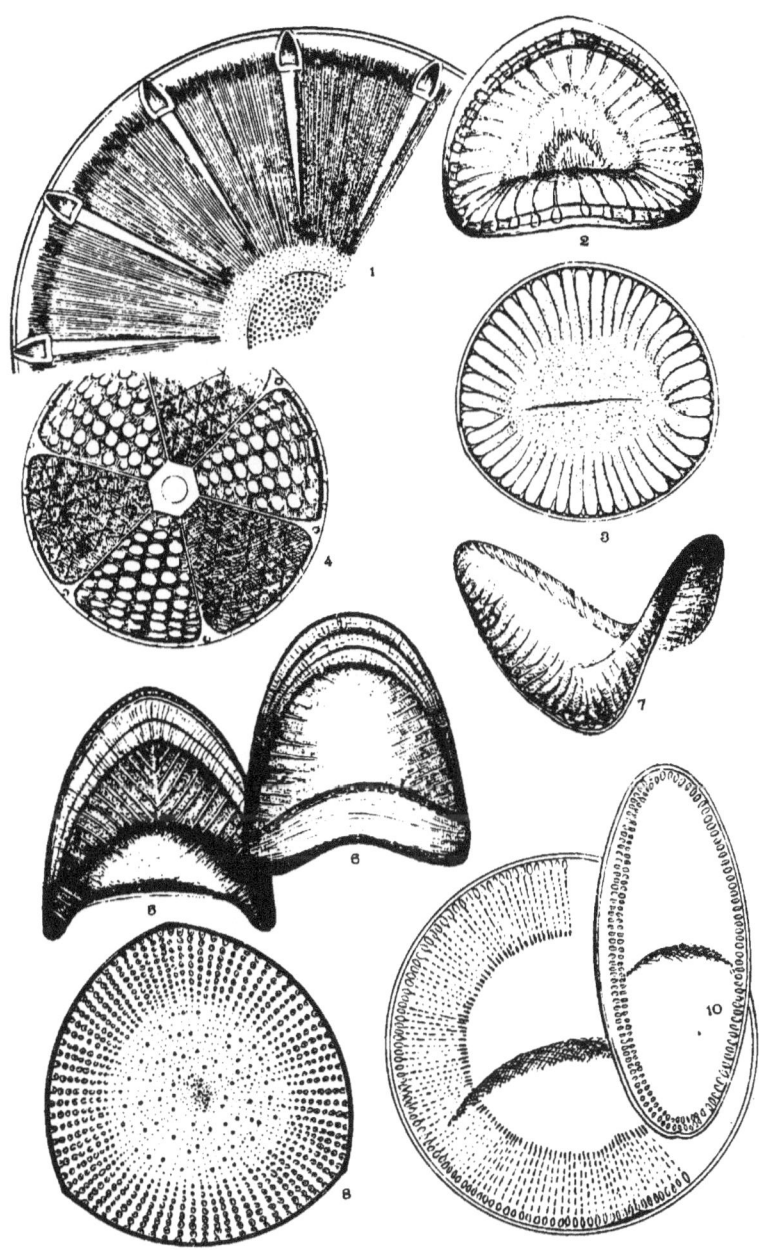

PLATE LXXII.

Figures magnified 500 diameters.

Fig. 1. CAMPYLODISCUS IMPERIALIS, Grev. T. M. S., 1860, p. 31, 1, f. 3; Schm. At., 17, f. 20; 52, f. 7; 53, f. 6, 7. Camp. Bay.
" 2. " BIFURCATAS, A. S.; Schm. At., 52, f. 8. Camp. Bay.
" 3. " ADORNATUS, A. S.; Schm. At., 51, f. 5; 52, f. 3. Campeachy Bay.
" 4. " RALFSII, S. B. D., 30, f. 257; Schm. At., 14, f. 1–3. Striae more decidedly radiate than shown.
" 5. " CONCINNUS, Grev., T. M. S., 1860, p. 30, 8, f. 2; Schm. At., 18, f. 16, 17.
" 6. " CONCINNUS, var. lineatus, Grun. May be a variety C. Hodgsoniis or imperialis. Camp. Bay.

Plate LXXII.

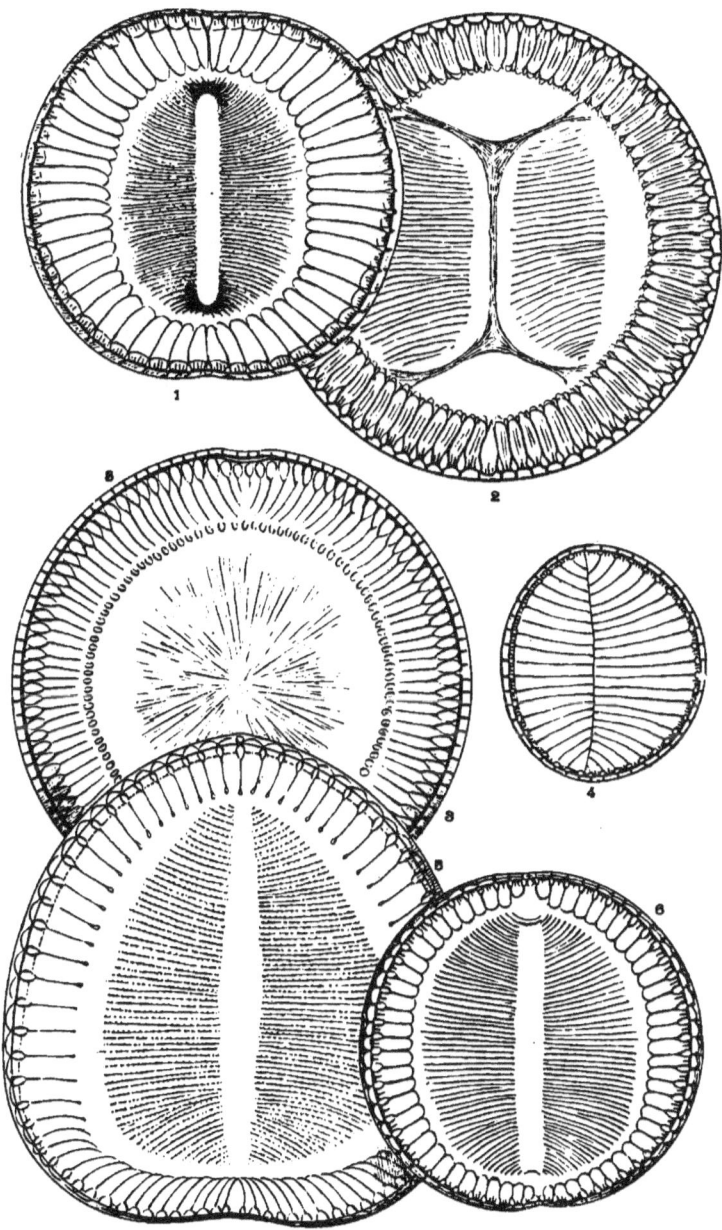

PLATE LXXIII.

Figures magnified 500 diameters.

Fig. 1.   CAMPYLODISCUS MULLERI, A. S., Schm. At., 14, f. 13.
"   2.   "   AMBIGUUS, Grev.; Schm. At., 18, f. 24.
"   3.   "   CLYPEUS, Ehrb. Mik., 10, 1, f. 1; 2, f. 21; Abh., 1869, 1, F, f. 1; K. B., 2 f. 5; V. H., 75, f. 1; Schm. At., 55, f. 1–3; 54, f. 7, 8.
"   4.   "   TABULATUS, A. S., Schm. At., 54, f. 4. Camp. Bay.
Figs. 5, 7.   "   ECHENEIS, Ehrb., Schm. At., 54, f. 3–6; V. H., 76, f. 1, 2.
Fig. 6.   "   SCHMIDTII, Grun.; Schm. At., 15, f. 12; 53, f. 10. Var. of C. imperialis?

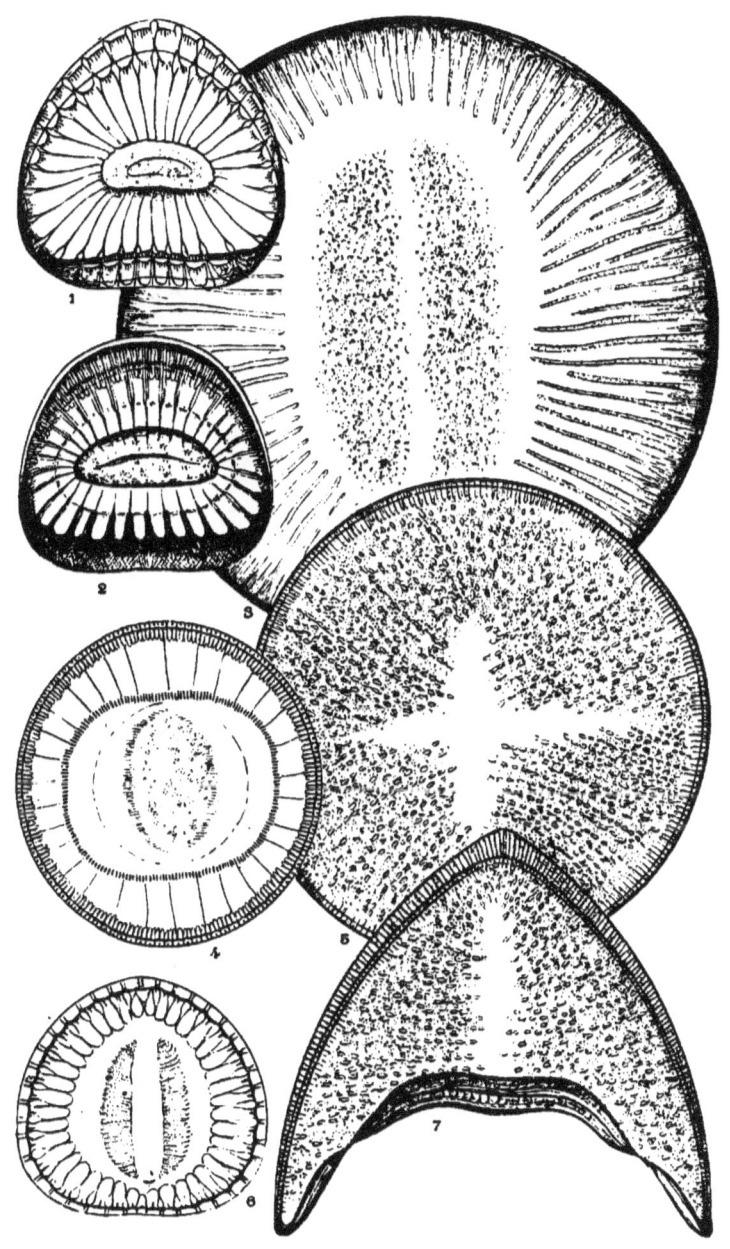

PLATE LXXIV.

Figures magnified 500 diameters.

Fig. 1. COSCINODISCUS CRASSUS, (Craspedodiscus) Bail.; Sill. J., 1856, p. 4; Schm. At., 61, f. 19. Virginia.
Figs. 2, 4. " SOL, Wallich.; T. M. C., 1860, p. 38, 2, f. 12; Schm. At., 58, f. 41, 42, 46, 47; V. II., 129, f. 6. Inserted under the impression the diatoms were found in the fossil deposits of Virginia. Are properly East India forms.
Fig. 3. CRASPEDODISCUS MICRODISCUS, Ehrb. Mik., 33, 17, f. 4; Prit., p. 832.
" 5. " ISOPORUS, (Coscinodiscus), Ehrb. Mik., 33, 17, f. 3. Virginia.
Figs. 6, 7. COSCINODISCUS PATINA, Kg., K. B., 1, f. 15; Sill. J., 1841, 2, f. 13, var.
Fig. 8. ACTINOPTYCHUS MINUTUS, Grev., T. M. S., 1866, p. 5, 1, f. 12.
Figs. 9, 10. MELOSIRA BAILEYI, (Cestodiscus) Am. Q. J. M., 1878, p. 19, 3, f. 9.
Fig. 11. ACTINOPTYCHUS GRUENDLERI, Schm. At., 100, f. 4. St. Monica.
" 12. " UNDULATUS, var. Crenulatus; Schm. At., 1. f. 1 4, 109, f. 1; M. D., 19, f. 7; V. II., 22 bis f. 14. Maryland.
" 13. COSCINODISCUS INTERMEDIUS, Ehrb. Mik., 33, 13, 3.
" 14. ACTINOPTYCHUS ELEGANS.
" 15. PERISTEPHANIA BAILEYI, Ehrb. Abh., 1870, p. 57, 3, 1, f. 13. Likely only a plate of Stephanodiscus Niagarae.
" 16. XANTHIOPYXIS UMBONATUS, Grev., T. M. S., 1866, p. 2, 1, f. 5.
" 17. DISCUS PORCELAINEOUS, Stodder. Stodder constituted this genus and species. He never figured it; mine is a figure from his description. A. J. M., 1879, f. 14.
" 18. PERISTEPHANIA EUTYCHIA, Ehrb. Mik., 35B, 4, f. 14. Cal.

Plate LXXIV.

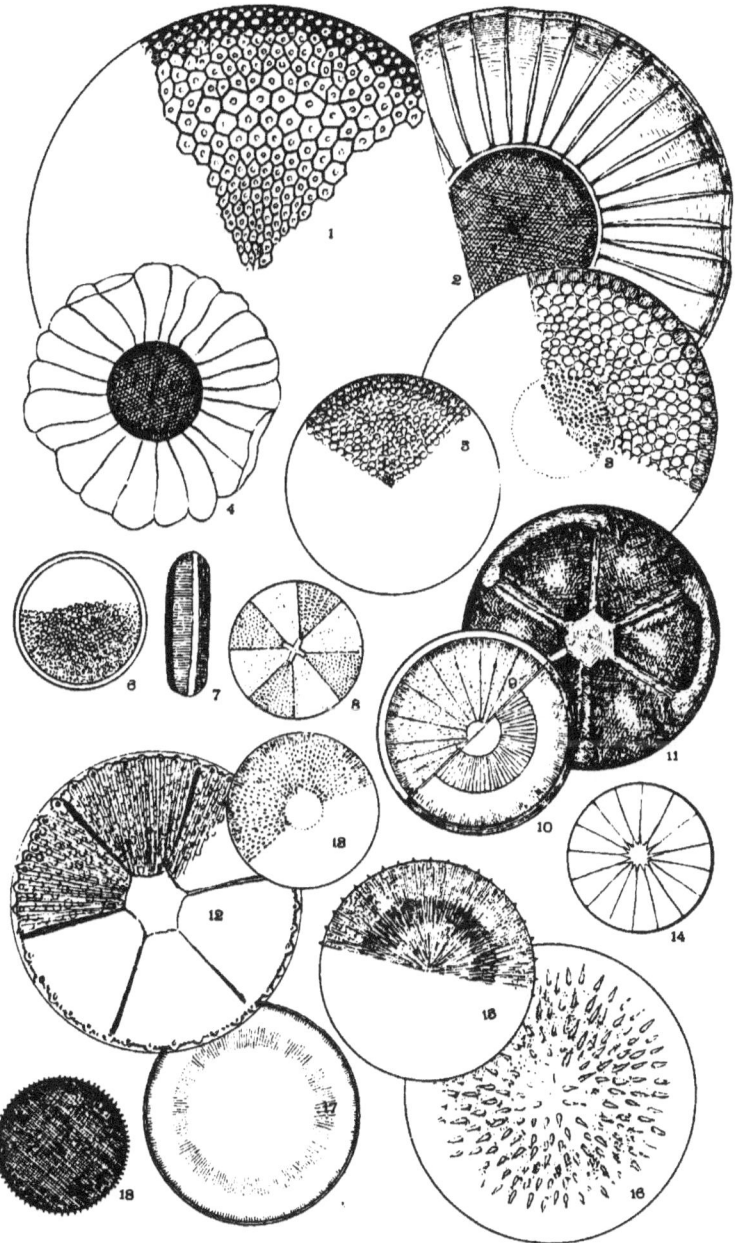

PLATE LXXV.

Figures magnified 500 diameters.

Fig. 1.   AULACODISCUS MOELLERI, A. S. Schm. At., 33, f. 14.   Nottingham, Md.
"   2.   "   DECORUS, (related to), Schm. At., 33, f. 9.   Monterey.
"   3.   STICTODISCUS HARDMANIANUS, forma minor.
"   4.   "   SIMPLEX, A. S., Sci. m. At., 74, f. 11.   San Francisco, Cal.   Appears very nearly related to Arachnodiscus.
Figs. 5, 6.   "   CALIFORNICUS, vars. Grev.; Schm. At., 74, f. 9.
Fig. 7.   "   var. ecostata, Grun.; Schm. At., 74, f. 6, 7.
"   8.   "   CALIFORNICUS, Grev.; Los Angeles, Schm. At., 74, f. 4.
Figs. 9, 10.   "   KITTONIANUS, Schm. At., 74, f. 16-18.   Nottingham, Md.
Fig. 11.   TRICERATIUM CINNAMOMIUM, (Cestodiscus) Grev. M. J., 1863, p. 232, 10, f. 12.
"   12.   COSMIODISCUS TENUIS, Grev.; V. H., 125, f. 13.
Figs. 13-15.   STICTODISCUS, A. Schm. determines these as inner plates of Stictodiscus, such as occur frequently with *Asteromphalus* and *Asterolampra* and have given rise to a number of spurious species. Grunow determines these three forms to be plates of Melosira clavigera.
Fig. 16.   EUNOTOGRAMMA DEBILIS, Grun.; Camp. Bay. V. H., 126, f. 17. May be with, or without the opening.
Figs. 17-19.   "   LAEVIS, Grun.; V. H., 126, f. 6, 7, 9.   N. C., Fla.

Plate LXXV.

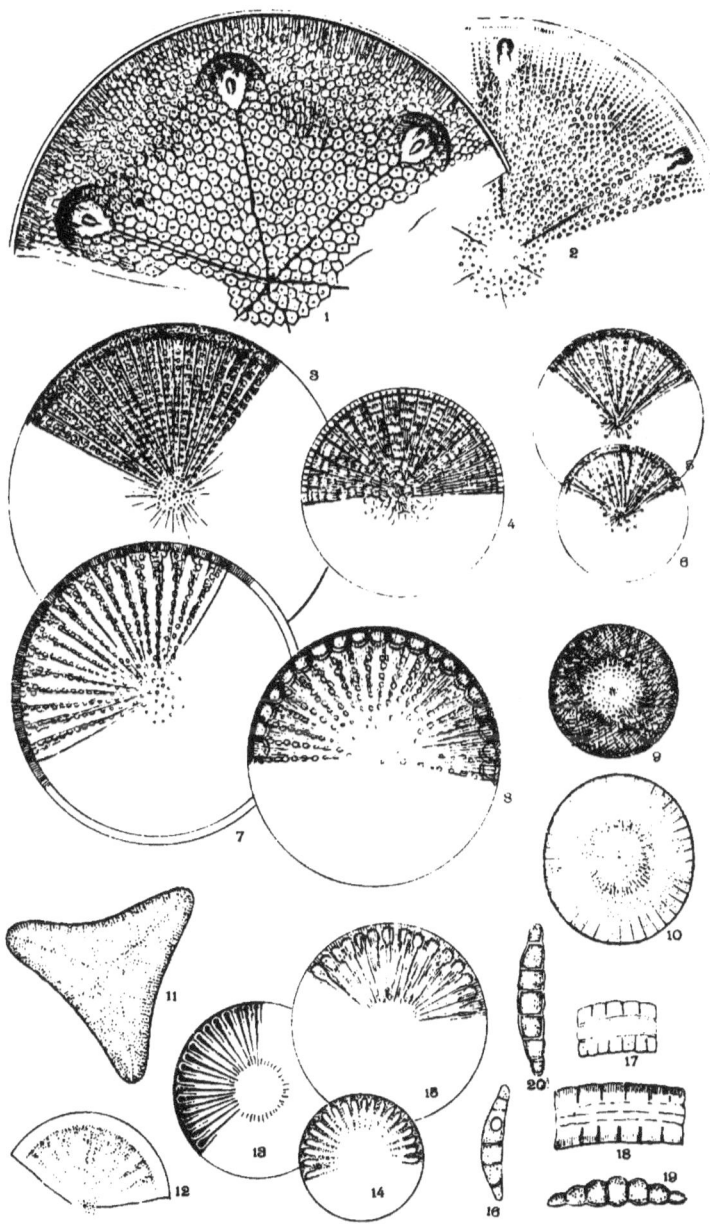

PLATE LXXVI.

Figures magnified 500 diameters.

Figs. 1, 2, 4.   EUPODISCUS ARGUS, Ehrb., S. B. D., 4, f. 39; Prit., 843, 6, f. 2; 11, f. 41; Lens, Vol. II, p. 29; V. H., 117, f. 3, 4, 5, 6. Florida, etc. Cells usually of more or less stellate appearance.
Fig. 3.   "   ROGERSII, Ehrb. Ber., 1844; Schm. At., 92. f. 5. Nottingham, Md. Cells are somewhat stellate as E. argus.
"   5   "   OCULATUS, Grev., T. M. S., 1862, p. 94, Pl. 9, f. 3. Monterey.
Figs. 6, (11).   "   RADIATUS, Bailey; V. H., 118, f. 1, 2; Bail., M. O., p. 39.
Fig. 7.   "   CALIFORNICUS, Grun.; (trioculatus, var.?) Gulf of California. V. H., 118, f. 8. This has the three spines just within the margin, H. L. S.
"   8.   GLYPHODISCUS STELLATUS, Grev.
"   9.   "   "   Grev.; var. major, Grun.; V. H., 118, f. 3. Santa Monica.
"   10.   "   GRUNOWII, A. S., Schm. At., 80, f. 6. May be a var. of G. stellatus. Crescent City.

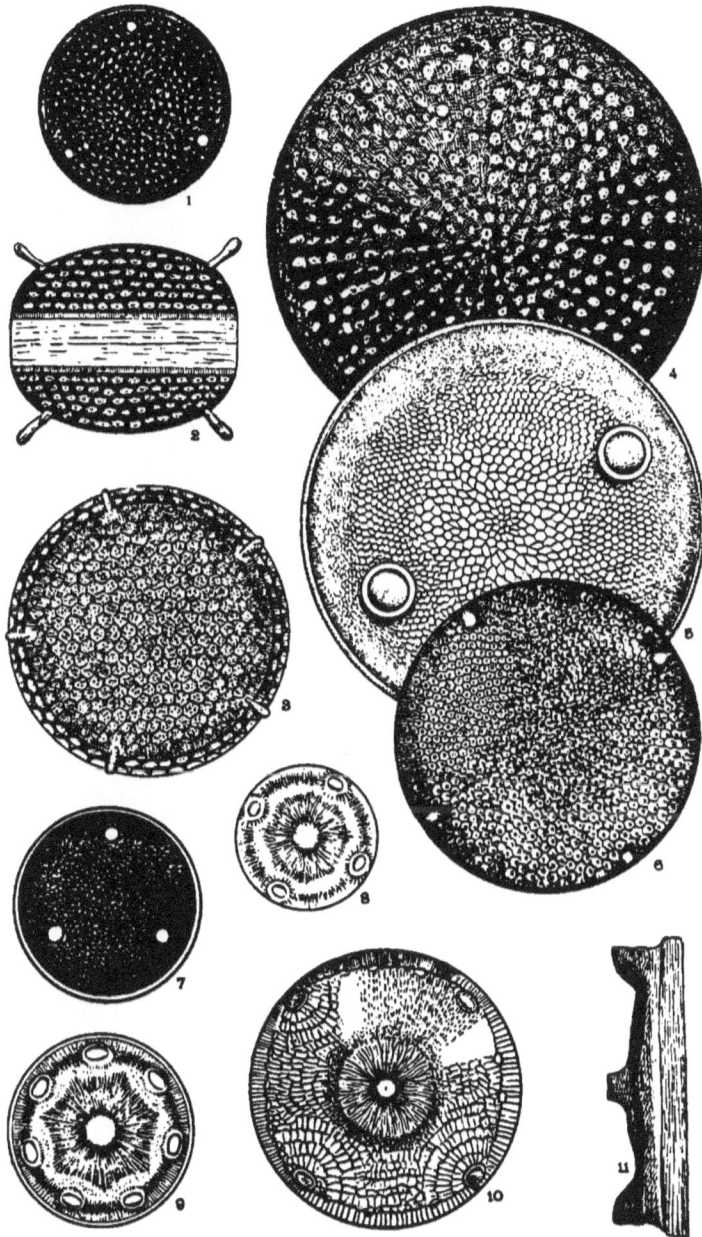

PLATE LXXVII.

Figures magnified 500 diameters.

Figs. 1, 2.   CRASPEDODISCUS ELEGANS, Ehrb. Ber., 1844, p. 266, f. 12; Sill. J., April, 1845, p. 324, 4, f. 1; Schm. At., 66, f. 1.   Nottingham, Md.
Fig. 3.   TRICERATIUM AMBLYCEROS, Ehrb. Ber., M. J., 1853, p. 250, 4. f. 14; Schm. At., 1, f. 25.   Richmond, Va.
"   4.   "   SPINOSUM, Bail.; Ried.; Schm. At., 87, f. 3.
Figs. 5, 6.   ACTINOPHYCHUS INTERPUNCTATUS, Shad., M. J., 1860, p. 94, 6. f. 17.   Richmond, Va.
Fig. 7.   ACTINOPTYCHUS RAEANUS, Challenger, 1887, 7, f. 4.   Santa Monica.
"   8.   EUPODISCUS RADIATUS, var. antiqua, J. D. Cox.   Richmond, Va., Atlantic City, artesian well; Bull. Tor. Bot. Club, March, 1889.
Figs. 9, 10.   TRICERATIUM SPINOSUM, vars., Schm. At., 76, f. 7.   Richmond, Va.

Plate LXXVII

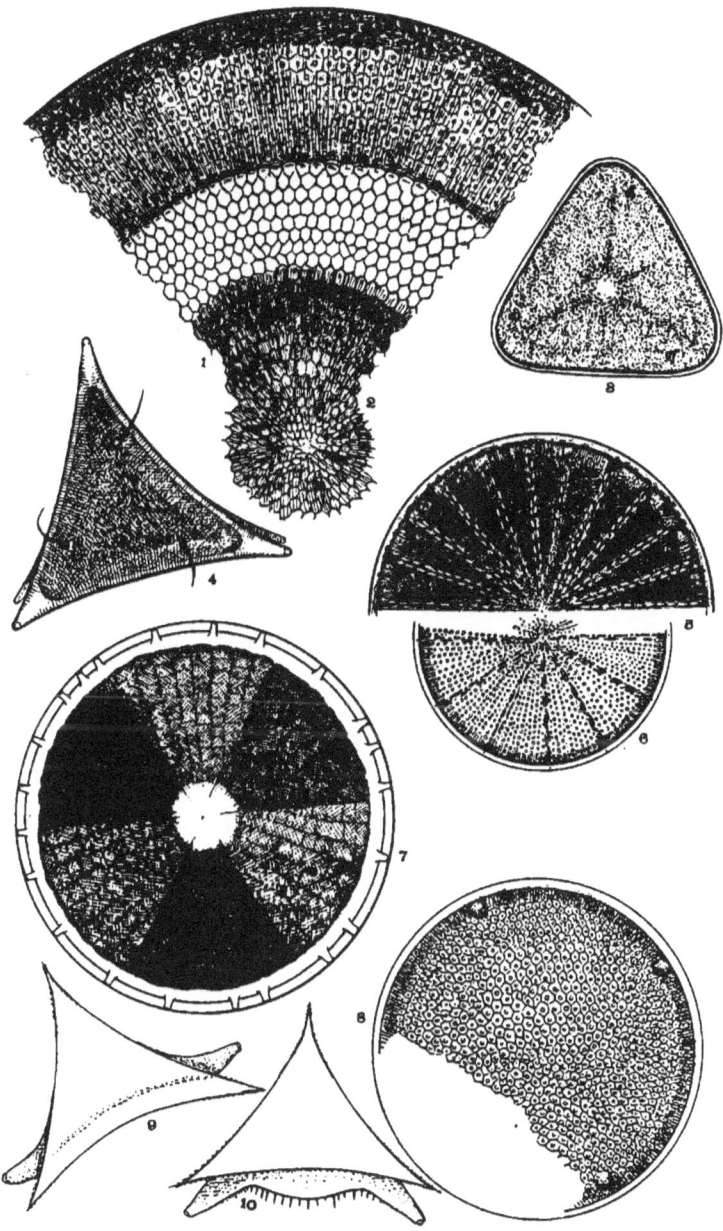

PLATE LXXVIII.

Figures magnified 500 diameters.

Figs. 1, 2. TERPSINOE INTERMEDIA, Grun.; var. Bull. Torrey Bot. Club, August, 1889, Plate 93, f. 2, 2a. Atlantic City.
Fig. 3-5. BIDDULPHIA WEISSFLOGII, Jan.; V. H., 100, f. 1, 2; Cast. 1887, p. 104, 26, f. 2. Doubtfully N. Amer. form.
" 6. " A questionable variety of this genus. Kain and Schultze, Bull. Tor. Bot. Club, Aug., 1880, Plate 93, f. 4. Atlantic City.
" 7. TRICERATIUM INDENTATUM, Kain and Schultze, Bull. Tor. Bot. Club, Aug., 1889, Pl., 92, f. 4. Artesian well, Atlantic City, N. J.
Figs. 8, 9. " HEILPRINIANUM, Kain and Schultze, artesian well, Atlantic City, N. J. Bull. Tor. Bot. Club, August, 1889, Pl. 93, f. 3, 3a. "Surface with central and angular elevations; the central elevation shaped like a truncated pyramid; punctae radiate and coarser at center."
Fig. 10. " KAINII, var. constrictum; Schmidt. Artesian well, Atlantic City. Bull. Tor. Bot. Club, 1889, Pl. 92, f. 3.
" 11. ACTINODISCUS ATLANTICUS, Kain and Schultze, artesian well, Atlantic City, N. J. Bull. Tor. Bot. Club, March, 1889, Pl. 92, f. 7. Resembles a form of A. splendens.
" 12. EUNOTIA AMERICANA, Kain and Schultze, artesian well, Atlantic City, N. J. Bull. Tor. Bot. Club, Aug., 1889, Pl. 93, f. 1.
" 13. ACTINOPTYCHUS UNDULATUS, var. verrucosa; W. and Chase, Pl. 2, f. 8.
" 14. SYMBOLOPHORA TRINITATUS, Ehrb. Ber., 1844, p. 88; Sill. J., April, 1845, 4, f. c; M. Dic., 19, f. 6. Prof. H. L. Smith thinks this is nothing more than a rounded fragment of Triceratium Marylandicum; hence, of no value.
" 15. DISCUS UNBENANT, Greg., T. M. S., 1857, Pl. 4, f. 48. Transf. to fill a vacant space.
" 16. CHAETOCEROS (didymus, Ehrb.)? Bull. Tor. Bot. Club, August, 1889, Plate 92, f. 6. Artesian well, Atlantic City.

Plate LXXVIII.

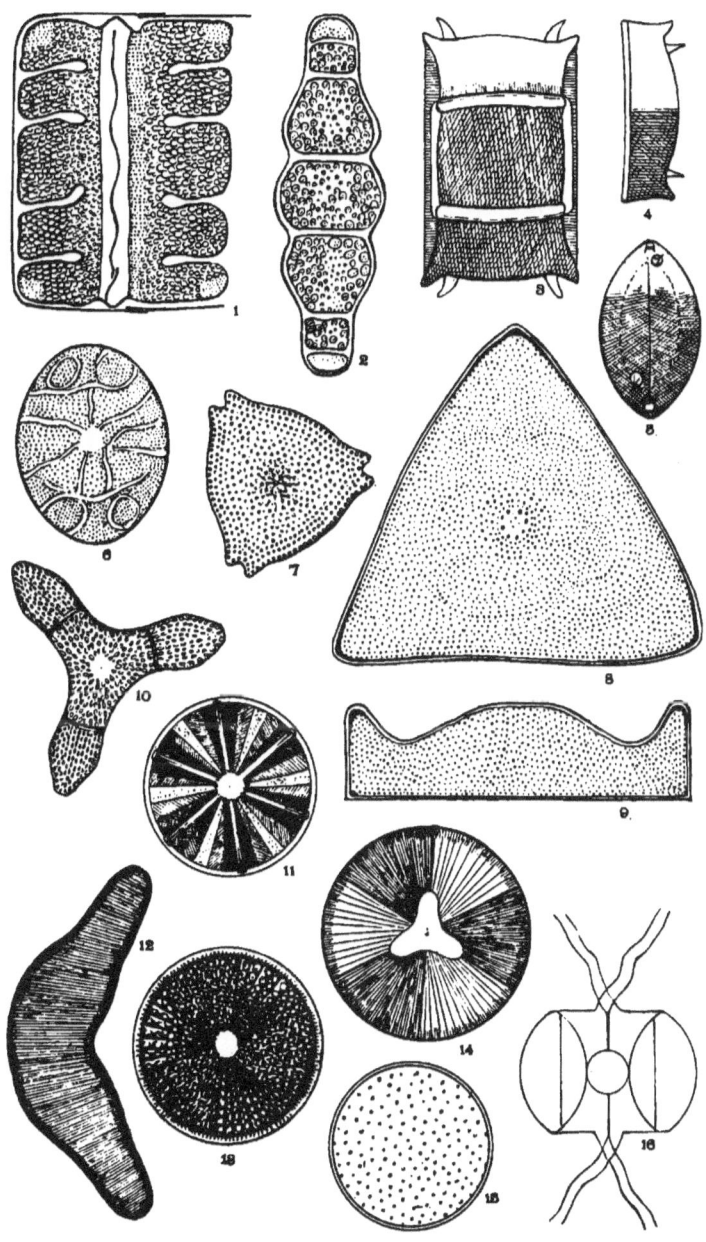

PLATE LXXIX.

Figures magnified 500 diameters.

Fig. 1. AULISCUS HARDMANIANUS, Grev.; T. M S., 1866, p. 6, 2, f. 17; Schm. At., 67, f. 1; 89, f. 4.
Figs. 2, 3. " PUNCTATUS, Grun.; Schm. At., 30, f. 10.
Fig. 4. " SPINOSUS, F. Christian; Schm. At., 125, f. 2; Bull. Torrey Bot. Club, March, 1889.
Figs. 5, 6, 7. " RECTICULATUS, Grev., T. M. S., 1863, p. 46, 2, f. 10; Schm., At., 30, f. 1-3.
" 8, 9, 10. " CLEVEI, Grun.; Schm. At., 31, f. 1-4.
" 11, 12. " GRUNOWII, var. Californica; A. S., Schm. At., 30, f. 14; 89, f. 7, 8.
" 13, 14. " PRUINOSUS, Bail.; Prit., p. 845, 6, f. 1; T. M. S., 1863, p. 48, 3, f. 13; Schm. At., 31, f. 6, 7.

Plate LXXIX.

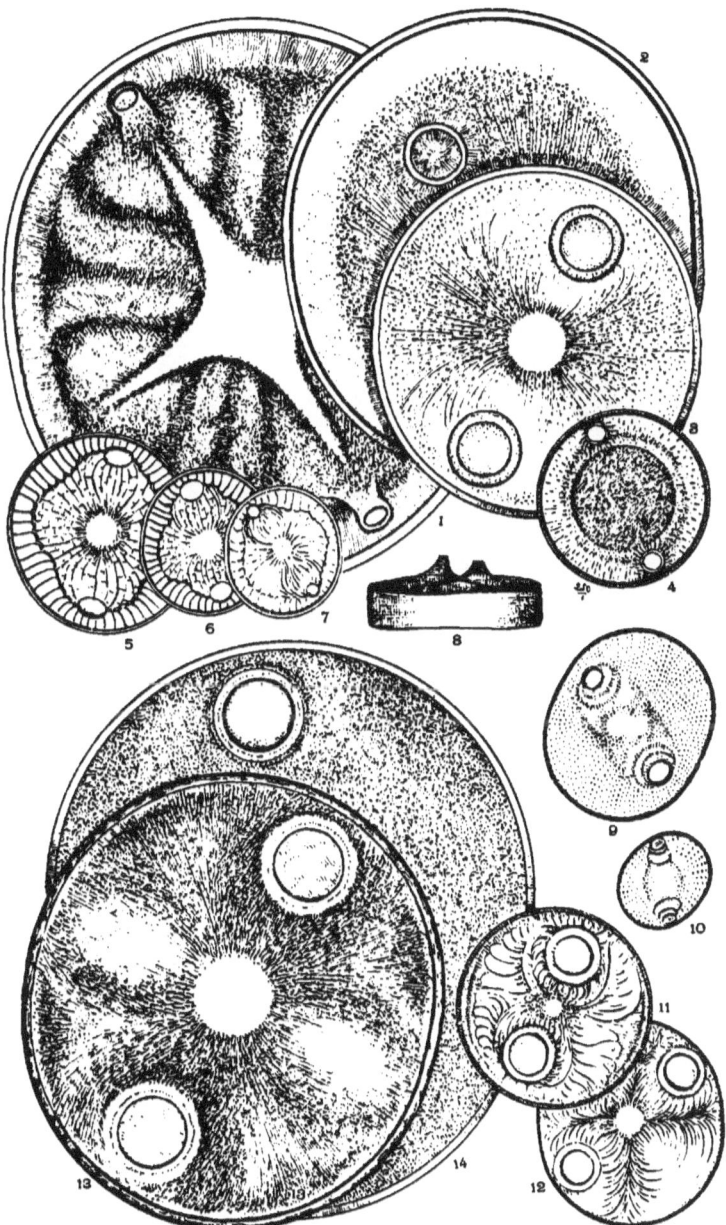

PLATE LXXX.

Figures magnified 500 diameters.

Fig. 1. AULISCUS JOHNSONII, A. S., Schm. At., 67, f. 2. Crescent City, Cal.
" 2. " SPECIOSUS, A. S., Schm. At., 108, f. 3. Santa Monica.
" 3. " BIDDULPHIA, var. Kitton; Schm. At., 89, f. 2. Santa Monica.
" 4. " BIDDULPHIA, Kitton; Schm. At., 67, f. 3.
" 5. " INTESTINALIS, A. S., Schm. At., 108, f. 2. Santa Monica.
" 6. " TEXTILIS, A. S., Schm. At., 89, f. 9.
" 7. " CARIBAEUS, Cleve.; Schm. At., 67, f. 9, 10.
" 8. " INCERTUS, Schm. At., 89, f. 19. Santa Monica.
" 9. " AMERICANUS, Ehrb. Mik., 33, f. 14, 2. Norwich, Conn. Probably a var. of A. sculptus.
" 10. AULACODISCUS ROGERSII, (Eupodiscus) Ehrb. Ber., 1844, p. 81; Schm. At., 92, f. 2-6; 107, f. 3. Maryland.
" 11. " ARGUS (Eupodiscus), Pensacola; Schm. At., 4.
" 12. " PROBABILIS, Schm. At., 36, f. 15, (13-16); 103, f. 3, 4. Rather exceptional figure. The normal form shows blank space from center to processes.

Plate LXXX.

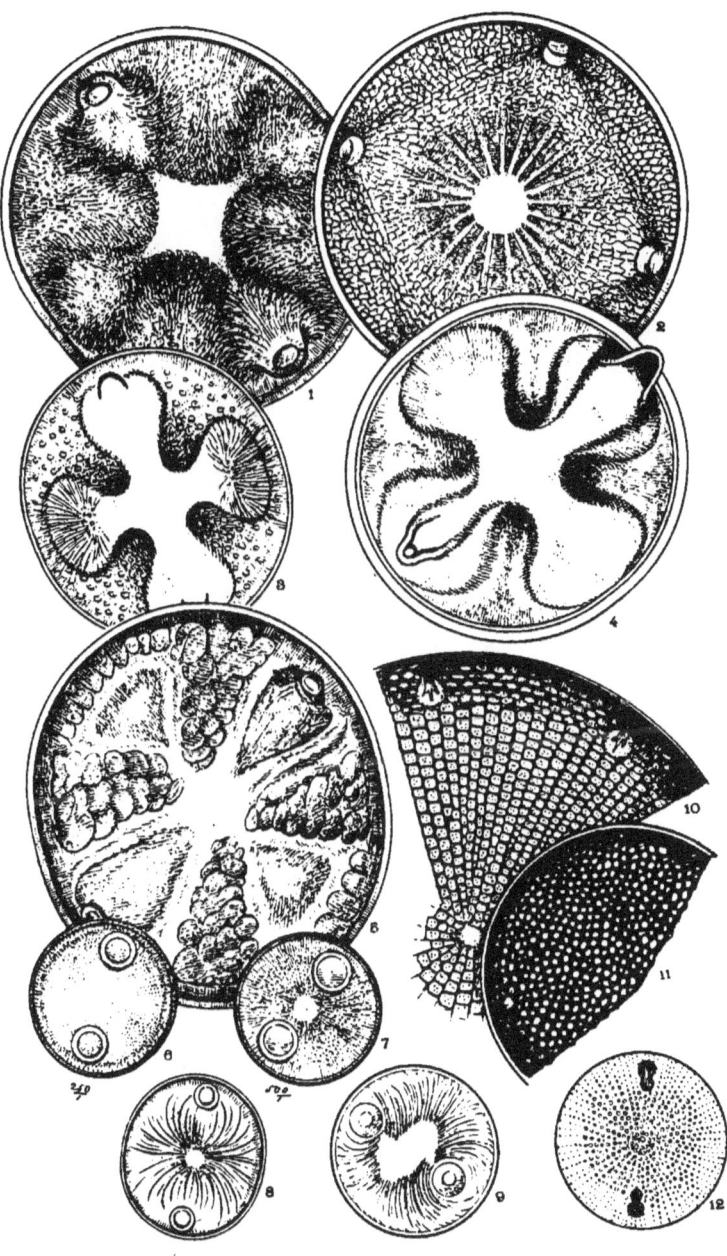

PLATE LXXXI.

Figures magnified 500 diameters.

Fig. 1. ASTEROLAMPRA HILTONIANA, (Asteromphalus), Grev., T. M. S., 1860, p. 111, 4, f. 15; Sm. Sp. T., No. 49.
" 2. " MORONIENS, Grev., M. J., 1863, p. 230, 7, f. 8; Schm. At., 38, f. 24.
" 3. " RALFSIANA, Grev., T. M. S., 1862, p. 50, 8, f. 31.
" 4. ASTEROMPHALUS ARACHNE, Breb., Schm. At., 38, f. 5. Figure should be somewhat ovoid.
" 5. ASTEROLAMPRA GREVILLEI, Grev., T. M. S., 1860, p. 113, 4, f. 21; V. H., 127, f. 12.
" 6. ASTEROMPHALUS FLABELLATUS, Grev., Cal. guano.
" 7. COSCINODISCUS RADIATUS, Ehrb. Mik., 39, 3, f. 17; Sill. J., 1842, p. 95, 2, f. 14; K. B., 1, f. 18; S. B. D., 3, f. 37; T. M. S., 1860, p. 48, 2, f. 22; V. H., 129, f. 5.
" 8. " ARGUS, var. Ehrb. Mik., 21, f. 2; 22, 5, f. 8; Schm. At., 61, f. 13; 113, f. 7.
" 9. " GIGAS, Ehrb. Mik., 18, f. 34; K. B., 1, f. 16; Schm. At., 64, f. 1. Cells not round, but somewhat angular.
" 10. " SUBCONCAVUS, Grun.; Schm. At., 59, f. 12, 13, 15; 62, f. 7.

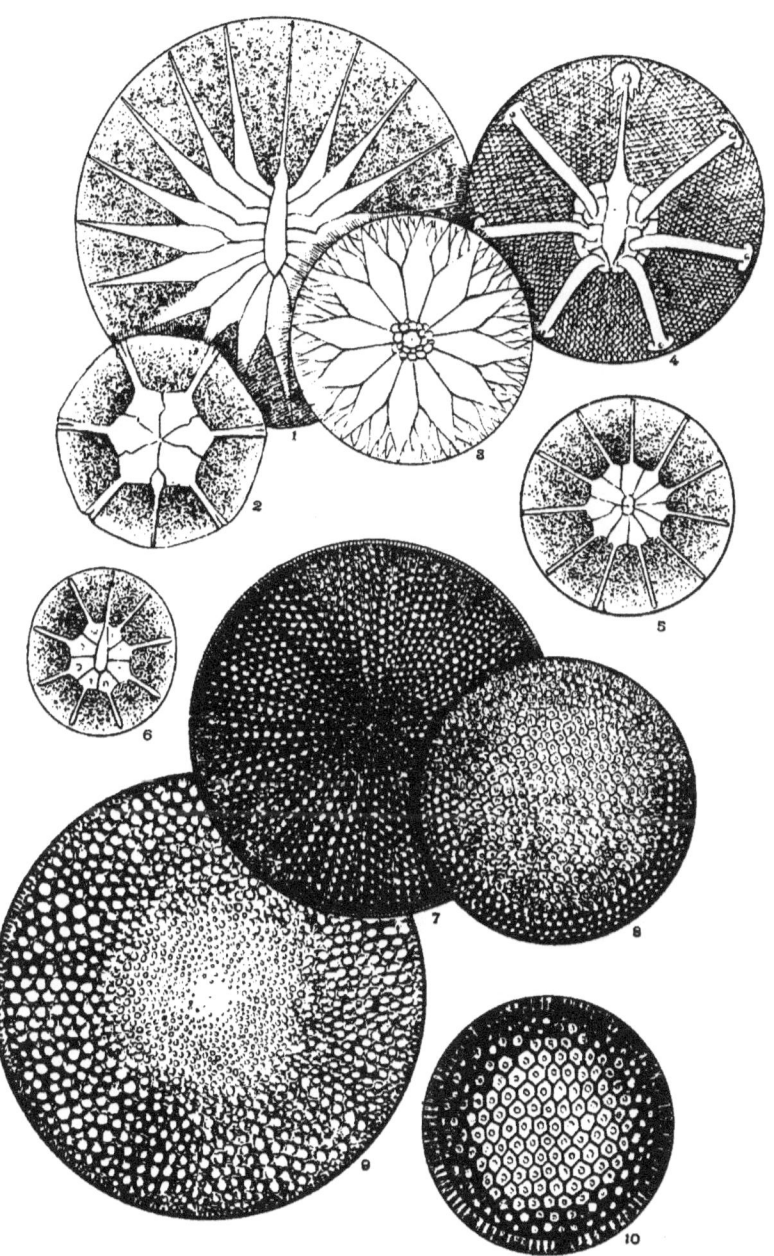

PLATE LXXXII.

Figures magnified 500 diameters.

Fig. 1. AULACODISCUS MARGARITACEUS, Ralfs.; Prit., p. 844; Schm. At., 37, f. 1-8. Cellules somewhat angular, too round in figure.
" 2. " MARGARITACEUS, smaller form, Cal. Schm. At., 37, f. 1-8. Rim too broad in figure.
" 3. " PETERSII, Ehrb.; Schm. At., 35, f. 1-4; 41, f. 1, 2. Side view. Processes really in the center of the elevated bosses.
" 4. " SPARSUS, Grev., T. M. S., 1866, p. 123, 11, f. 6; Schm. At., 36, f. 12. Cal.
" 5. " KITTONII, Arn.; Prit., p. 844, 8, f. 24; Schm. At., 36, f. 5-7; 41, f. 6; M. J., 1860, p. 95, 6, f. 13. Areolation like Coscinodiscus, more or less hexagonal.
Figs. 6, 7. " PETERSII, Ehrb.; Figs. 3, 6, 7, are varieties of the same species.

Plate LXXXII.

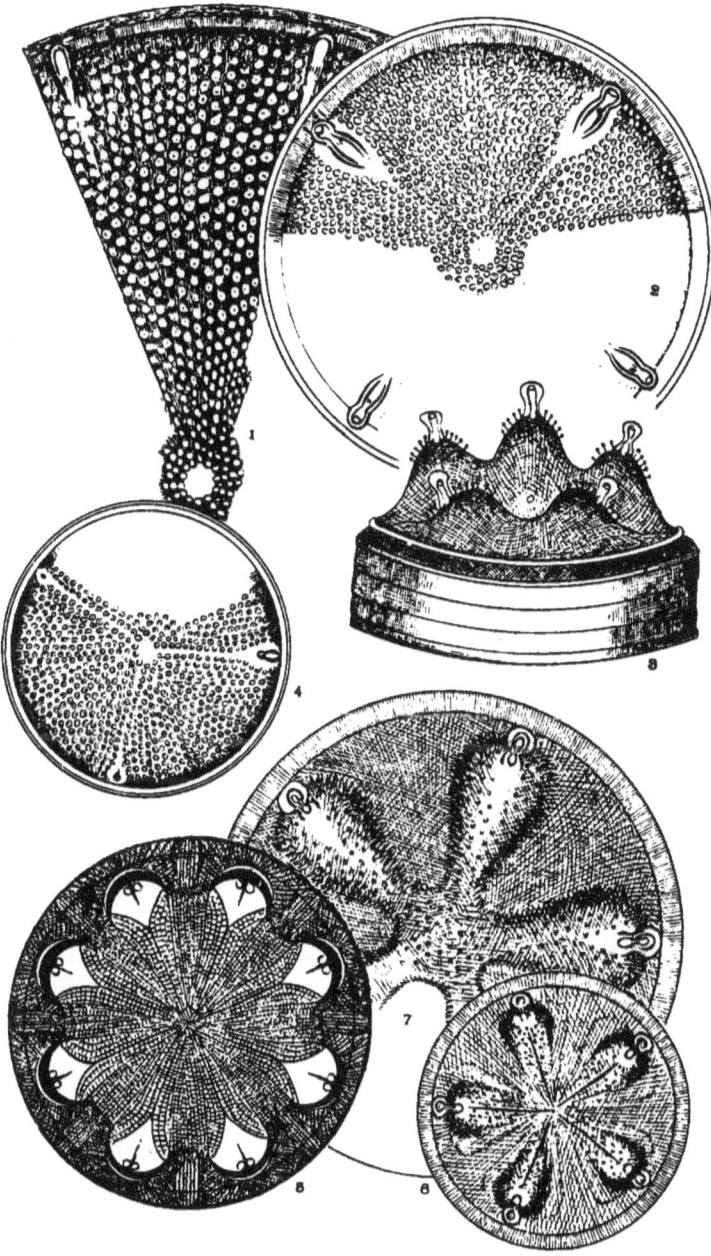

PLATE LXXXIII.

Figures magnified 500 diameters.

Fig. 1.   CAMPYLODISCUS GREENLEAFIANUS, Grun.; Schm. At., 15, f. 3.
"   2.   "   STELLATUS, Grev., M. J., 1859, p. 157, 7, f. 3; Prit., p. 796.
"   3.   "   TRIUMPHANS, A. S., Schm. At., 15, f. 4, 5.
"   4.   "   GRUENDLERII, Grun.; Schm. At., 15, f. 1, 2, not 51, f. 13. Camp. Bay.
"   5.   "   PUNCTULATUS, Grun.; Schm. At., 17, f. 4. Form not quite circular.
"   6.   "   INTERMEDIUS, Grun.; Gulf of Mexico. Schm. At., 14, f. 30.
"   7.   "   ECCLESIANUS, Grev.; Camp. Bay. Schm. At., 16, f. 9.
"   8.   "   BIMARGINATUS, A. S., Schm. At., 16, f. 7. Camp. Bay.
"   9.   "   CREBRESTRIATUS, Grev.; Schm. At., 14, f. 28; 53, f. 18.
"   10.   "   RALFSII, W. S., S. B. D., 30, f. 257; Schm. At., 14, f. 1-4. Gulf of Mexico, Camp. Bay.
"   11.   "   ROTULA, Grun.; Schm. At., 14, f. 10. Camp. Bay.
Figs. 12, 13.   "   SAUERBECKII, Gründ.; Schm. At., 52, f. 6; 53, f. 3, 4.
"   14, 15,   "   PHALANGIUM, A. S., Schm. At., 14, f. 11, 12. Camp. Bay.
Fig. 16.   "   SIMULANS, Greg.; T. M. S., 1857, p. 71, 1, f. 41. Schm. At., 17, f. 12-14.

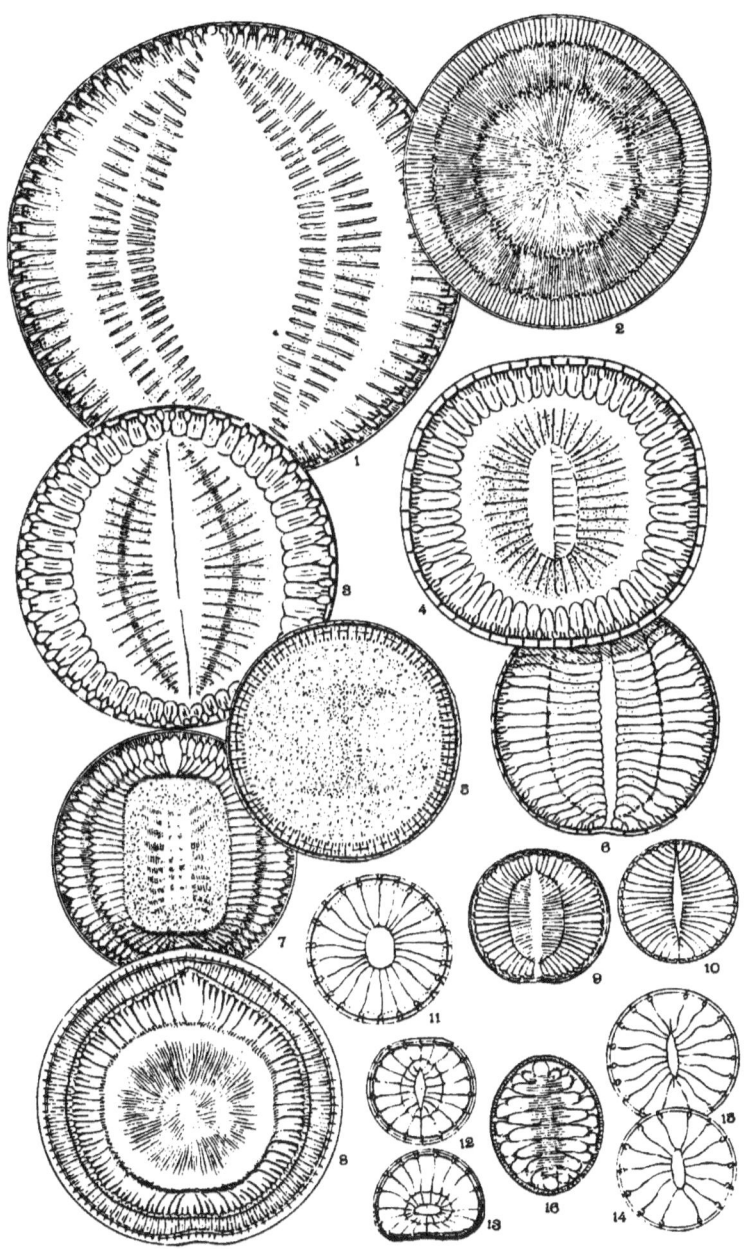

PLATE LXXXIV.

Figures magnified 500 diameters.

Fig. 1. RAPHIDODISCUS MARYLANDICA, T. C.—*Microscope*, May, 1889.
" 2. " CHRISTIANII, Gascoigne.—*Microscope*, May, 1889.
Figs. 3, 4. " FEBIGERII, T. C.—*Microscope*, May, 1889.

These three described species must be reduced to a single one. The impression conveyed by the examination of a photograph of *R. Marylandica*, is that of two different diatoms accidentally wedged together; the markings of MILOSIRA can be plainly made out when those of Raphidodiscus are out of focus; and R. Christianii is simply Marylandica without the enveloping MILOSIRA; or more correctly, *Marylandica* and *Christianii* lodged in a Milosira. R. Febigerii is simply *Christianii* with its marginal rim broken away. C. M. Vorce, Cleveland, O.—*The Microscope*, May, 1889.

Fig. 5. STEPHANOPYXIS SPINOSISSIMA, Grun.; Schm. At., 123, f. 18. St. Monica.
" 6. PSEUDOAULISCUS RADIATUS, Schm. At., 32, f. 28.
" 7. ODONTODISCUS SUBTILIS, Grun. Do not see why this should be called Odontodiscus without any evidence of teeth, (Odontos.)
Figs. 8, 9. STEPHANOPYXIS SUPERBA, Schm. At., 123, f. 4, 5. Large and small forms.
Fig. 10. XANTHIOPSIS UMBONATUS, Grev., T. M. S., 1866, p. 2, 1, f. 5.

Plate LXXXIV.

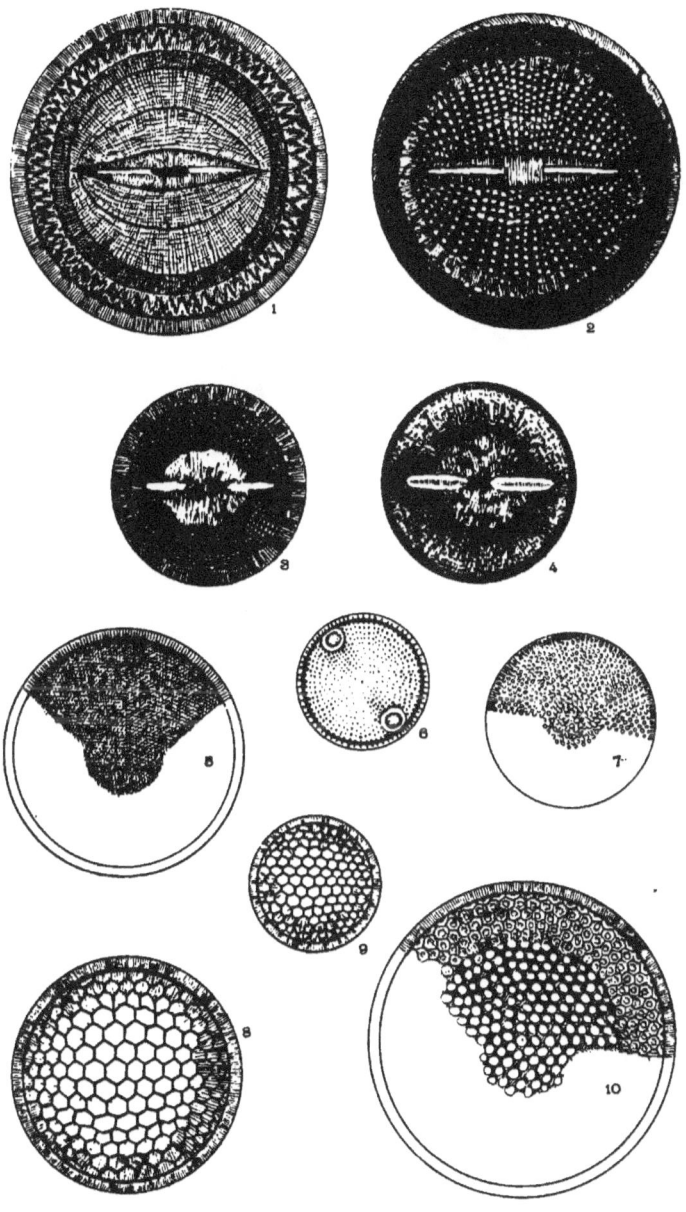

PLATE LXXXV.

Figures magnified 500 diameters.

Fig. 1. ACTINOCYCLUS ELLIPTICUS, Grun.; V. H., 126, f. 10. Richmond, Va.
" 2. ACTINOPTYCHUS SOCIUS, A. S., Schm. At., 1, f. 11.
" 3. " PRAETER, A. S., Schm. At., 100, f. 5. Maryland deposit.
" 4. " AREOLATUS, Schm. At., 1, f. 9. California.
" 5. " AMBLYOCEROS, (Triceratium amblyoceros, Ehrb.) Schm. At., 1, f. 25. Richmond, Va.
" 6. " NITIDUS, Schm. At., 1, f. 7. Mexico.
" 7. " SUBTILIS, Greg., forma major; V. H., 125, f. 11.
" 8. " INCERTUS, Grun.; St. Monica. V. H., 125, f. 4.
" 9. ACTINOCYCLUS EHRENBERGII, Ralfs.; St. Monica. V. H., 125, f. 1.
" 10. ACTINOPTYCHUS LAEVIGATUS, Grun.; Depos. Monterey, Cal. V. H., 122, f. 7.
" 11. " TRIGONUS, A. S., Schm. At., 1, f. 24. Camp. Bay.
" 12. ACTINOCYCLUS EHRENBERGII, var. intermedia, Grun.; V. H., 124, f. 5. Depos. St. Monica.
" 13. " RALFSII, var. Monica, Grun., Cal. V. H., 124, f. 3. Punctae of border are quincunx.
" 14. " ALIENUS, var. Californica, Grun.; St. Monica. V. H., 125, f. 10.
" 15. ACTINOPTYCHUS VULGARIS, var. Virginica, Grun.; Richmond, Va. V. H., 121, f. 7.
" 16. " GLABRATUS, var. Angelorum, Grun.; St. Monica, Cal. V. H., 120, f. 9.
" 17. " GLABRATUS, var. Montereyi, Grun.; Depos. Monterey. V. H., 120, f. 7.
" 18. " SPINIFERUS, (vulgaris var.) Grun.; St. Monica, V. H., 121, f. 5, 6.
" 19. " GLABRATUS, var. incisa, Grun.; St. Monica. V. H., 120, f. 8.
" 20. " VULGARIS, var. Monicae, Grun.; Cal. V. H., 121, f. 9.

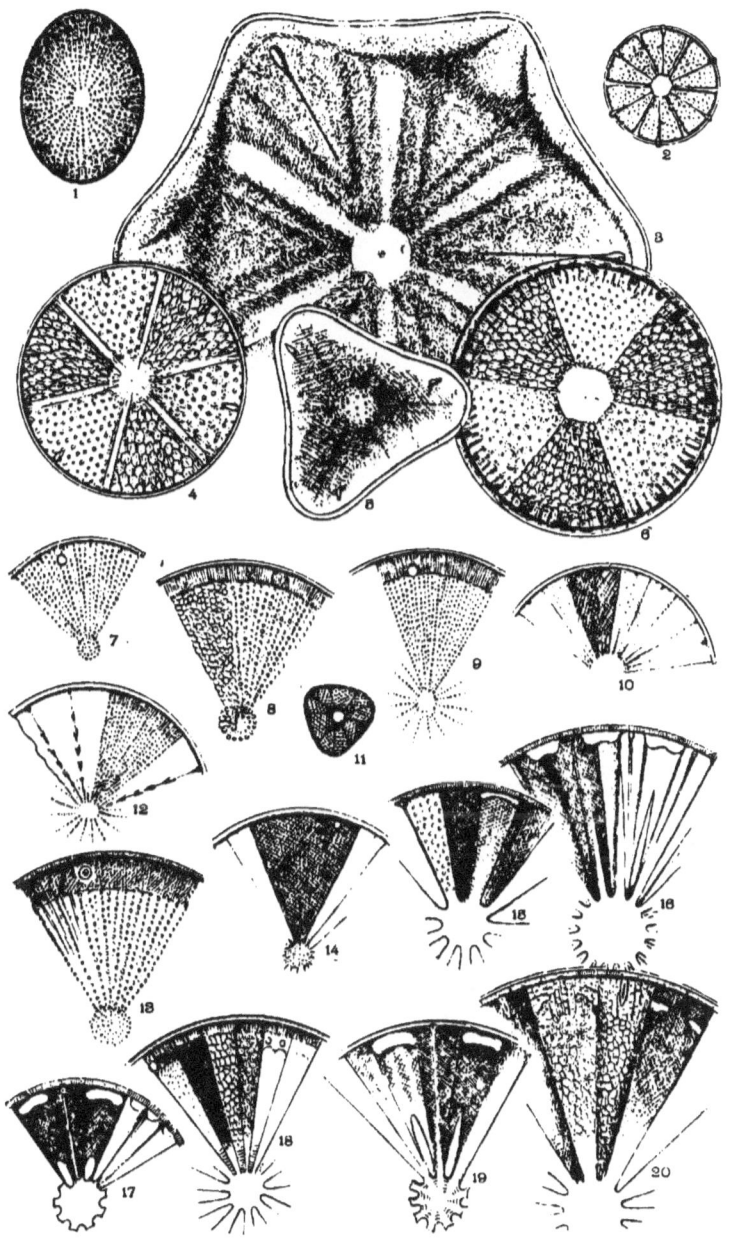

PLATE LXXXVI.

Figures magnified 500 diameters.

Fig. 1. COSCINODISCUS BIANGULATUS, A. S., Schm. At., 63, f. 13. Nottingham, Md.
" 2. " ARMATUS, Grev.; Schm. At., 67, f. 4. Richmond, Va.
" 3. CRASPIDODISCUS COSCINODISCUS, Grun.; Schm. At., 66, f. 4. Monterey.
" 4. " OCULUS IRIDIS, Ehrb.; Schm. At., 63, f. 3-7.
" 5. CRASPIDODISCUS RHOMBICUS, Grun.; Schm. At., 66, f. 13. Monterey. "More likely a valve of Biddulphia than a Craspidodiscus," H. L. S.
" 6. COSCINODISCUS HETEROPORUS, var. Grun.; Monterey. Schm. At., 61, f. 5, 4.
" 7. CRASPIDODISCUS ELEGANS, Ehrb.; Schm. At., 66, f. 1. Nottingham, Md.
" 8. " COSCINODISCUS, Ehrb.; Schm. At., 66, f. 3. Nottingham, Md.
" 9. COSCINODISCUS APICULATUS, var. Ehrb.; Schm. At., 64, f. 7-9. Richmond, Va.

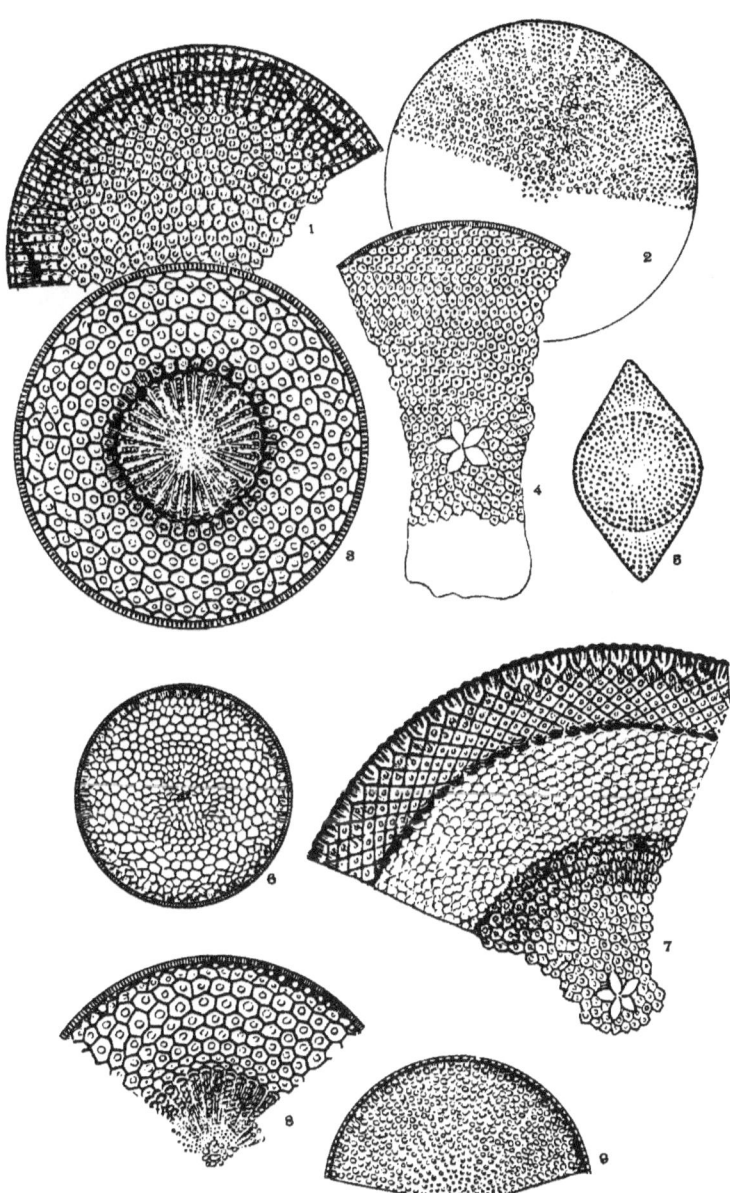

PLATE LXXXVII.

Figures magnified 500 diameters.

Fig. 1. COSCINODISCUS BOREALIS, Bail.; Sill. J., July, 1856, p. 3; Schm. At., 63, f. 11.
" 2. " CRASSUS, Bail.; Sill, J., 1856, p. 4; Schm. At., 61, f. 19.
" 3. " DIORAMA, Schm. At., 64, f. 2, = C. gigas, Ehrb.
" 4. " ASTEROMPHALUS, Ehrb.; with large cells. St. Monica. Schm. At., 63, f. 12.
" 5. " ROBUSTUS, Grev.; St. Monica. Schm. At., 62, f. 6.
" 6. " CENTRALIS, Ehrb.; Cal. Schm. At., 63, f. 1; 60, f. 12.
" 7. " SUSPECTUS, Janisch. Cal. Schm. At., 59, f. 2. The meshes only slightly smaller at margin than in center.
" 8. " OMPHALANTHUS, Ehrb. Margin arched. Monterey. Schm. At., 63, f. 2.
" 9. " PERFORATUS, Ehrb.; Richmond, Va. Series of puncta radiating; Schm. At., 64, f. 12.
" 10. " LINEATUS, Ehrb.; Schm. At., 59, f. 26.
" 11. " ROBUSTUS, Grev.; St. Monica. Smaller form of Fig. 5. Schm. At., 62, f. 5.

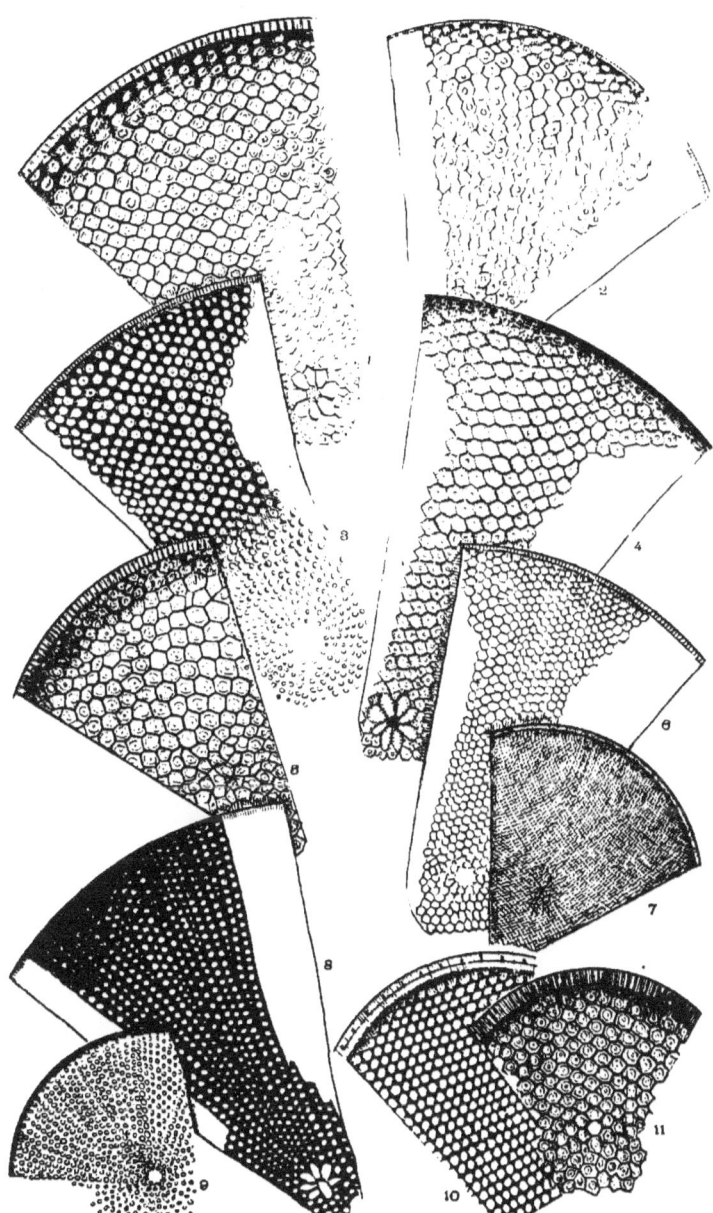

PLATE LXXXVIII.

Figures magnified 500 diameters.

Fig. 1. AULACODISCUS CRUX, Ehrb.; Richmond, Va. Areolation (cellules) *hexagonal*. Schm. At., 124, f. 1.
" 2. " KITTONII, Arnott. San Francisco. Schm. At., 36, f. 3 = A. Ehrenbergii, Janisch. Monterey.
" 3. " THUMII, A. S. The cellules are not simply round and smooth; the diatom is composed of two plates, the upper one is perforated and the other dotted all over with granules. St. Monica. Schm. At., 108, f. 8.
" 4. " OREGONUS, Bailey. Areolation more or less hexagonal.
" 5. " CALIFORNICA, Schm. At., 34, f. 4, 5.
" 6. " OREGONUS, smaller var. of Fig. 4.
" 7. " CIRCUMDATUS, A. S., Schm. At. 35, f. 5. California.
" 8. " PROBABILIS, A. S., Schm. At., 36, f. 15. Monterey. Cellules arranged more or less in circles around the middle. Nearly akin to A. Brownii.
" 9. " SOLLITIANUS. Nor.; Schm. At., 33, f. 10. Nottingham, Md. Surface dotted with very small prickles.
" 10. " BROWNEI, Norm. Monterey. Schm. At., 105, f. 6; hardly separable from *A. probabilis*, Fig. 8.
" 11. " KINKERI, A. S., Schm. At., 106, f. 45. St. Monica. Possibly a var. of A. crux.

Plate LXXXVIII.

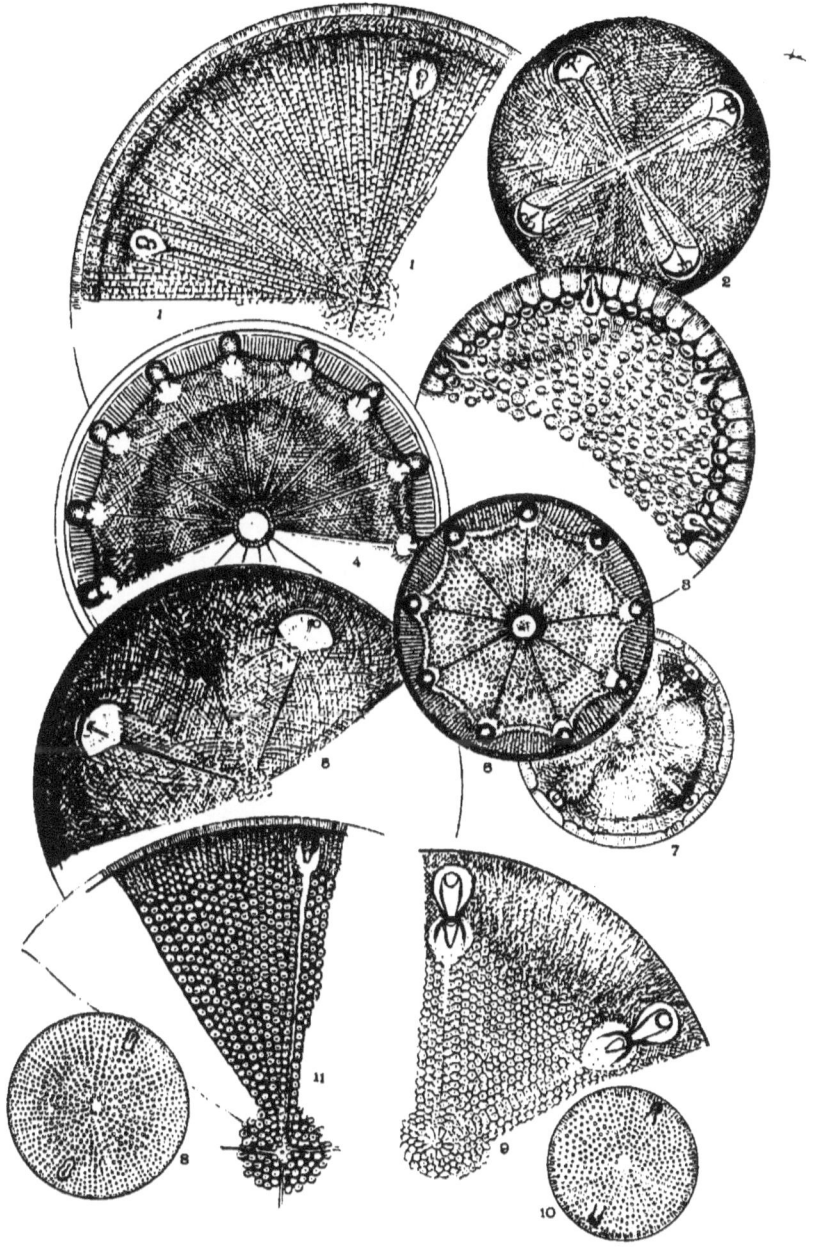

PLATE LXXXIX.

Figures magnified 500 diameters.

Fig. 1. AULISCUS SCULPTUS, Ralfs.; Prit., p. 845, 6, f. 3; Schm. At., 30, f. 8; V. 11., 107, f. 1, 2.
" 2. " MACRAEANUS, Grev.; T. M. S., 1863, p. 51, 3, f. 18; Schm. At., 31, f. 5.
" 3. " RADIATUS, Bail.; T. M. S., 1863, p. 49, 3, f. 14.
" 4. " PERUVIANUS, Grev.; T. M. S., 1862, p. 25, 2, f. 6; Schm. At., 32, f. 29. Cal.
" 5. " SCHMIDTII, Grun.; Camp. Bay. Schm. At., 30, f. 7.
" 6. " CAELATUS, Bail., N. Sp., p. 6, f. 3, 4; Schm. At., 32, f. 14, 15. Mont.
" 7. " RACEMOSUS, Ralf.; Schm. At., 30, f. 12, 13.
Figs. 8, (11). " CONFLUENS, Grun.; Schm. At., 31, f. 16.
Fig. 9. " CAELATUS, var. Bailey; Camp. Bay. Schm. At., 32, f. 17.
" 10. " " var. latecostata, Schm. At., 32, f. 18. Fig. too round, ought to be more oval.
" 11. " CONFLUENS, *vide* Fig. 8.
Figs. 12, 13. " MIRABILIS, a small and larger form, after Schm. At., 89, f. 10-13.
Fig. 14. " STOECKHARDTII, Janisch; Schm. At., 30, f. 11. Same as A. racemosus, *vide* Fig. 7.

Plate LXXXIX.

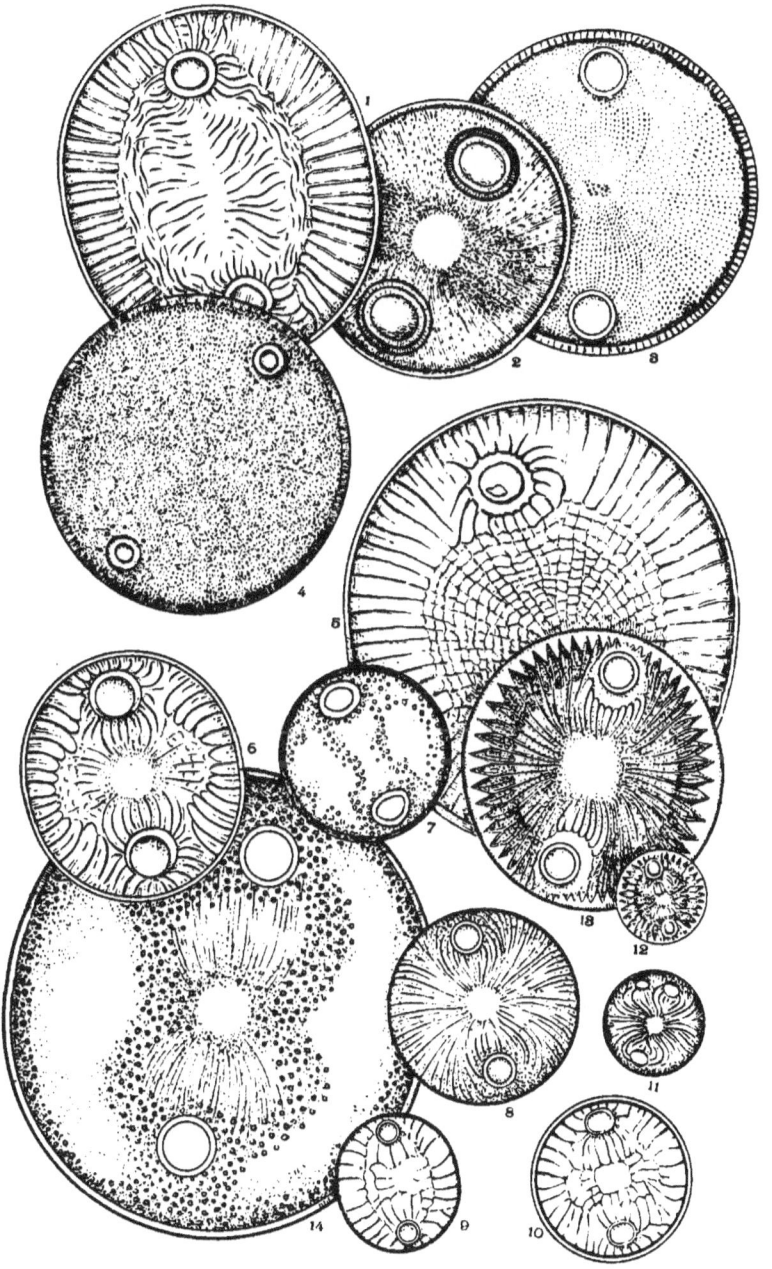

PLATE XC.

Figures magnified 500 diameters.

Fig. 1. COSCINODISCUS EXCAVATUS, Grev.; Prit., p. 829, S. f. 26; Schm. At., 65, f. 1. North Carolina, New Jersey, etc.
" 2. " SECERNENDUS, (Secernendus), A. S., Schm. At., 114, f. 1. Maryland.
" 3. " RADIATUS, Ehrb. Abh., 1839, p. 148; Mik., 39, 3. f. 17; 35 A, 17, f. 6; Sill. J., 1842, p. 95, 2, f. 14; S. B. D., 3, f. 57; Schm. At., 60, f. 5, 9, 6, 7, 10.
" 4. " WOODWARDIA, Eulens. Doubtfully this species and doubtfully N. Amer., Schm. At., 61, f. 2; 65, f. 2.
" 5. " FLORIDULUS, A. S., Schm. At., 113, f. 16. St. Monica. Rays not straight, but more or less curved, in concentric and excentric series of cellules. Hardly separable from C. Radiatus.
" 6. " LEPTOPUS, Grun. C. lineatus, var. V. H., 131, f. 5. California.
Figs. 7, 8. " VELATUS, Schm. At., 62, f. 10-12, places these doubtfully to C. velatus. Virginia. May be varieties of C. Marginatus.
Fig. 9. " IMPRESSUS, Grun. St. Monica. Center of cell somewhat depressed.
" 10. " EXCENTRICUS, Ehrb. California. Schm. At., 58, f. 49. Figure rather finely lined. Might be more in hexagonal cells.
" 11. " SUBVELATUS, so named provisionally by A. S.; Schm. At., 65, f. 9. Monterey.
" 12. " HETEROPORUS, var. Grun., Monterey; Schm. At., 61, f. 4, 5.

Plate XC.

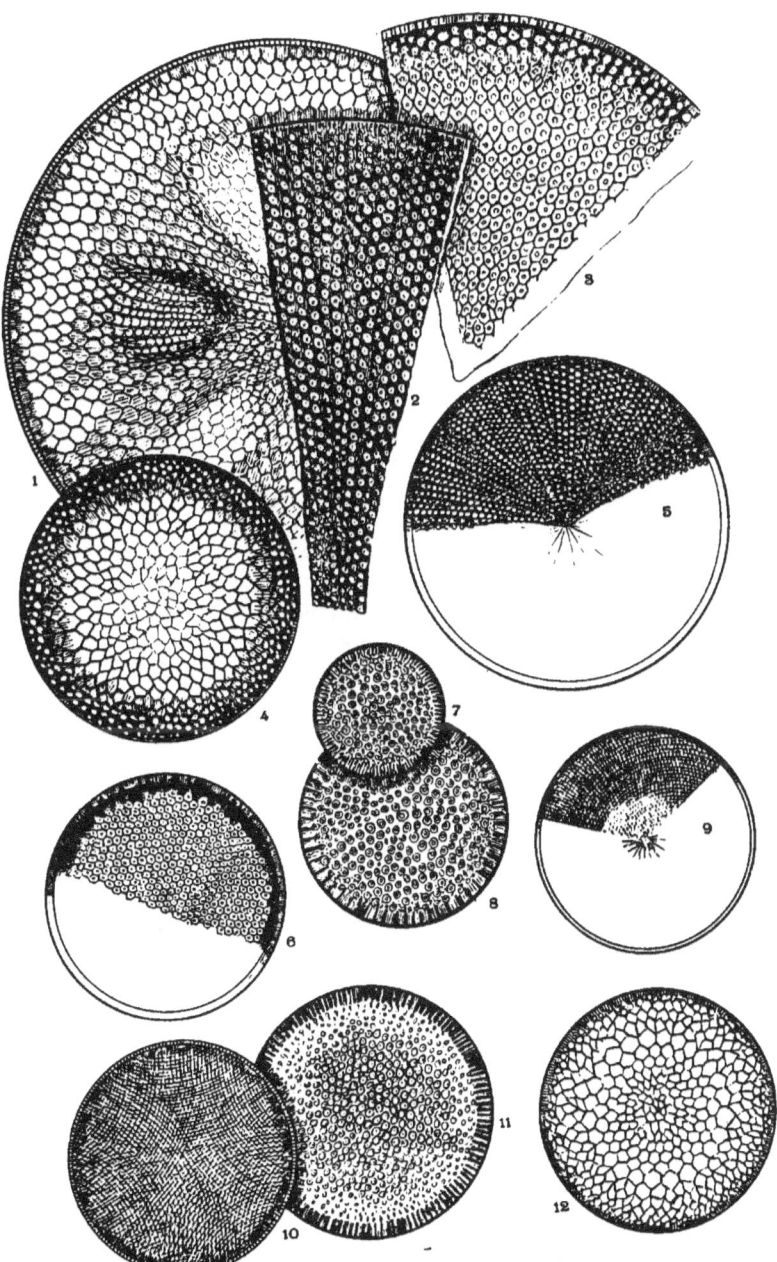

PLATE XCI.

Figures magnified 500 diameters.

Fig. 1.   ARACHNOIDISCUS EHRENBERGII, Bail., Wilkes Ex., p. 174, 9, f. 9; S. B. D., 31, f. 256; Schm. At., 68, f. 1; 73, f. 1.
"   2.   "   EHRENBERGII, var. Montereyana, Grun.; Schm. At., 68, f. 2.
"   3.   "   EHRENBERGII, var. Californica, A. S., Schm. At., 73, f. 1.
"   4.   "   EHRENBERGII, var. Californica, Weissfl.; Schm. At., 68, f. 3.
"   5.   "   GREVILLEANUS, Hardm.; Schm. At., 68, f. 5. Monterey.
"   6.   "   INDICUS, Ehrb., var.; Schm. At., 73, f. 2; Mik., 36, f. 24.
Figs. 7, (10, 11).   "   INDICUS, vars.; Schm. At., 68, f. 6, 9, 10. Center less filled; hardly a rosette as shown in Figs. 10 and 11.
"   8, 9.   "   ORNATUS, var. Montereyana; A. S., Schm. At. 73, f. 7, 9. Rosette does not show blank center large enough in either figure; it is quite apparent on the valve, H. L. S. MSS.

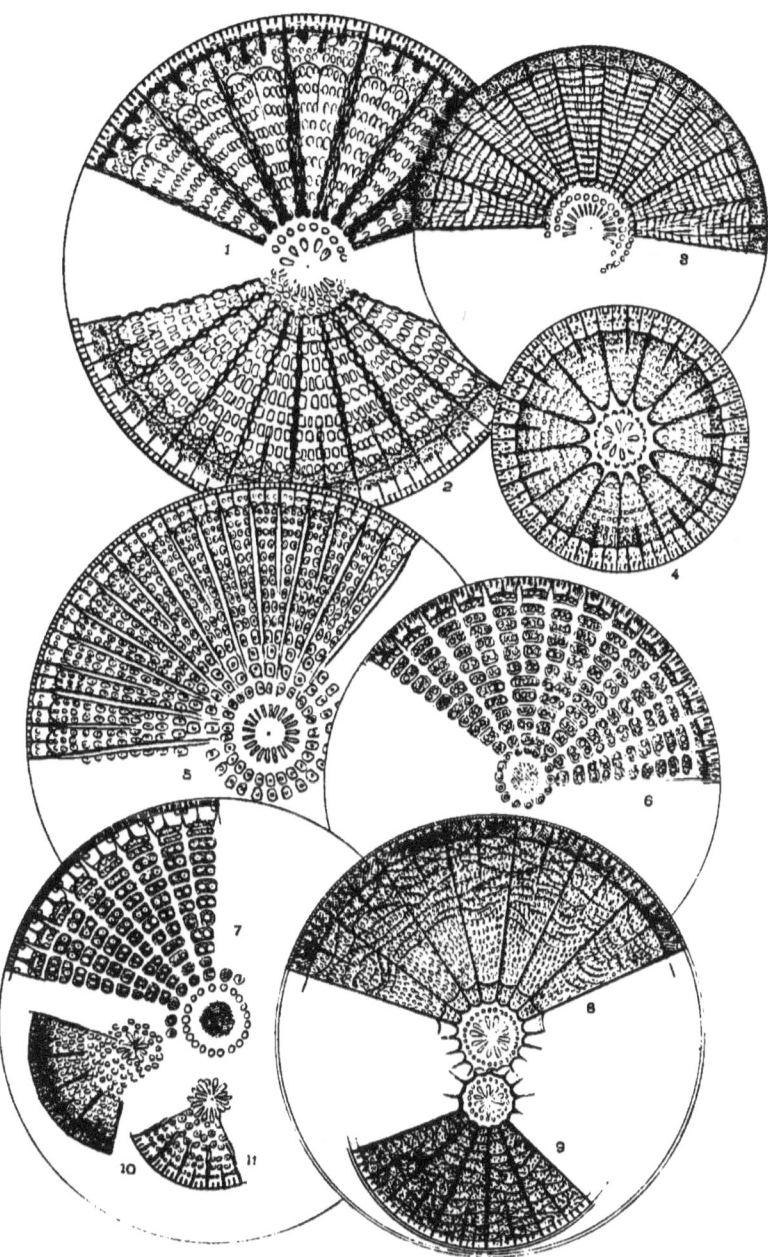

PLATE XCII.

Figures magnified 500 diameters.

Fig. 1.   ACTINOPTYCHUS HELIOPELTA, Grunow, lower plate; Schm. At., 100, f. 2; V. H., 123, f. 3. Maryland. Edge-wise the disc is undulating.
" 2.   "   PFITZERI, Gründ.; Schm. At., 29, f. 1. Cal.
Figs. 3, (7).   "   GRUENDLERI, A. S.; St. Monica and Monterey. Schm. At., 1, f. 22; 100, f. 4.
" 4, 5, 6.   "   UNDULATUS, Ehrb.; Monterey. Schm. At., 1, f. 6; V. H., 122, f. 3.
Fig. 8.   "   EXCELLENS, Schum., 1867, p. 64; Schm. At., 1, f. 14.
" 9.   "   SPLENDENS, Schad.; V. H., 119, f. 1, 2, 4.
" 10.   "   " var. crucifera; Cal., St. Monica. V. H., 120, f. 2.
" 11.   "   SPLENDENS, var. Californica; Schm. At., 1, f. 26; V. H., 120, f. 1.
" 12.   "   SPLENDENS, var. Halionyx; V. H., 119, f. 3.
" 13.   "   TRIANGULUS, A. S.; Piscataway, Me. Probably Tric. Marylandicum; Schm. At., 1, f. 26.

Plate XCII.

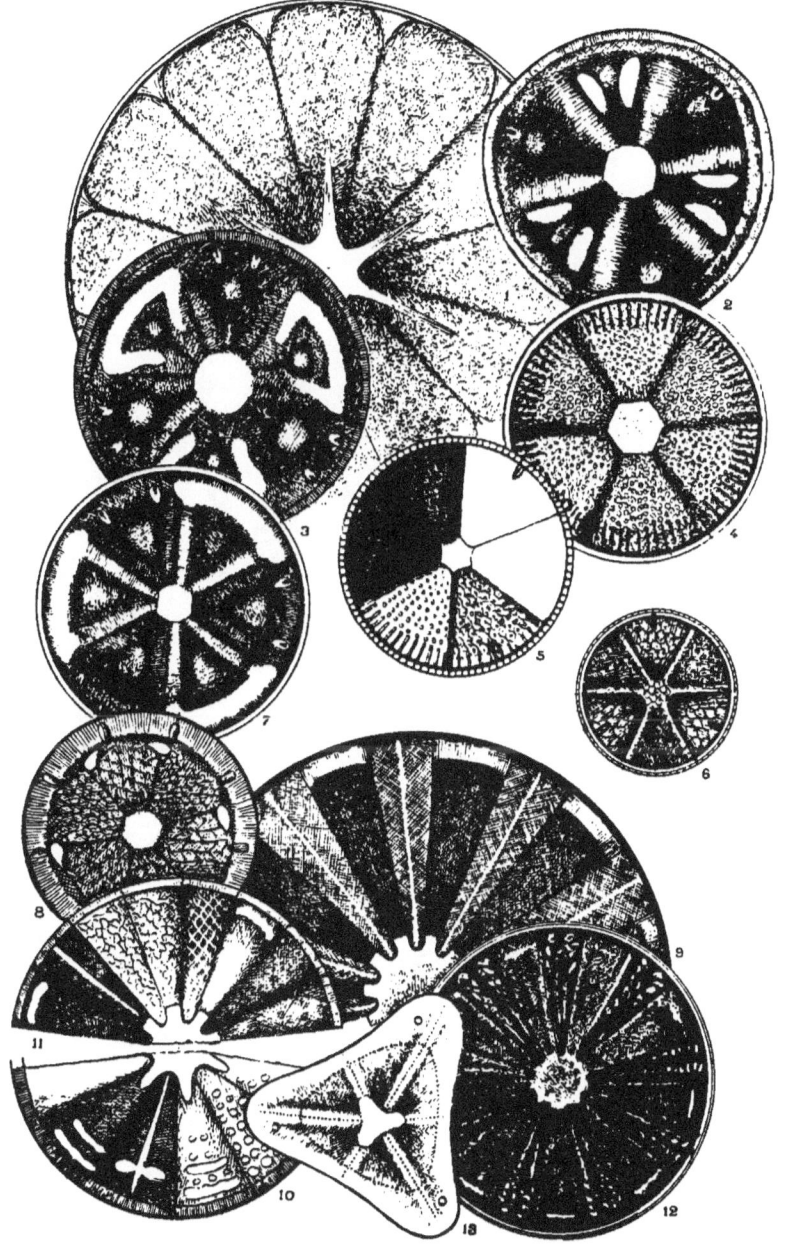

PLATE XCIII.

Figures magnified 500 diameters.

Fig. 1.   ASTEROLAMPRA MARYLANDICA, Ehrb.
Figs. 4, 5, 6.   "   "   varieties, M. J., 1860, p. 94, 5, f. 3; T. M. S., 1860, p. 47, 2, f. 13, 14; p. 108, 3, f. 1-4; 1862, p. 44, 7, f. 1-3.
Fig. 2.   "   VARIABILIS, Grev.; T. M. S., 1860, p. 111, 3, f. 6, 8.
"   3.   "   BROOKII, Bail.; Sill. J., July, 1856, p. 2, 1, f. 1; Prit., 837, 5, f. 79; Schm. At., 38, f. 9, 21, 23.
"   7.   "   BREBISSONIANA, Grev.; T. M. S., 1860, p. 114. 3, f. 9. Monterey.
Figs. 8, 9.   "   DARWINII, Grev.; T. M. S., 1860, p. 116, 4, f. 12, 13. Monterey.
Fig. 10.   "   ROTULA, Grev.; T. M. S., 1860, p. 120, 3, f. 5. Monterey.
" 11.   "   ELEGANS, Grev.; M. J., 1859, p. 161, 1, f, 6; T. M. S., 1860, p. 118, 4, f. 16.

Plate XCIII

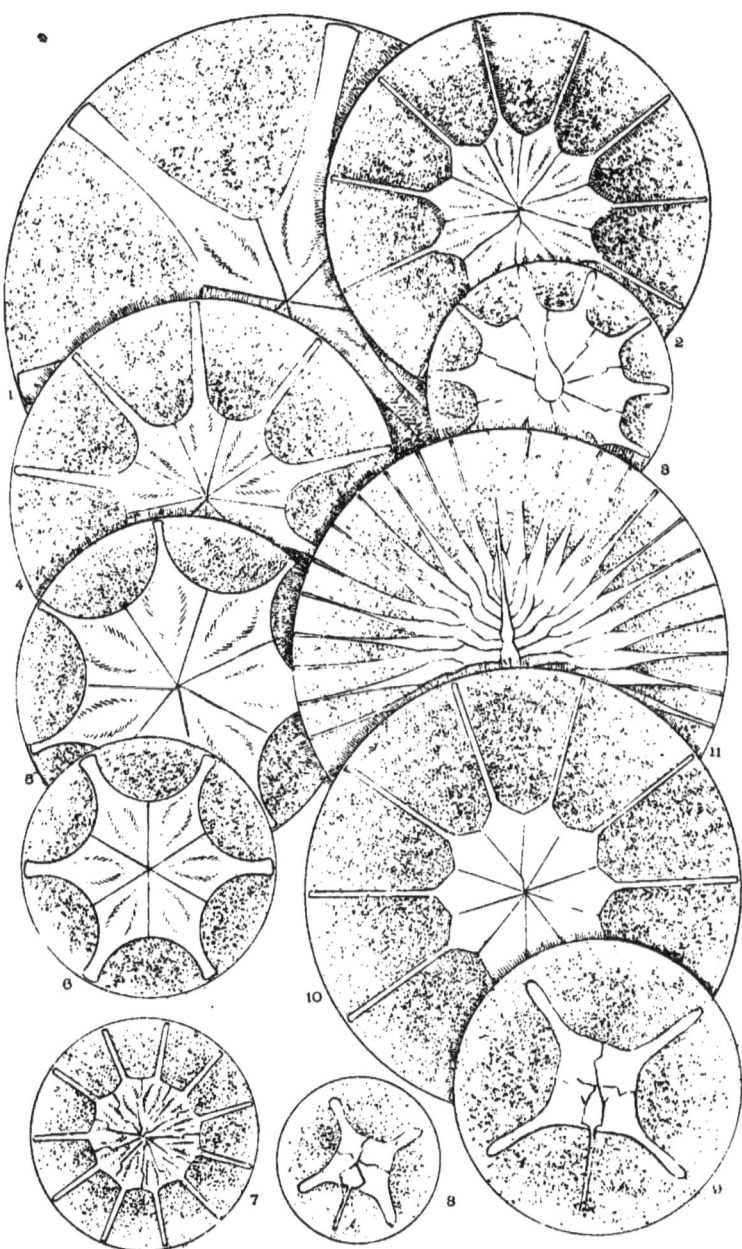

PLATE XCIV.

Figures magnified 500 diameters.

Fig. 1. COSCINODISCUS ELEGANS, Grev., T. M. S., 1866, p. 3, 1, f. 6; Schm. At., 58, f. 7. Piscataway, Mo.
" 2. " SYMMETRICUS, Grev., T. M. S., 1861, p. 68, 8, f. 2; Schm. At., 57, f. 25-27. Areolation, hexagonal.
" 3. " CONCAVUS, Ehrb. Mik., 21, f. 4; 18, f. 38; Schm. At., 59, f. 16. Monterey.
Figs. 4, 5. " COCCONEIFORMIS, A. S. Monterey. Schm. At., 58, f. 25, 26.
Fig. 6. " RADIOLATUS, Ehrb. Mik., 39, 3, f. 18; 22, f. 4; 18, f. 36; K. B., 29, f. 91; Schm. At., 60, f. 11. Cellules rather too irregular and too radiate, H. L. S.
" 7. " NODULIFER, Jan., or better C. RADIOLUS, var. nodulifer. Schm. At., 59, f. 20. Camp. Bay, Cal.
Figs. 8, 9. " SUBTILIS, Ehrb. Richmond, Virginia. 57, f. 11, and 15, 16 vars. The latter with areolation somewhat larger.
Fig. 10. " NITIDULUS, Grun. Camp. Bay. Schm. At., 58, f. 20.
" 11. " RADIOLUS, Grun. Monterey. V. H., 132, f. 7.
" 12. " CENTRALIS, Grun.; not Ehrenberg's figure, which is very indefinite. Schm. At., 60, f. 12.
" 13. " APICULATUS, (= perforatus) Ehrb.; Schm. At., 64, f. 5, 6. Richmond, Va.
" 14. ACTINOCYCLUS CURVATULUS, Janisch; Schm. At., 57, f. 31.
" 15. COSCINODISCUS, NITIDULUS, Grun. Camp. Bay. Schm. At., 58, f. 21.
" 16. " MARGINATO-LINEATUS, A. S., Schm. At., 59, f. 33. Camp. Bank.
" 17. " ROTULA, Grunow; Schm. At., 57, f. 6, 7; Camp. Bay.
" 18. " LEWISIANUS, Grev.; Schm. At., 66, f. 12; T. M. S., 1866, p. 78, 8, f. 8-10. Rappahannock, Va. According to Greville the granules have a distinct radiate character.
" 19. " STELLIGER, Grun. Camp. Bay. Schm. At., 58, f. 10.
" 20. " MINOR, Ehrb. Abh., 1839, p. 147, 3, f. 2; K. B., 1, f. 12, 13; Schm. At., 58, f. 39, 40.
Figs. 21, 26. " MARGINULATUS, var. curvato-striata; V. H., 94, f. 32.
" 22. " NITIDULUS, Greg. Camp. Bay. Schm. At., 58, f. 17.
" 23. " " var., Schm. At., 58, f. 19.
" 24. " NOTTINGHAMENSIS, Grun. Maryland. V. H., 129, f. 2.
" 25. " MARGINULATUS, var. Campeachiana, Grun.; V. H., 94, f. 33.
" 27. " MARGINULATUS, Grun.; var. Stellulifera, Grun. Camp. Bay. V. H., 94, f. 34.

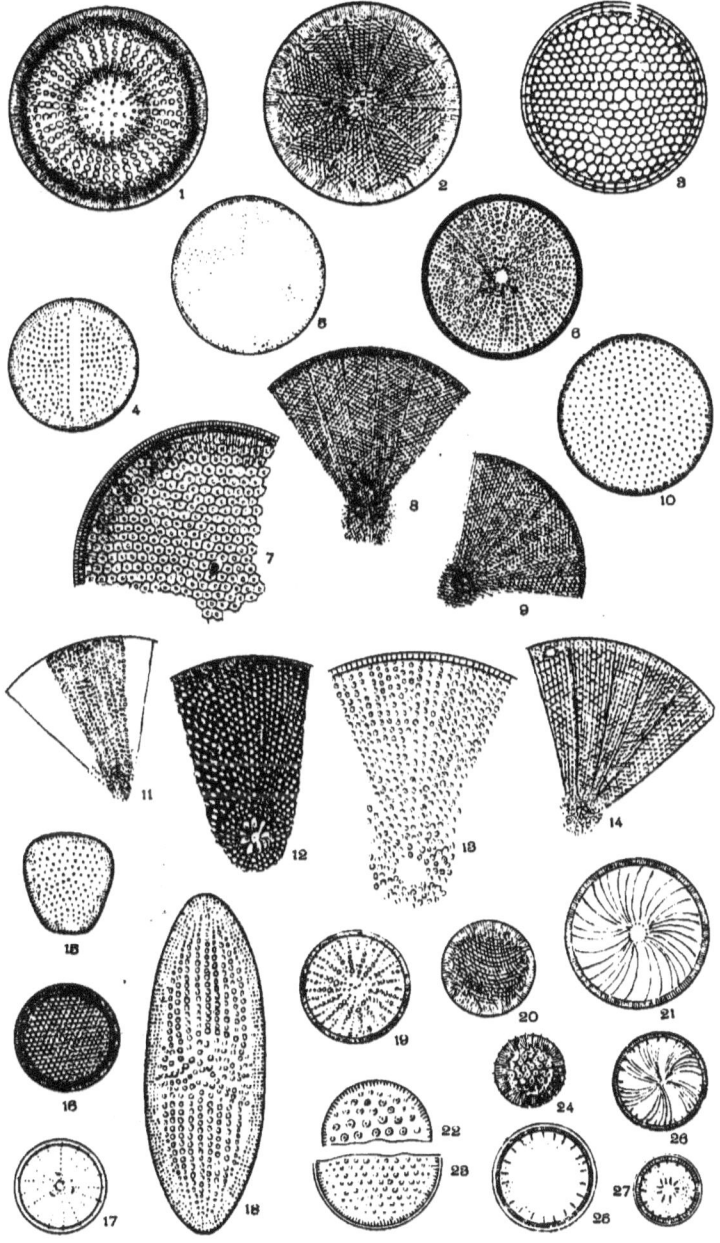

PLATE XCV.

Figures magnified 500 diameters.

Figs. 1–4.   BIDDULPHIA TUOMEYII, Bail. Camp. Bay. Schm. At., 119, f. 1–8; T. M. S., 1859, p. 8, 1, f. 1, 2; V. H., 98, f. 2, 3. The granules of central part of Fig. 4, arranged radially.
" 5, 6.   "   SUBORBICULARIS, Grun.; V. H., 100, f. 15, 16. Nottingham, Maryland.
" 7, (10).   "   RHOMBUS, (Ehrb.), W. Smith, (*Zygoceros Ehrb., Denticella Ehrb., Odontella, Kuetz*).
" 8, 11–13.   "   ROPERIANA, Grev., V. H., 99, f. 4–6; M. J., 1859, p. 163, 7, f. 11–13. The granules are quincunx, on all the hoops.
Fig. 9.   "   MOBILENSIS, (Bailey), Grun. *Zygoceros, Bailey, Biddulphia, Baileyi, W. Smith.* Sill. J., 1845, p. 336, 4, f. 2, 4; V. H., 101, f. 4–6.

Plate XCV.

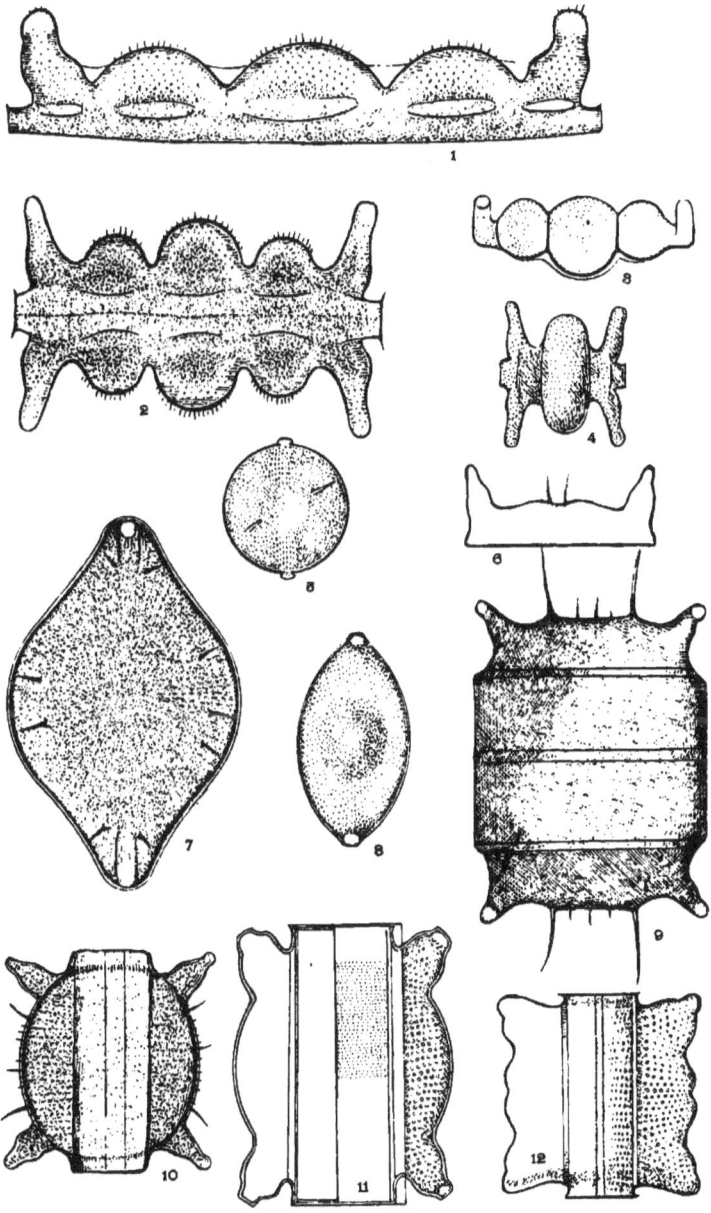

PLATE XCVI.

Figures magnified 500 diameters.

Figs. 1, 2. BIDDULPHIA PULCHELLA, Gray. A front and a lateral view. (*Denticella Biddulphia*, Ehrb.; *Biddulphia Locularis*, Kütz.) Camp. Bank. V. H., 97, f. 1; Schm. At., 118, f. 26-29.
Fig. 3. " PULCHELLA, Gray.; a lateral view.
Figs. 4, 5. " EDWARDSII, (front and lateral view); V. H., 100, f. 9, 10. Only a prickly form of B. Roperiana, Grev.; M. J., 1859, p. 163, 7, f. 11-13.
Fig. 6. " LONGICRUCIS, Grev.; M. J., 1859, p. 163, 7. f. 10.
Figs. 7, 8. " SETICULOSA, Grun.; Virginia. V. H., 101, f. 7, 8.
" 9, 10, 11. " AURITA, Lyngb.; V. H., 98, f. 4-9; S. B. D., 45, f. 319.
Fig. 12. " LONGISPINA, Grun.; St. Monica. V. H., 102, f. 6.
Figs. 13, 14. " RETICULATA, V. H., 102, f. 1, 2; Roper, T. M. S., 1859, p. 14, 2, f. 13, 17; Schm. At., 78, f. 21-23; 84, f. 9-16; 85, f. 8.

Plate XCVI.

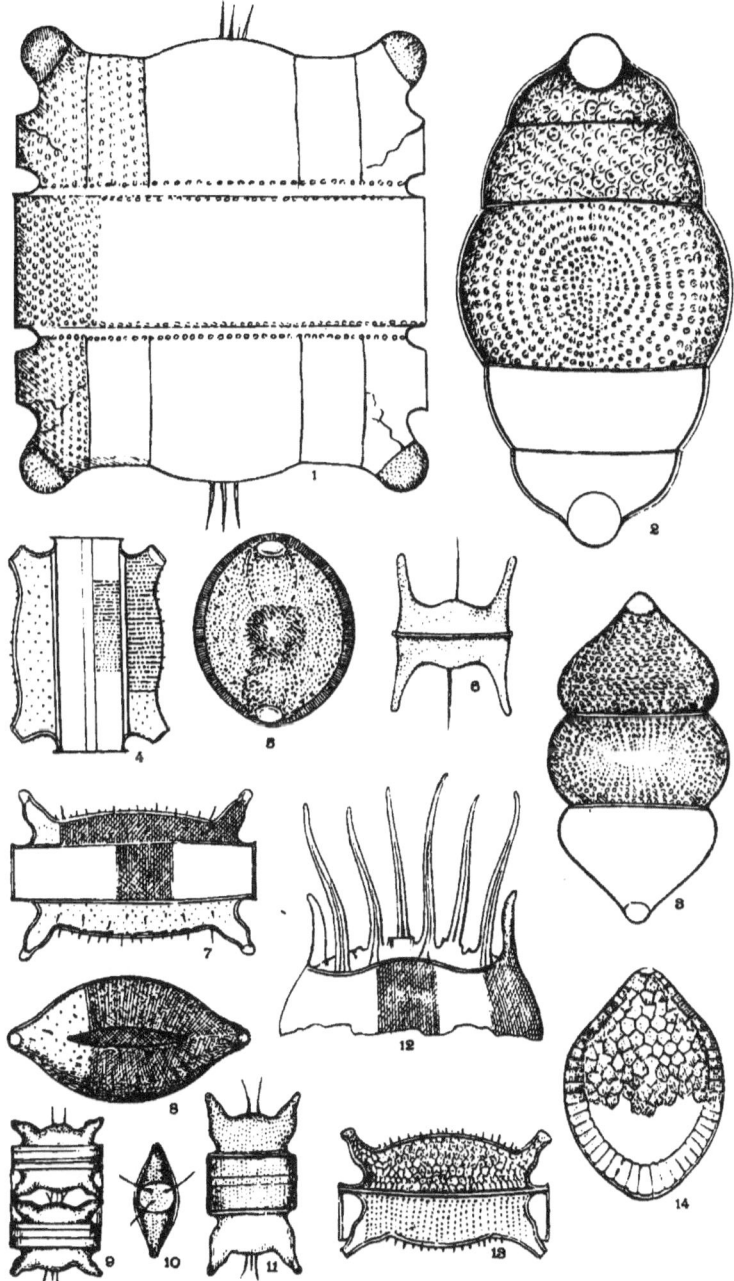

PLATE XCVII.

Figures magnified 500 diameters

Figs. 1, 2, (5, 11). BIDDULPHIA LAEVIS, Ehrb. Mik., 33, 15, f. 16; T. M. S., 1859, p. 18, 2, f. 24–26, Sill. J., 1842, 2, f. 8. A few of the many varieties widely distributed; frequent along the sea board, but also inland, salt marshes of Kansas, Utah, etc.

" 3, 4. " TURGIDUS, W. S.; *Ceratanlus: Odontella, Ehrb.* Spines of the valves wavy, not rigid. V. H., 104, f. 1, 2; S. B. D., 62, f. 384; T. M. S., 1859, p. 17, 2, f. 23.

" 5, (11). " POLYMORPHUS (Cerataulus), V. H., 104, f. 4, 3. Odontella, polymorphus, Ehrb. Biddulphia laevis, W. S.

Fig. 6. " JOHNSONIANUS, Var. Grev.; Schm. At., 115, f. 15.

" 7. " TURGIDUS, Ehrb., Schm. At., 116, f. 2. *Vide* Figs. 3, 4, above.

Figs. 8, (10). " OVALIS, *Cerataulus*, A. S., Schm. At., 115, f. 7, and 115, f. 5. This diatom is not only finely dotted but areolated like a Coscinodiscus.

Fig. 9. " CALIFORNICUS (*Cerataulus*), A. S.

" 11. " *Vide* Fig. 5.

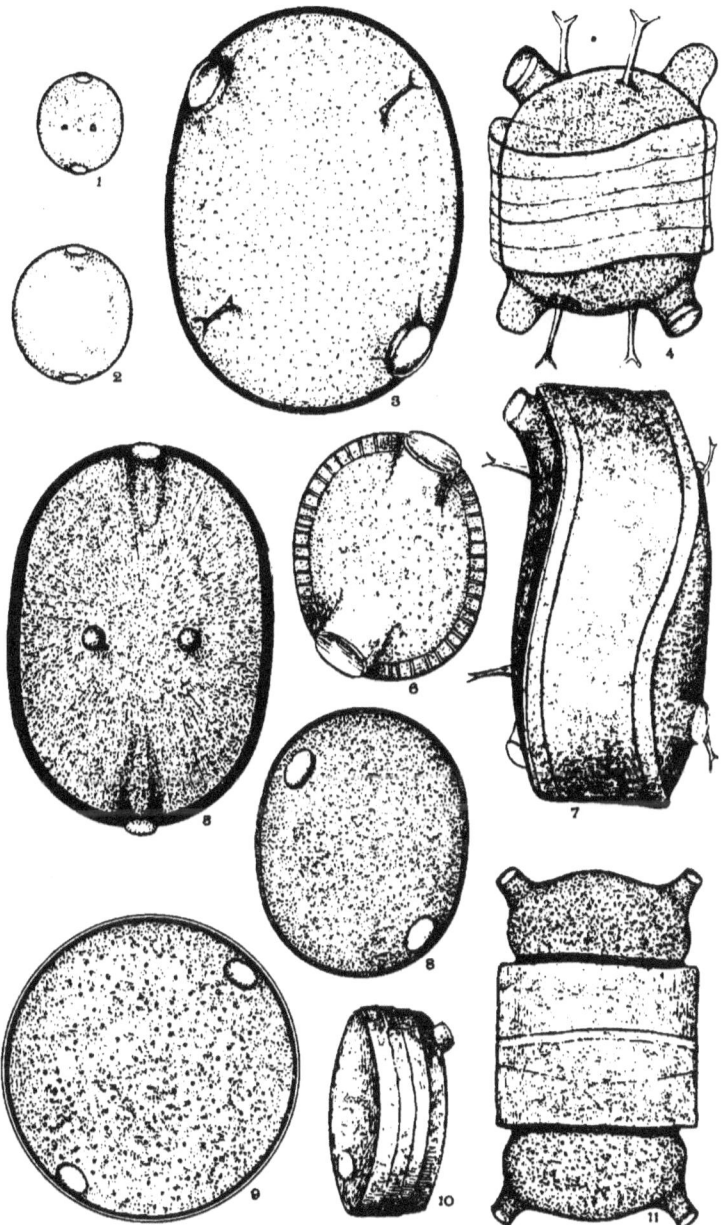

PLATE XCVIII.

Figures magnified 500 diameters.

Figs. 1, 2, 3.   BIDDULPHIA BAILEYII, W. S.; S. B. D., 45, f. 322; 62, f. 322; T. M. S., 1859, p. 12, 1, f. 59; V. H., 101, f. 4, 5. Nearly akin to B. mobiliensis.

Fig. 4.   "   WOOLMANII, Kain and Schultze; Bull. Torrey Bot. Club, March, '89. Atlantic City. Artesian well.

Figs. 5, 6.   "   DECIPIENS, Grun.; V. H., 100, f. 3, 4. Atlantic City.

" 7, 8, 9.   "   ELEGANTULA, var. of Tuomeyii, Grev.; Schm. At., 119, f. 9. 10, 11.

Fig. 10.   "   COOKIANA, Kain and Schultze. Artesian well, Atlantic City; Bull. Torrey Bot. Club, March, 1889. Var. of B. Ornata, Cast., 1887, p. 105, 23, f. 9.

Figs. 11, 12.   "   TENUIS, Bail., B. J. N. H., vol. vii, 1 p. f. 25. Probably identical with B. Baileyi. *Vide* Figs. 1–3.

" 13, 14.   "   TRINACRIA, Bail.; B. J. N. H., p. 338, 1, f. 34, 35. A a variety of Bid. Baileyi.

" 15, 16, 17.   "   PORPEIA QUADRATA, Grev., T. M. S., 1863, p. 65, 6, f. 20; V. H., 92, bis f. 15.

" 18, 19.   "   BRITTONIANA, Kain and Schultze, Bull. Torrey Bot. Club, Aug., 1889, p. 208, 92, f. 1, a. b, c. Artesian well, Atlantic City.

Plate XCVIII.

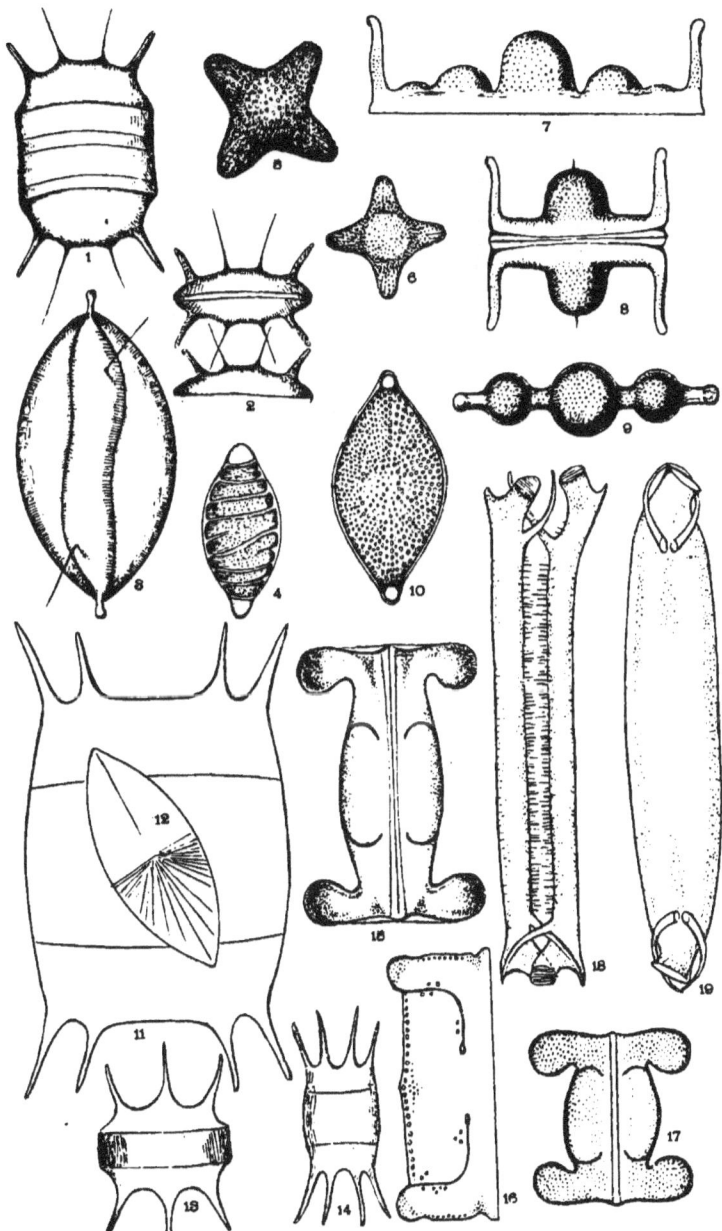

PLATE XCIX.

Figures magnified 500 diameters.

Figs. 1, 2. TRICERATIUM FAVUS, Ehrb. Mik., 19, f. 17; K. B., 18, f. 11; S. B. D., 5, f. 44; 30, f. 44; V. II., 107, f. 1-4; Schm. At., 82, f. 13, 14.
" 3, 4. " FIMBRIATUM, Wall., M. J., 1858, p. 247, 12, f. 4-9; Lens, I, p. 100; II, p. 105; Schm. At., 82, f. 6, 7. Florida.
" 5, 6. " FIMBRIATUM, Grun.; V. II., 109, f. 1, 3; Schm. At., 99, f. 10-13; forma major, and forma pusilla. St. Monica.
Fig. 7. " CALIFORNICUM, Grun. St. Monica. V. II., 108, f. 11.
Figs. 8, 9. " INELEGANS, Grev. St. Monica. Fig. 8, var. micropora, Grun.; Fig. 9, var. Yucatensis, Grun.; V. II., 90, f. 3, 4. (*Odontella*).
Fig. 10. " CONSIMILE, Grun., (*Odontella*). St. Monica. V. II., 108, f. 2.
" 11. " IRREGULARE, var. Grun. Petersburg, Va. V. II., 111, f. 10.
" 12. " INELEGANS, Biddulphia, var. aracopora. St. Monica. V. II., 110, f. 2.

Plate XCIX.

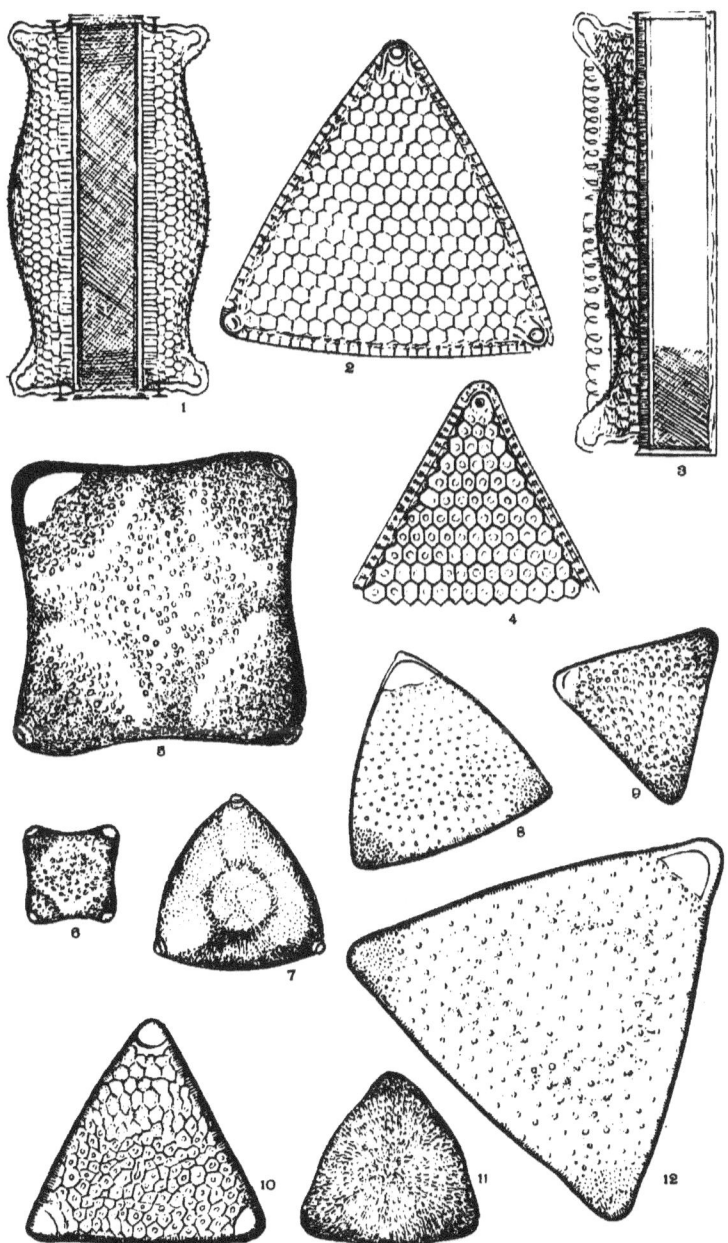

PLATE C.

Figures magnified 500 diameters.

Fig. 1.	TRICERATIUM	TABELLARIUM, Brightw.,var. deplosticta, Grun.; Camp. Bay. Schm. At., 77, f. 1.
" 2.	"	VENULOSUM, Grev.; Gulf of Mexico. Schm. At., 77, f. 8.
" 3.	"	BULLOSUM, Witt.; Camp. Bay; Schm. At., 78, f. 33.
Figs. 1, (7).	"	VENULOSUM, Grev.; Camp. Bay. Schm. At., 77, f. 9, 7.
Fig. 5.	"	TRIPARTITUM. Grun.; Camp. Bay. V. II., 110, f. 8. (Biddulphia.)
" 6.	"	HETEROPORUM, Grun.; St. Monica. (Biddulphia), V. II., 112, f. 2.
" 8.	"	INTERPUNCTATUM, Grun.; Nottingham, Md. Schm. At., 76, f. 7. A very variable species often apparently closely related to T. elegans, Grev.
Figs. 9, 10.	"	TRISULCUM, Bailey; Camp. Bay. Schm. At., 78, f. 5 8, 112, f. 11-18.
" 11, 12.	"	PARALLELUM, Grev.; Monterey. Schm. At., 75, f. 11, 12.
" 13, 14.	"	HARRISONIANUM, Nor. and Grev.; Camp. Bay. Schm. At., 75, f. 14-16; T. M. S., 1861, p. 76, 9, f. 9.
Fig. 15.	"	PARALLELUM, forma trigona, Grev.; Schm. At., 76, f. 14.
Figs. 16, 18.	"	ALTERNANS, Ehrb.; Schm. At., 78, t. 10, 12, (9-20). Camp. Bay, etc.
Fig. 17.	"	SECERNENDUM, A. S., Schm. At., 76, f. 8. Nottingham, Md.

Plate C.

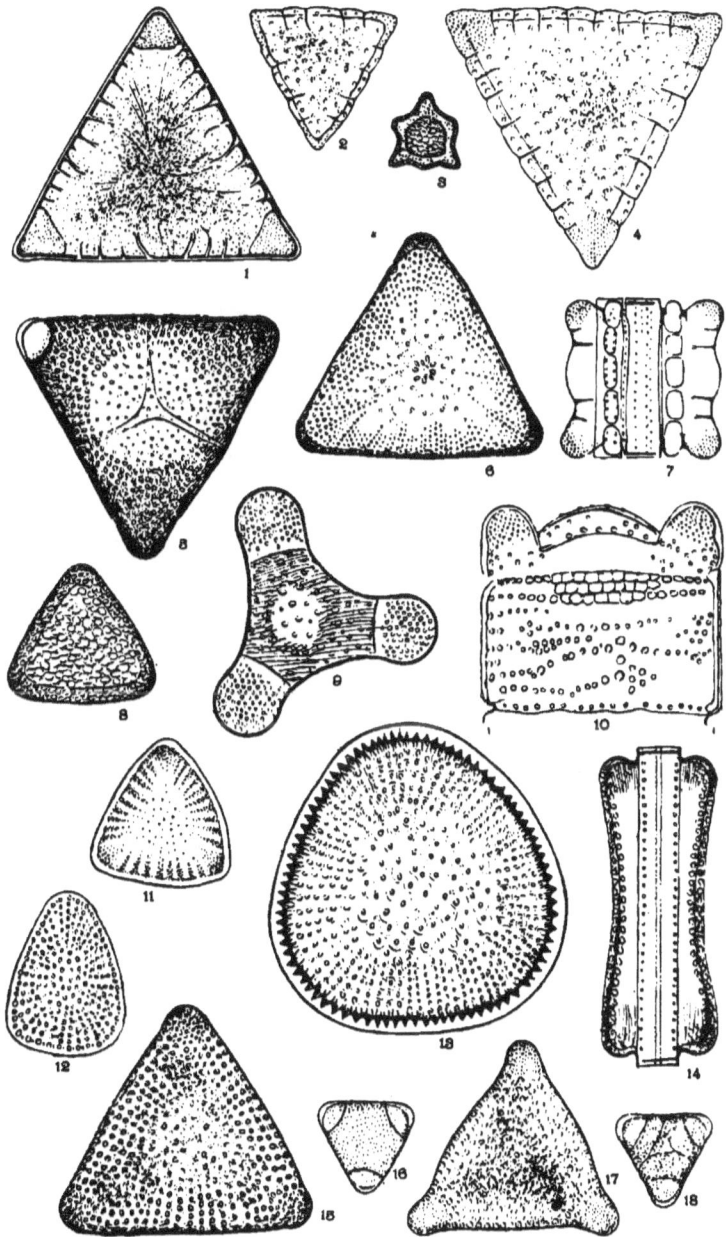

PLATE CI.

Figures magnified 500 diameters.

Fig. 1.   TRICERATIUM ANTILLARUM, Cleve., 1878, Schm. At., 99, f. 14.
" 2.   "   KITTONIANUM, Grev. Maryland. T. M. S., 1865, p. 8, 2, f. 18.
" 3.   "   SOLINOCEROS, Ehrb., Sill. J., April, 1845, p. 329, 1, f. 23; M. J., 1853, p. 248, 4, f. 1; Schm. At., 96, f. 11; 77, f. 21.
Figs. 4, 5.   "   SUBCORNUTUM, Grun.; Camp. Bay. Schm. At., 99, f. 15, 18. Granules in radiate series.
" 6, 7.   "   ELEGANS, Grev. Camp. Bay. Schm. At., 99, f. 11.
Fig. 8.   "   MARYLANDICUM, Bright. Maryland. M. J., 1856, Pl. 17, f. 17; Sm. Sp. T., No. 601.
Figs. 9, (12).   "   PUNCTATUM, Bright., M. J., 1856, p. 275, 17, f. 18; Schm. At., 76, f. 19, 20.
Fig. 10.   "   BROWNIANUM, Grev., T. M. S., 1861, p. 72, 8, f. 16. Maryland.
" 11.   "   ROBUSTUM, Grev. T. M. S., 1861, p. 71, 8, f. 15. Maryland.
" 13.   "   RECEPTUM, A. S.; Schm. At., 81, f. 10. Santa Monica; related to T. Shadboldtii, Bail.; and T. acceptum, Grev.
" 14.   "   AMOENUM, Grev.; T. M. S., 1861, p. 75, 9, f. 7. Maryland.
" 15.   "   ORNATUM, Grun. Camp. Bay. Schm. At., 98, f. 18.
" 18.   "   PENTACRINUS, Wallach. Camp. Bank. Schm. At., 98, f. 7. Previously figured as Amphitetras Ornata.

Plate CI.

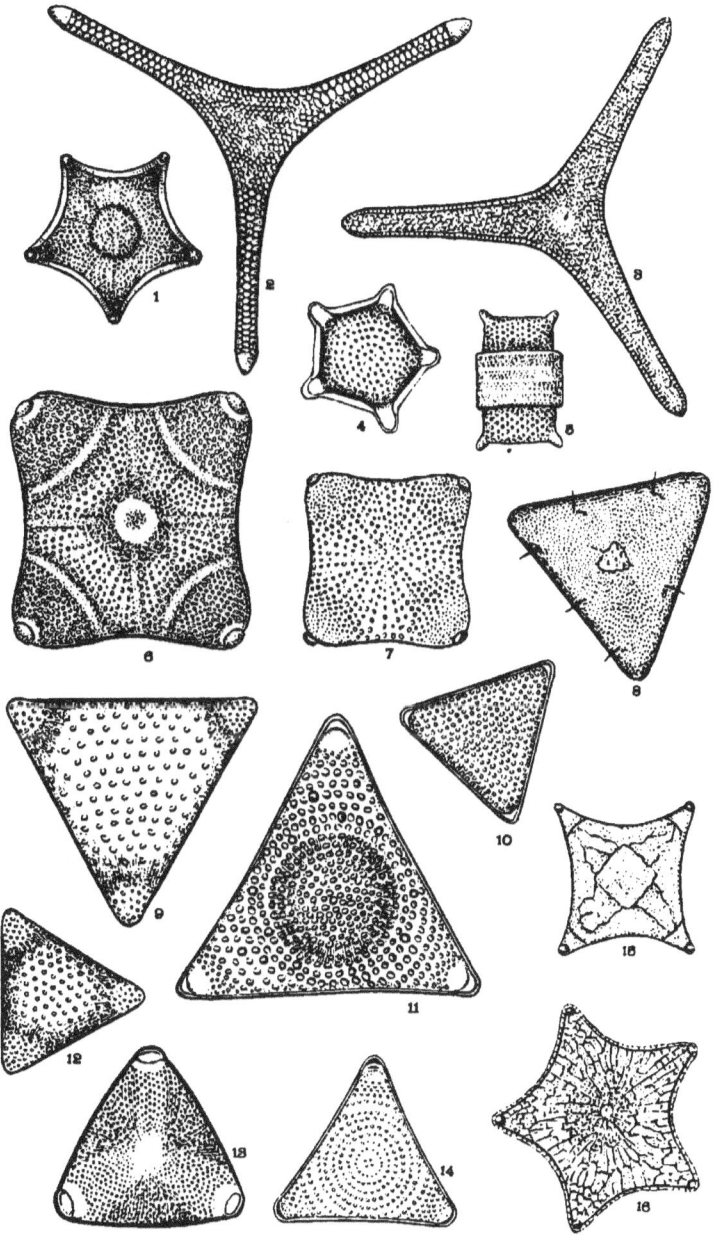

PLATE CII.

Figures magnified 500 diameters.

Fig. 1. TRICERATIUM MONTEREYI, Bright.; Santa Monica. Schm. At., 94, f. 1, 2, 3.
Figs. 2, (5). " SPINOSUM, Bail.; Camp. Bank. Schm. At., 87, f. 2, 3.
Fig. 3. " SUBROTUNDATUM, A. S.; Nottingham, Md. Schm. At., 93, f. 1.
" 4. " JUCATENSE, Grun.; Camp. Bay. Schm. At., 76, f. 13.
" 6. " CONDECORUM, Bright.; Nottingham, Md. Schm. At., 76, f. 27.
" 7. " MURICATUM, Bright.; Camp. Bank. Schm. At., 83, f. 8, 9.
" 8. " FISCHERII, A. S.; Patuxet River. Schm. At., 76, f. 34.
" 9. " TESSELATUM, Grev. var.; Nottingham, Md. Schm. At., 76, f. 33.
" 10. " OBSCURUM, forma minor, Grev.; Piscataway, Me. Schm. At., 76, f. 5.
" 11. " VALIDUM, Grun.; Santa Monica. Schm. At., 94, f. 5.
" 12. " UNCINATUM, Grun.; Pacific Coast. Schm. At., 94, f. 4.
" 13. " DURIUM, Bright.; Camp. Bay. An abnormal form; Schm. At., 78, f. 26.

Plate CII.

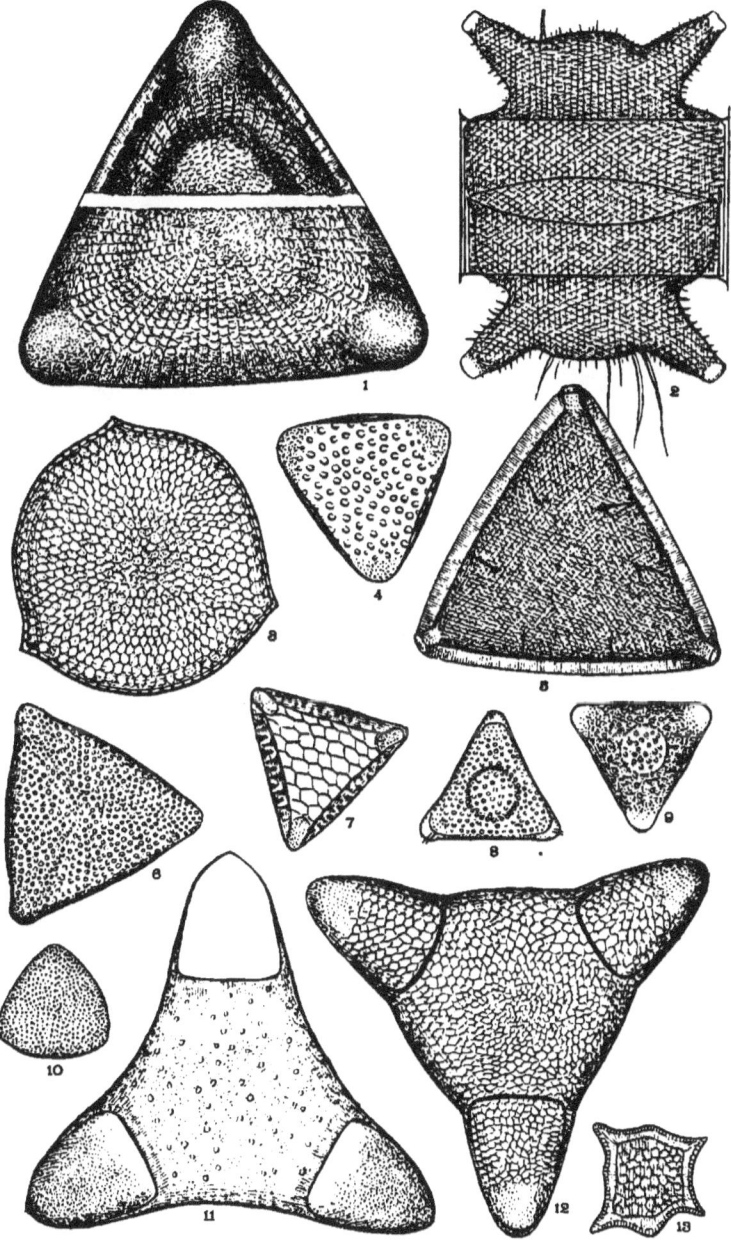

PLATE CIII.

Figures magnified 500 diameters.

Fig. 1. ACTINOCYCLUS CRASSUS, W. S.; V. H., 125, f. 8. — Eupodiscus crassus; S. B. D., I, p. 24, 4, f. 41; (actinocyclus octonarius and crassus).
" 2. ACTINOPTYCHUS HELIOPELTA, Grun. Nottingham, Md. V. II., 123, f. 3; V. II., 109, f. 2.
" 3. " BISMARKII, A. S., Schm. At., 91, f. 4. Santa Monica.
" 4. ACTINOCYCLUS NIAGARAE, H. L. S., A. Q. J. M., 1878, p. 17, 3, f. 10; J. M., 1879, 135, 6, f. 10.
" 5. " MONILIFORMIS, Ralfs.; V. H., 124, f. 9, N. a? Allied to Eupodiscus crassus, W. S., compare Fig. 1.
" 6. " TENUISSIMUS, Cleve., 1878, p. 21, 5, f. 34; J. M., 1879, p. 136, 5, f. 134; V. H., 125, f. 2.
" 7. ACTINOPTYCHUS PULCHELLUS, Grun.; V. II., 123, f. 5.
Figs. 8, 9, 10. BIDDULPHIA LAEVIS, Ehrb. Mik., 33, 15, f. 16; Sill. J., 1842, 2, f. 8; Bail., Gailionella. Present specimen from Nebraska.
Fig. 11. " GIGAS, Ehrb. Mik., 33, 12, f. 11. Fossil. Va. Of doubtful value.
" 12. " GIGAS, Ehrb. Oregon. Mik., 33, no Biddulphia, but Melosira.
" 13. HUTTONIA RICHARDTII, var. Grun. Virginia. Schm. At., 116, f. 4.

Plate CIII

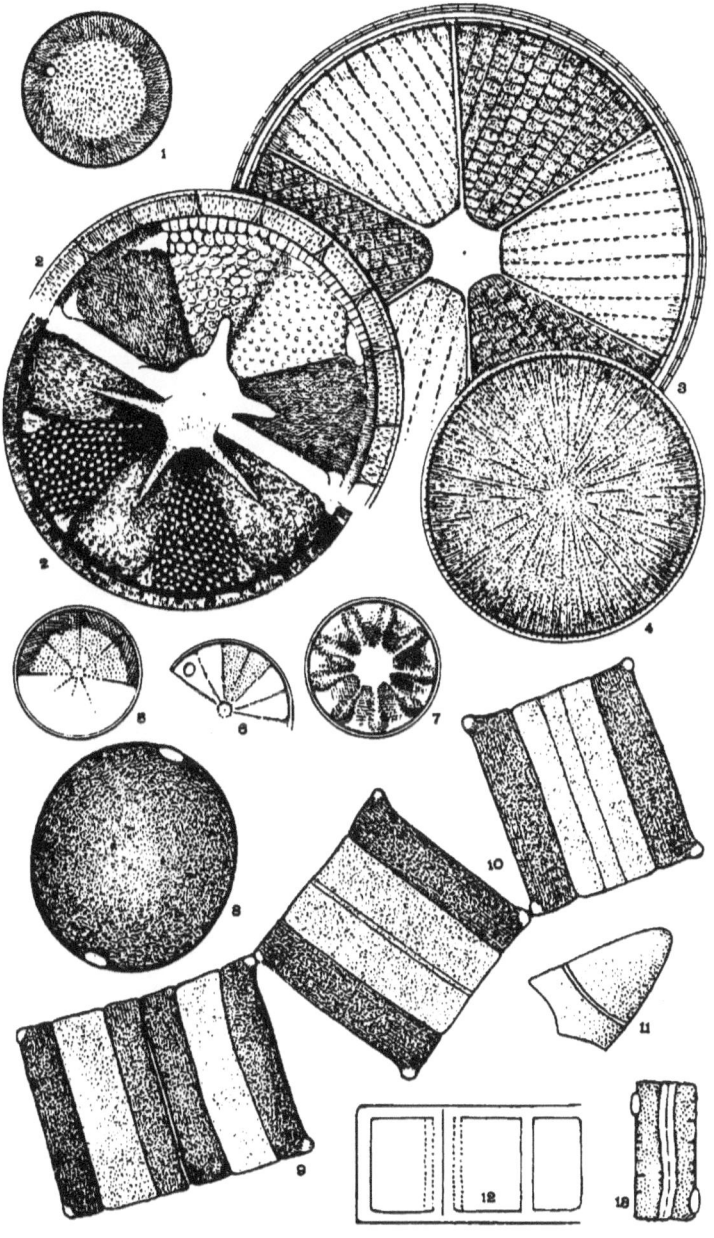

PLATE CIV.

Figures magnified 500 diameters.

Fig. 1. TRICERATIUM GRANDE, Bright., var. of Tr. favus fossil, Petersburg, Virginia; M. J., 1853, p. 249, 4, f. 8; Schm. At., 82, f. 5. *T. favus*, 82, f. 13, 14.
" 2. " ANTEDILUVIANUM (Amphitetras), V. II., 109, f. 4, 5.
" 3. " UNDULATUM, (Ehrb.) Bright., M. J., 1853, p. 248; V. H., 116, f. 7. Ditylium, Nottingham deposit, Md.
" 4. " INELEGANS, Grev.; T. M. S., 1866, p. 8, 2, f. 21; Schm. At., 81, f. 16. Santa Monica, Cal.
Figs. 5, 6. " RETICULUM, Ehrb. Mik., 33, 16, f. 13; 18, f. 50; M. J., 1853, p. 251, 4, f. 17.
Fig. 7. " OBTUSUM, Ehrb. Mik., 18, f. 48, 49; T. M. S., 1866, p. 8. Fossil, Richmond, Virginia.
" 8. " KAINII, Schultze; Bull. Tor. Bot. Club, Pl. 89, f. 6, March, 1889. Artesian well, Atlantic City.
" 9. " UNDULATUM, Ehrb., (Virg.), M. J., 1853, p. 250, 4, f. 13; V. H., 116, f. 13, var. Deposit, Petersburg, Virginia.
Figs. 10, (17). " BIQUADRATUM, Janisch, Schm. At., 98, f. 4 and f. 5. Leton Bank.
" 11, 12, 14. " STRIOLATUM, Ehrb. Attached to smaller forms of algae. Front views (11, 14) show a fine but distinct hexagonal areolation. M. J., 1853, p. 250, 4, f. 10; T. M. S., 1854, 6, f. 3; S. B. D., 5, f. 4, 6.
" 13, (16). " EHRENBERGII, Grün.; (Discoplea undulata), Ehrb. Fossil deposit, Nottingham, Maryland; V. H., 115, f. 7, 8.
" 18, 19. " LITHODESMIUM (Minusculum), Grun. Deposit, San Diego, California; V. H., 116, f. 2-4.

Plate CIV

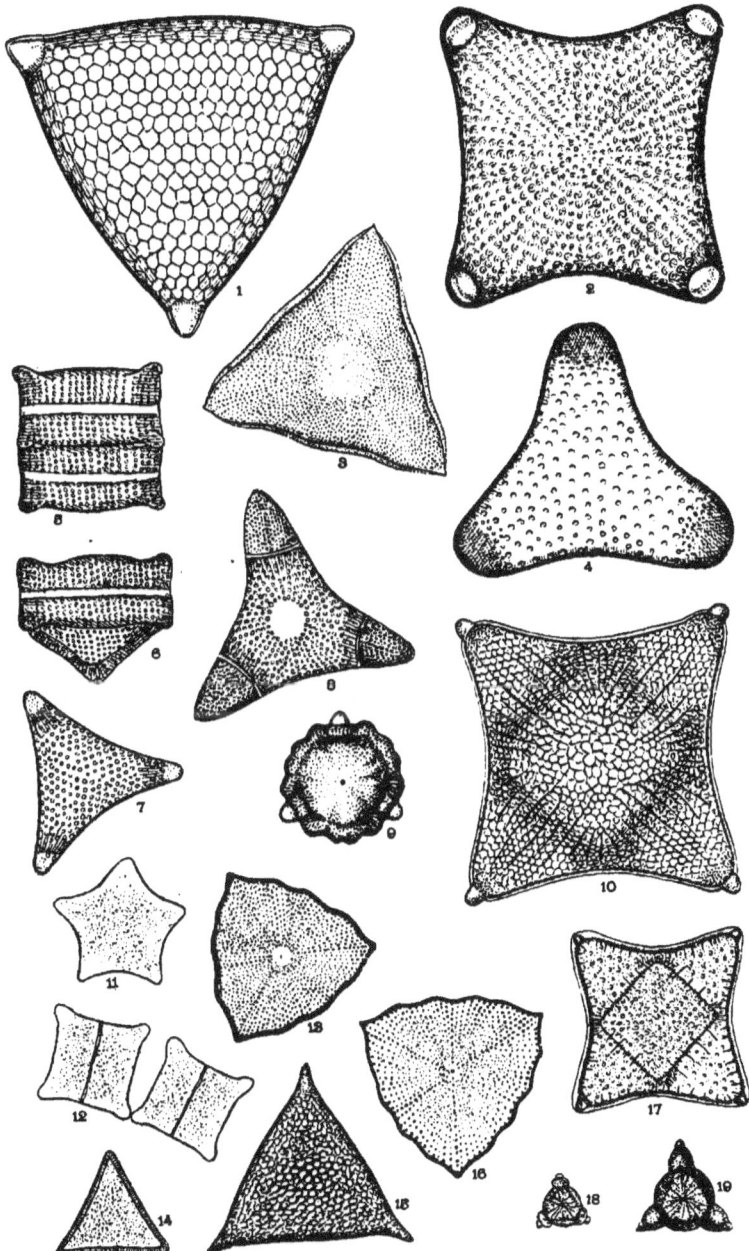

PLATE CV.

Figures magnified 500 diameters.

Fig. 1. TRICERATIUM TRIDACTYLUM, Bright.; M. J., 1853, p. 248, 4, f. 3. Schm. At., 87, f. 12. Fossil, Petersburg, Va.
" 2. " SEMICIRCULARE, (Euodia Brightwellii), M. J., 1853, p. 252, 4, f. 21.
Figs. 3, 4. " SEMICIRCULARE, Artesian well, Atlantic City. Bull. Tor. Bot. Club, March, 1889.
Fig. 5. " ARCTICUM, var. Californica, Grun. forma tetragone.
" 6. " MEMBRANACEUM, Bright.; M. J., 1853, p. 251, 4, f. 15. Virginia.
" 7. " AMERICANUM, Ralfs.; Schm. At., 76, f. 3, 28. Nottingham, Md.
" 8. " ARCTICUM, Bright. var. Californicum, Grun.; Schm. At., 79, f. 5.
Figs. 9, 10. HYDROSERA TRIQUETRA, Wallich; Schm. At., 78, f. 36, 37.
Fig. 11. " " according to Pritch.
" 12. " " Slide, Santa Cruz, Cal.
" 13. TRICERATIUM CALIFORNICUM, (Lithodesmium) V. H., 115, f. 9.
" 14. " UNDULATUM (Lithodesmium).
Figs. 15, 16. " KAINII, Schultze, Artesian well, Atlantic City, N. J. Bull. Tor. Bot Club, March, 1889.
Fig. 17. " REGINA, Heib., Schm. At., 97, f. 3-5. Atlantic City, N. J., artesian well.
" 18. " CINNAMOMEUM, Grev. Nearly allied to *Cestodiscus;* with marginal spines, about 5 on a side it would be *Cestodiscus;* without the spines it is Triceratium. Camp. Bay. V. H., 126, f. 1.
Figs.19,20,21. EUODIA JANSCHII, Grun.; Santa Monica. V. H., 127, f. 1-4.

Plate C V.

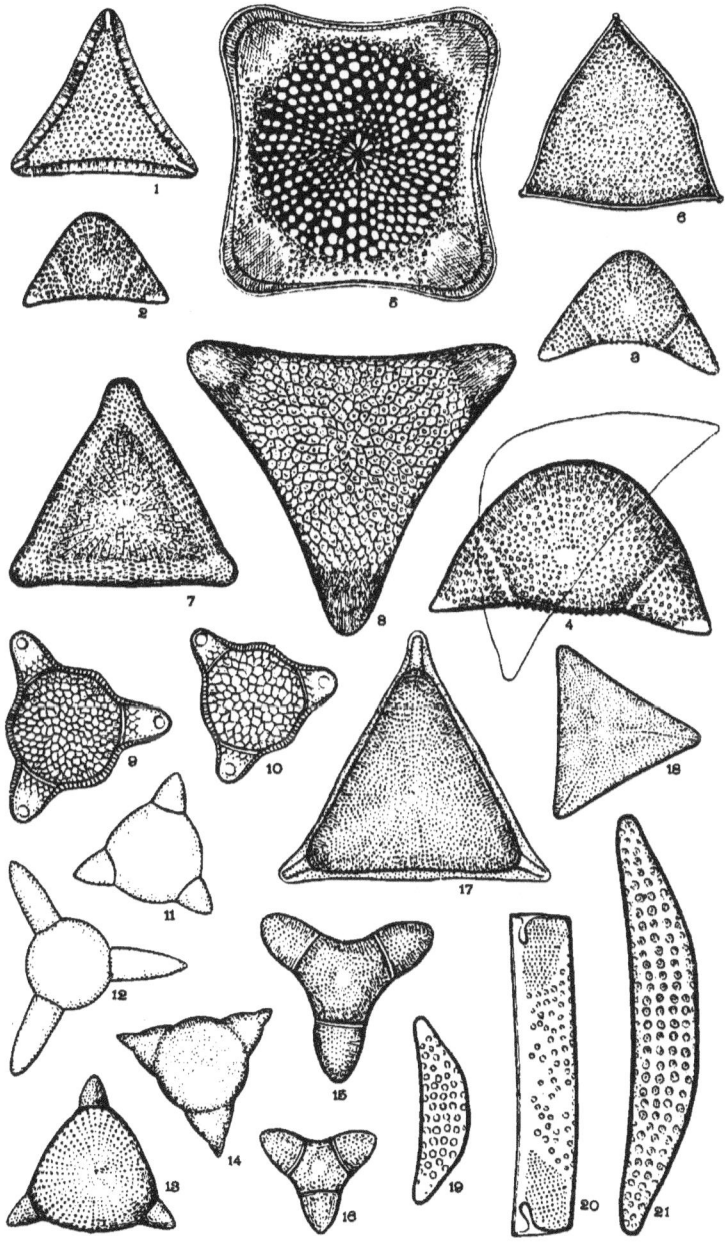

PLATE CVI.

Figures magnified 500 diameters.

Fig. 1.   TRICERATIUM SCITULUM, Bright, M. J., 1853, p. 250, 4, f. 9; Schm. At., 83, f. 11-16. Gulf of Mexico.

" 2.   "   ALTERNANS, Ehrb. Camp. Bay. S. B. D., 5, f. 45; V. H., 113, f. 6; Schm. At., 78, f. 9-20.

Figs. 3-5.   "   SCULPTUM, Shadb., T. M. S., 1854, p. 15, 2, f. 4; Schm. At., 76, f. 9-12; V. H., 109, f. 7, 8. Camp. Bay. Granules hardly prominent enough.

" 6.   "   CONSIMILE, Grun.; V. H., 108, f. 2; Schm. At., 84, f. 13, 14. Santa Monica.

" 7.   "   ARCTICUM, forma Campeachiana. Camp. Bay. V. H., 112, f. 1. Cellules hexangular, not so rectangular.

" 8.   "   QUADRANGULARE, Grev. Santa Monica, Cal. Probably only a quadrangular form of T. arcticum. The cellulae are really radiate.

Figs. 9, 11.   "   FRAGLICH? Schm. At., 86, f. 9, 10. Crescent City, Oregon. Figures only slightly magnified.

Fig. 10.   "   OBLIQUUM, Grun.; V. H., 110, f. 11. Cal.

Plate CVI.

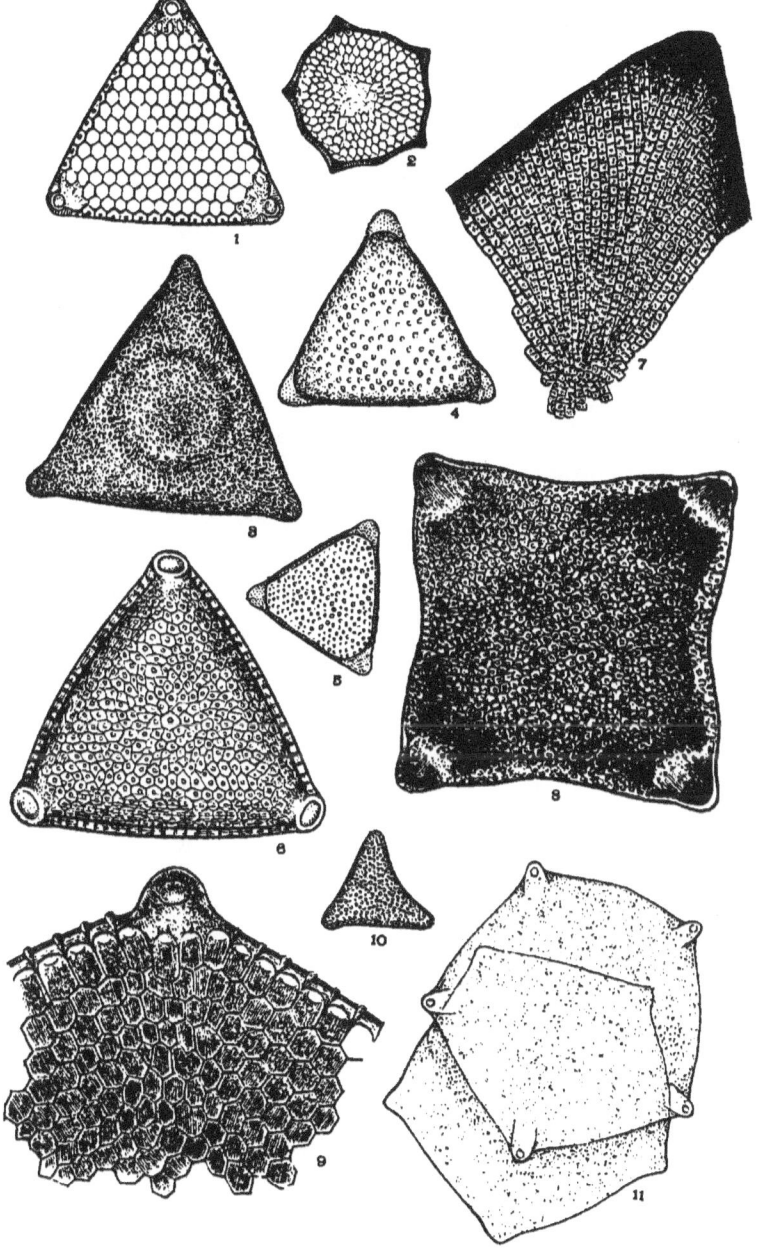

PLATE CVII.

Figures magnified 500 diameters.

Figs. 1, 2.   AMPHIPRARA NEREIS, Lewis. Atlantic sea-board, Narragansett. Figures too large, H. L. S. Length of valve .002″, or .0045″. N. and R. Sp., 6, 1, f. 6.
"   3, 4.   "   CONSERTA, Lewis N. and R., 1, f. 5. Prof. Smith writes: "I believe I am the only one who has specimens of this diatom; it is scarcely siliceous.
Fig. 5.   "   COSTATA. O'Meary, Irish; erroneously enrolled as N. American. M. J., Vol. IX, N. S., 12, f. 6.
Figs. 6, (9).   TRICERATUM MONTEREYI, Brightw. Copied from fossil diatoms of New Zealand, (Pl. XI, f. 25,) to show the extreme convexity of the terminal valve of a frustule.
Fig. 7.   SURIRELLA PULCHRA, Lewis N. and R., p. 4, 1, f. 1. Allied to *S. fastuosa* and *S. eximia*. St. Mary's River, Ga., salt marsh, Florida. Length of valve .005″-.009″.
"   8.   MASTOGLOIA ELEGANS, Lewis, p. 346; Proc. A. N. Sc., Pl. 3, f. 9.
"   9.   TRICERATIUM MONTEREYI, *vide* above Fig. 6.
"   10.   MASTOGLOIA ANGULATA, Lewis N. and R., Pl. 2, f. 4. The granules are really arranged in regular quincunx order. Atlantic City, Rockaway, etc.
"   11.   "   SUBMARGINATA, Cleve. and Gr., N., and S. K. D., 1, f. 2. The figure of this diatom is differently represented in size by different authors; it is probably only about half as large as here represented.

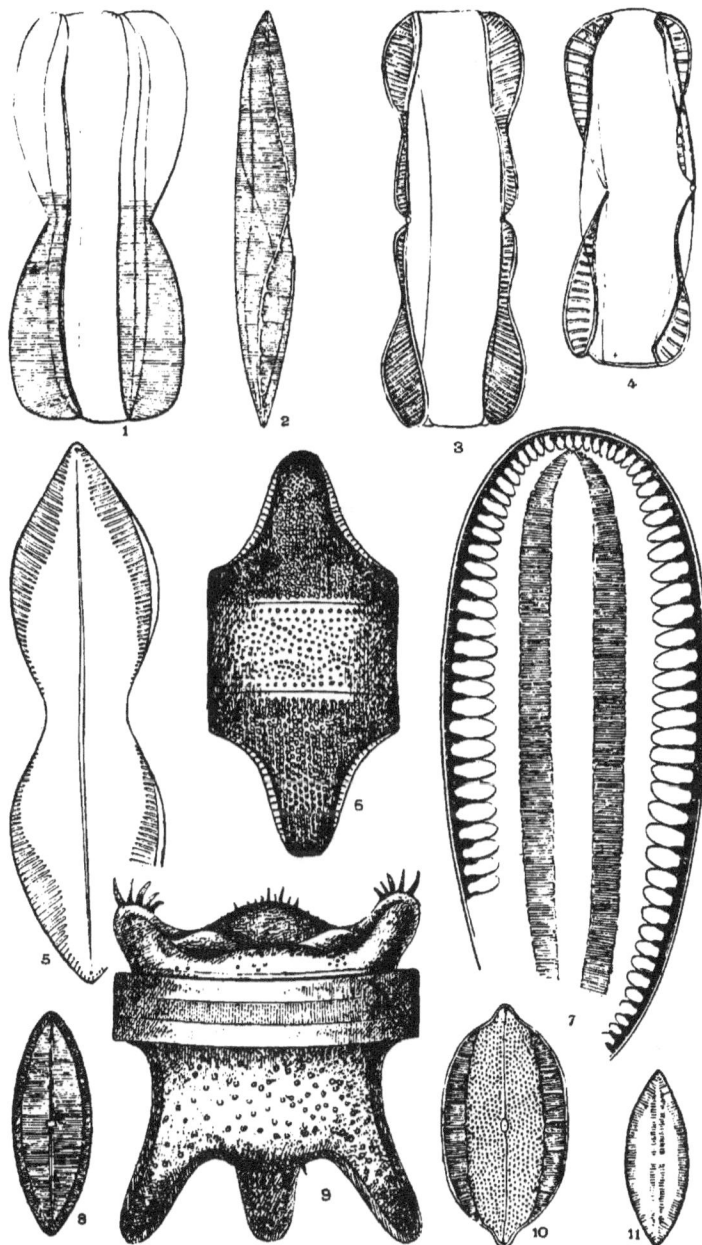

PLATE CVIII.

Figures magnified 500 diameters.

Fig. 1. TRICERATIUM PILEOLUS, Ehrb. Mik., 35a, 21, f. 17.
"   2.   "   PILEUS, Ehrb. Mik., 19, f. 18.
"   3.   "   ONTUSUM, Bright, M. J., 1853, p. 251, 4. f. 20. Virginia.
"   4. AMPHITETRAS ORNATA, Grev.; Tric. pentacrinum, Wallach, M. J., 1858, 12, f. 10-12; T. M. S., 1861, p. 45; Schm. At., 98, f. 18.
Figs. 5, 6. CYMATOPLEURA ELLIPTICA, S. B. D,, 10, f. 80. Normal and twisted form, or forma spiralis.
Fig. 7. COSCINODISCUS CALIFORNICA. Do not recognize this.
"   8. AULISCUS MUTABILIS, Grev. Monterey. T. M. S., 1863, p. 44, Pl. 2, f. 11.
"   9. ACTINOPTYCHUS BISEPTINARIUM, Ehrb. Mik., 33, 16, f. 5. One of the forms of A. Ehrenbergii.
"   10.   "   QUINARIUS, Ehrb., K. B., 1, f. 20; Mik., 33, f. 16.
Figs. 11, 12. COSCINODISCUS PUNCTATUS, Ehrb. Mik., 18, f. 40, 41; Sm. Sp. T., No. 97.
"   13-15. SURIRELLA CAMPYLODISCUS, K. B., 28, f. 26; Rab., 3, f. 4. The same as Campylodiscus spiralis, which is also better proportioned. See Plate.
Fig. 16. TRICERATIUM RETICULUM, Schm. At., 76, f. 26. Compare Pl. 104, f. 5, 6.

Plate CVIII.

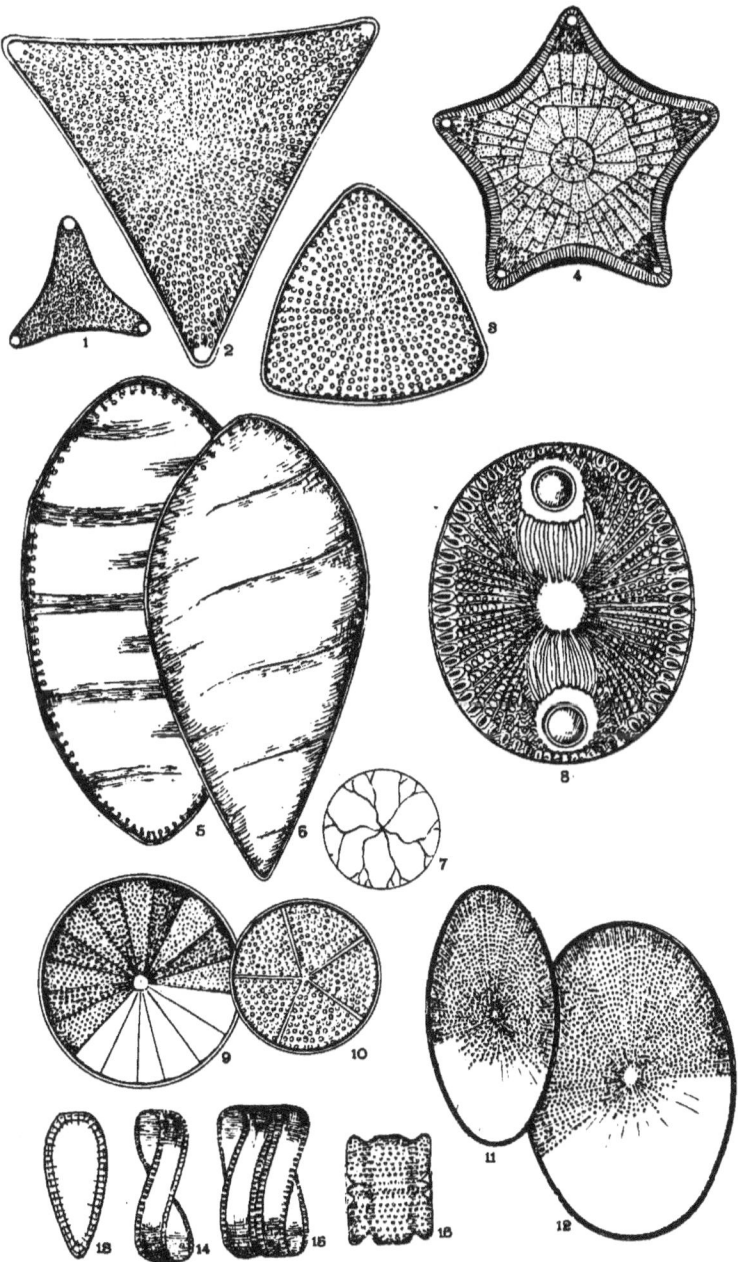

PLATE CIX.

Figs. 1-5. ISTHMIA NERVOSA, S. B. D., Pl. 47. Figs. 1 and 2 magnified 500 diameters. Figs. 3-5 various positions magnified 250 diameters, attached to stem of Polysiphonia.

"   6-9. ISTHMIA ENERVIS, S. B. D., Pl. 48. Figures 8, 9 magnified 500 diameters. Figures 6, 7 magnified 250 diameters.

Plate CIX.

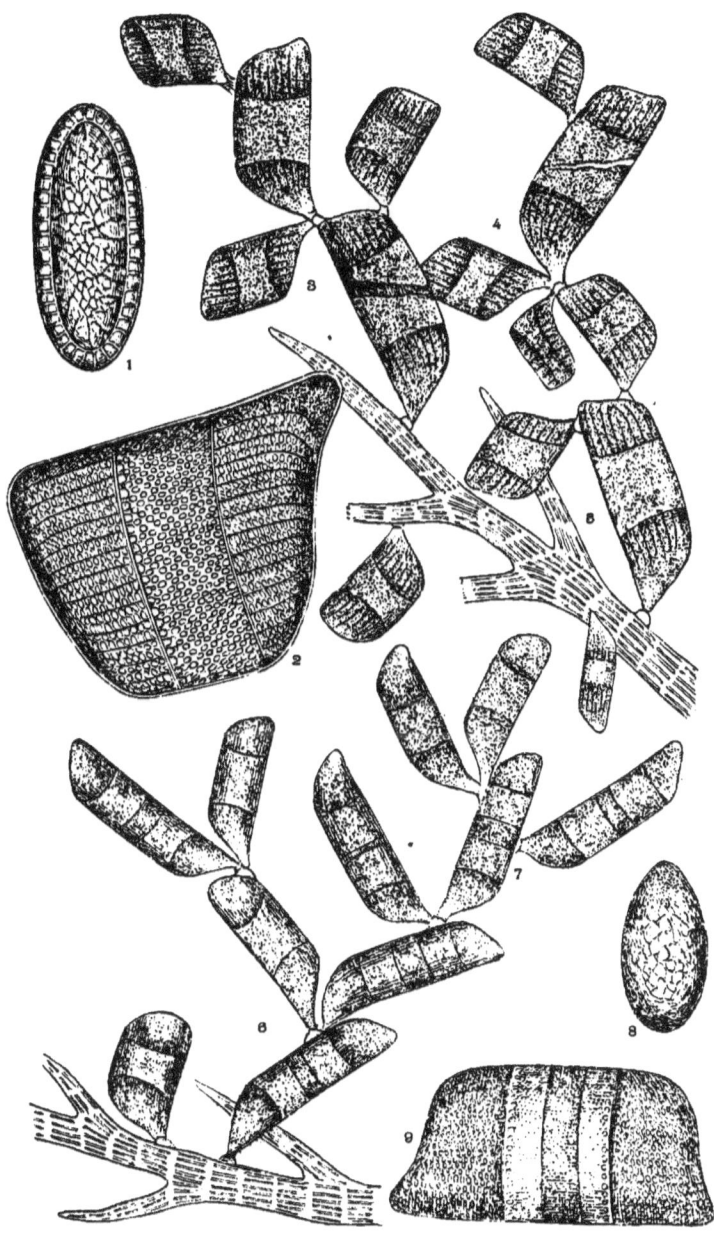

PLATE CX.

Figures magnified 500 diameters.

Fig. 1. COSCINODISCUS INCRETUS, A. S., Schm. At., 139, f. 1. Santa Monica.
" 2. TRICERATIUM CAMPEACHIANUM.
Figs. 3, 4. CAMPYLODISCUS BICOSTATA, W. S.; T. M. S., 1854, p. 75, 6, f. 4; V. H., 75, f. 2; Schm. At., 55, f. 4, 5.
Fig. 5. CYCLOTELLA COMPTA, var. Bodonica, Kg.; V. H., 92, f. 16-22; V. H., 93, f. 21, etc. Br. Columbia.
Figs. 6, 7. BIDDULPHIA ANGULATA, A. S.; Nottingham, Md. Schm. At., 141, 7, 8. Valves are marked somewhat like *Coscinodiscus*.
Fig. 8. HYALODISCUS RECTICULATUS, A. S., Schm. At., 140, f. 7. Monterey.
Figs. 9, 10. BIDDULPHIA MEMBRANACEA, Cleve.; W. Ind. diatom; 5. f. 33. Near *Zygoceros Balaena* of Ehrb.

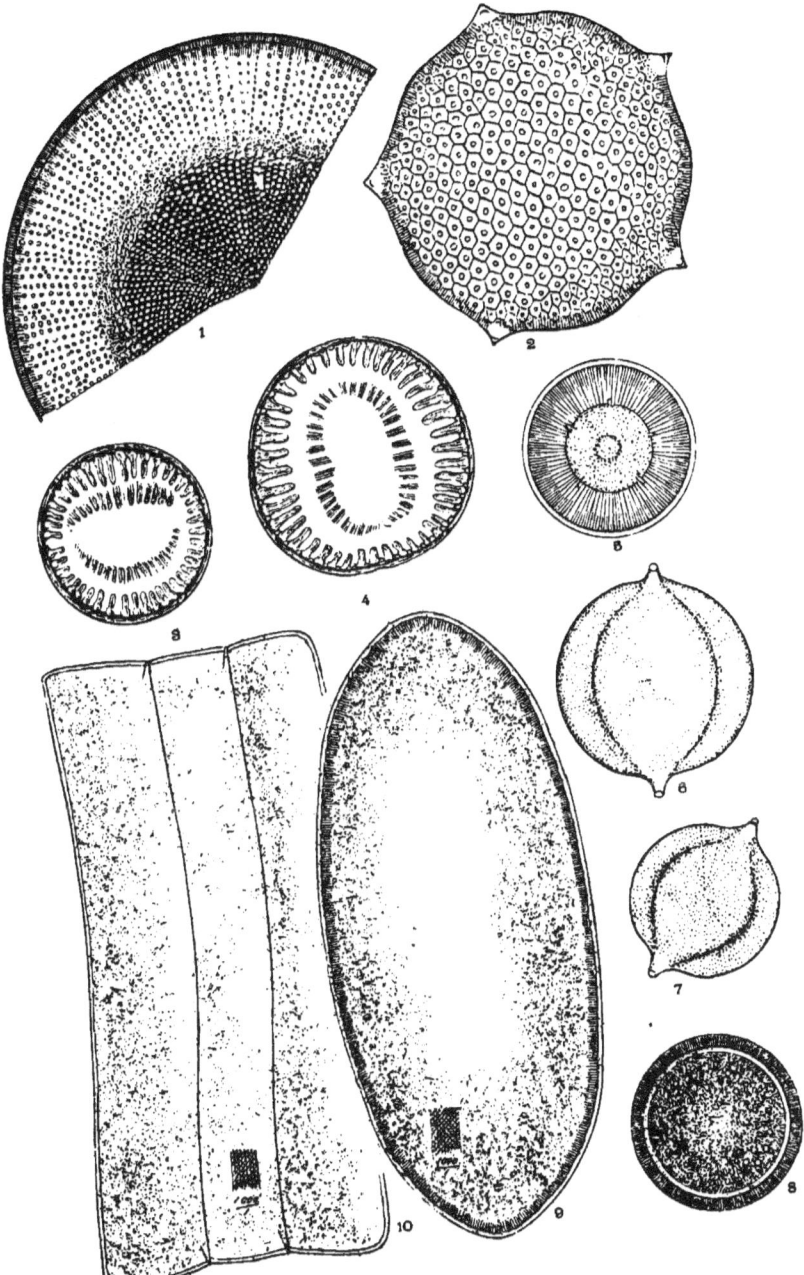

PLATE CXI.

Figures magnified 500 diameters.

Figs. 1, 2. NAVICULA MACULATA. Edwards, Stauroneis maculata, Bailey. M. J., 1860, p. 129; Schm. At., 62, f. 10; Sm. Sp. T., No. 293. Near latissima, vide Schultze's fig., Bul. Torrey Bot. Club, 1887, April.
Fig. 3. " PRETEXTA, var. (Ehrb.), Grundl. Camp. Bay.
" 4. TRICERATIUM SCULPTUM, var. Grundl. Atlantic City, Nottingham, etc., Schm. At., 76, f. 12.
" 5. AULISCUS PRUINOSUS, Bail. Gulf of Mexico. T. M. S., 1863. p. 19. Pl. 3, f. 13; Schm. At., 31, f. 6, 7; 108, f. 10.
" 6. " SPINOSUS, T. Christian, closely allied to *Glyphodiscus*. Atlantic City, Bul. Torrey Bot. Club, March, 1889, p. 73.
" 7. NAVICULA MARGINATA (Mastogloia), Lewis N. and R. Sp., p. 26, 2, f. 1. 'from photo. Charleston, South Carolina.
Figs. 8, 9. " GRUENDLERI. A. S.; Schm. At., 12, f. 35-37. Cal.
Fig. 10. " VULPINA, (V. H., 9, f. 18), a var. of N. acuta, K. B., 3, f. 69; Sm. Sp., No. 240. Lake Michigan.
" 11. SURIRELLA NOBILIS, N. S. Sporangial form, S. B. D., 8, f. 63. — Sur. splendida, var. Lake Michigan.
" 12. COSCINODISCUS VELATUS? From photo., S. C.; Schm. At., 62. f. 10.
Figs. 13, 14. BIBLARIUM ELLIPTICUM, vars., Ehrb. Mik., 33, 2, f. 5; 33, 12, f. 2; M. D., 13, f. 38.
" 15, 16. MERIDION CIRCULARE, var. contorta, Lake Michigan, frequent. For. Mcr. constrictum, compare Ralfs. A. N. H., 1843, p. 458, 18, f. 2; Rab. S. D., 1, f. 2; and S. B. D., 32, f. 278.
Fig. 17. EUPODISCUS ROGERSII, (Argus, var). Artesian well, Atlantic City, Schm. At. 92, f. 2–6. From photo.
" 18. TRICERATIUM AMERICANUM, Ralfs., Schm. At., 76, f. 3, 20. Artesian well, Atlantic City.
" 19. CYMBELLA ROTUNDATA, H. H. C. An extreme form of this is called by Bailey Cymb. gibba. Grunow calls the same diatoms *Encyonema triangulare*.

Plate CXI.

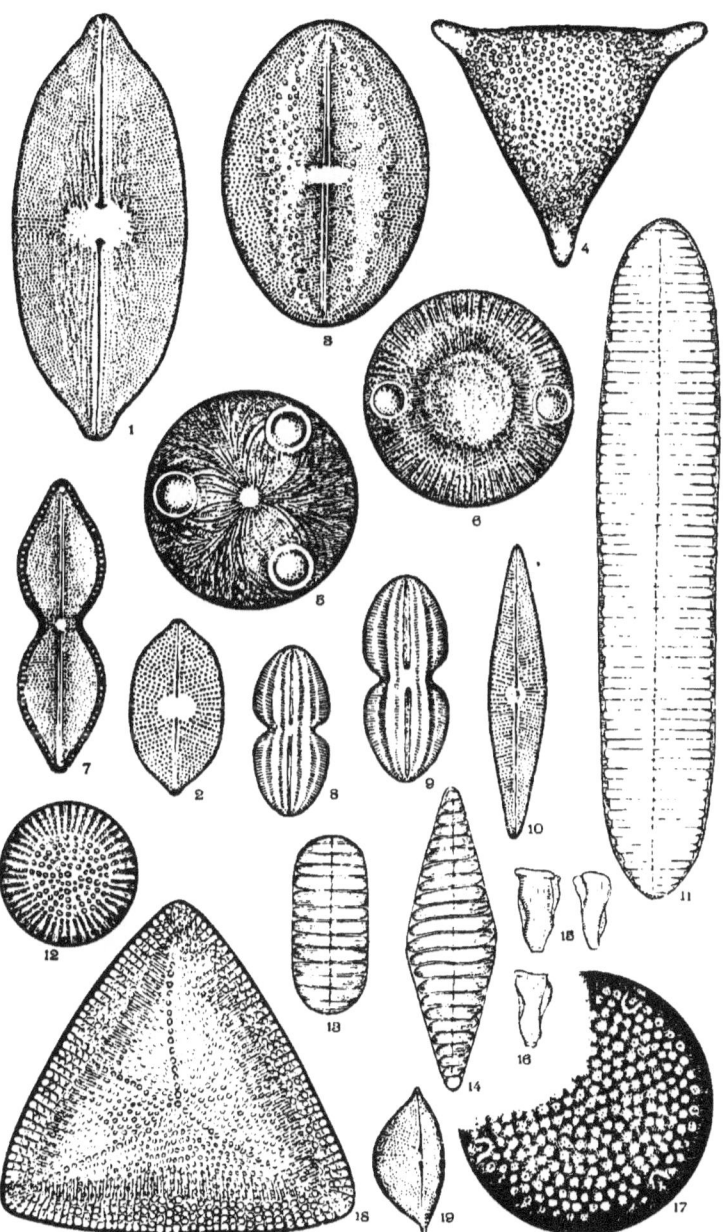

PLATE CXII.

Figures magnified 500 diameters.

Fig. 1. AMPHIPLEURA MAXIMA, W. and Chase, 2, f. 5. Oregon.
" 2. MELOSIRA ARENARIA, Rab. 2, f. 5, etc.
" 3. ACTINOPTYCHUS IRREGULARE, V. H., 132, f. 11.
Figs. 4, 5. MELOSIRA UNDULATA, Slide and S. B., 2, f. 9.
Fig. 6. COSCINODISCUS DENARIUS, Schm. At., 57, f. 24. Atlantic City.
" 7. TRICERATIUM SUBROTUNDATUM, Schm. At., 93. Atlantic City.
" 8. COSCINODISCUS MARGINATUS, K. B., 1, f. 7. Amer.
" 9. STAURONEIS ICOSTAURUM. K. B., 29, f. 10.
" 10. " PHYLLODES, K. B., 29, and Rab., 9.
" 11. " LINEATA, K. B., 29, f. 5.
" 12. AMPHORA FLEXUOSA, Lens, Vol. 2, No. 2, Pl., 1, f. 3.
" 13. STEPHANODISCUS ASTRÆA, V. H., 95, f. 9, var. minutiala.
" 14. DISCLOPEA OREGONICA.
" 15. NITZSCHIA BILOBATA, not half but whole frustule. Compare Plate 43, f. 5, 6; S. B. D., Vol. 1, p. 42, Pl. 15, f. 113.

Plate CXII.

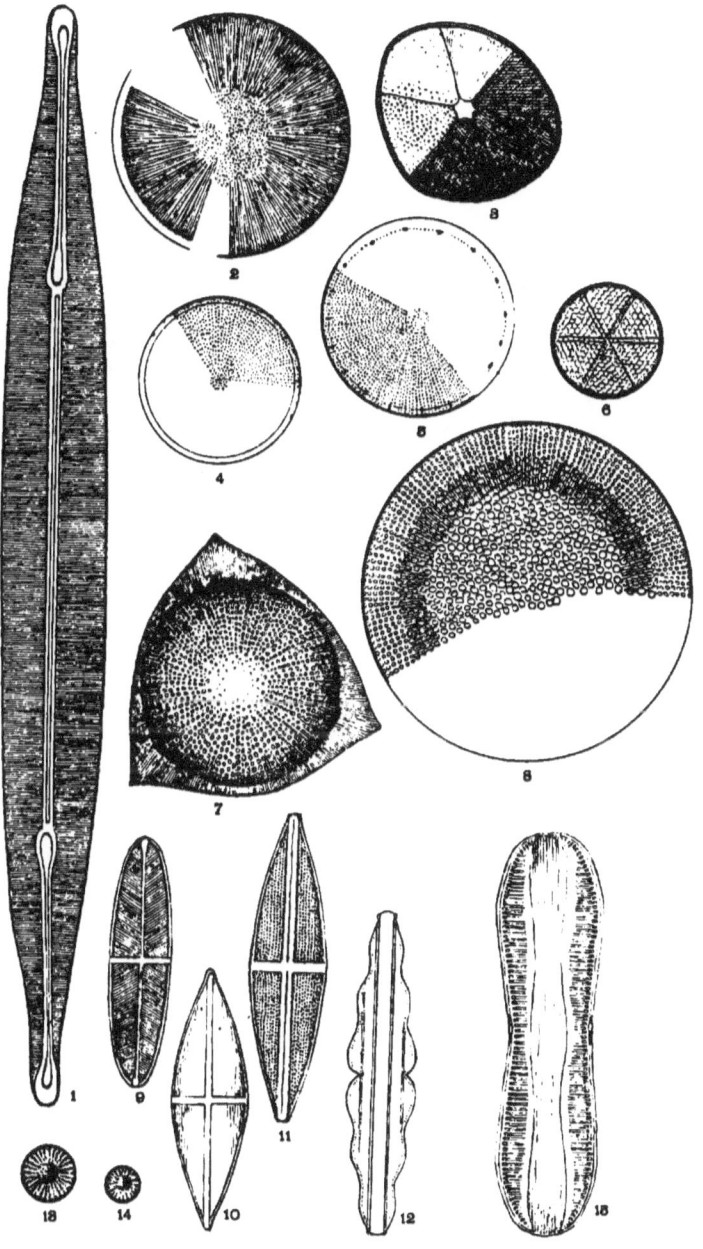

www.ingramcontent.com/pod-product-compliance
Lightning Source LLC
Chambersburg PA
CBHW020831020526
44114CB00040B/535